# Energy Materials

# Energy Materials

Edited by

M Eswaramoorthy
Subi J George
A Sundaresan
C N R Rao

School of Advanced Materials
Jawaharlal Nehru Centre for Advanced Scientific Research (JNCASR), India

World Scientific

NEW JERSEY • LONDON • SINGAPORE • BEIJING • SHANGHAI • HONG KONG • TAIPEI • CHENNAI • TOKYO

*Published by*

World Scientific Publishing Co. Pte. Ltd.
5 Toh Tuck Link, Singapore 596224
*USA office:* 27 Warren Street, Suite 401-402, Hackensack, NJ 07601
*UK office:* 57 Shelton Street, Covent Garden, London WC2H 9HE

Library of Congress Control Number: 2023000831

**British Library Cataloguing-in-Publication Data**
A catalogue record for this book is available from the British Library.

**ENERGY MATERIALS**



ISBN 978-981-127-094-9 (hardcover)
ISBN 978-981-127-095-6 (ebook for institutions)
ISBN 978-981-127-096-3 (ebook for individuals)

For any available supplementary material, please visit
https://www.worldscientific.com/worldscibooks/10.1142/13268#t=suppl

https://doi.org/10.1142/9789811270956_fmatter

## Contents

https://doi.org/10.1142/9789811270956_fmatter

## Preface

The world is in short supply of energy. Along with environmental factors, it has become obligatory for science to assist with the problem. An important component of material science is the field of Energy Materials. The various aspects of energy include electrical power — batteries, supercapacitors, thermoelectric energy conversion, photovoltaics and so on. Hydrogen is available in abundance, but catalysts are required for the catalysis, and are important for providing hydrogen. Catalysts or porous solids therefore have universal appeal in terms of usage and applications. Then, there exist nuclear energy materials too.

So overall, energy materials are extremely important in the world today. And in this book, we have covered the various topics dealing with energy. Such a book is possible with the help of experts in the School of Advanced Materials at the Jawaharlal Nehru Centre for Advanced Scientific Research. The School has faculty with expertise in most areas of energy materials and they have helped us in bringing out this volume.

M. Eswaramoorthy
Subi J. George
A. Sundaresan
C.N.R. Rao
*Editors*

https://doi.org/10.1142/9789811270956_0001

# Chapter 1

## Solution-Processed Photovoltaics

Abdul Azeez and K.S. Narayan*

*Chemistry and Physics of Materials Unit (CPMU), School of Advanced Materials (SAMat), Jawaharlal Nehru Centre for Advanced Scientific Research, Jakkur, P.O., Jakkur, Bengaluru 560064, India*

**narayan@jncasr.ac.in*

Increasing energy demand and the deepening climate change crisis require renewable energy technologies. In this regard, solar photovoltaic technologies based on inorganic, organic and hybrid materials are prominent choices. Third-generation solar cells based on organic semiconductors and hybrid perovskites are promising due to their unique features of bandgap tunability, low-cost, and low-temperature solution processability. The remarkable improvements in the performance of non-fullerene acceptors-based single junction and tandem organic solar cells present a promising way forward in solar energy harvesting in various applications involving flexible, semi-transparent, and wearable devices. Material design and device engineering strategies to reduce the high open-circuit voltage loss by suppressing the non-radiative recombination loss are needed to improve its viability. Solar cells based on hybrid perovskite are on the verge of widespread commercialization, with power conversion efficiency exceeding the 25% limit. Combining organic solar cells and hybrid perovskite in tandem structures also offers attractive options for high efficiency. A critical barrier for both of these solution-processed technologies are issues related to stability and degradation. This chapter discusses the fundamentals, current status and challenges of both technologies in general and a survey of non-fullerene acceptors-based organic solar cells.

## 1. Introduction

State-of-the-art photovoltaic (PV) technologies are largely based on single crystalline (Si, GaAs, GaInP), polycrystalline (Si, CIGS, CdTe), and amorphous silicon. Silicon enjoys the domination of active material choice in the solar PV landscape. Since economics, durability and performance dictate the choice, it is hard to see an alternative to Si in the near future. But, with increasing demand for renewable energy sources due to the deepening climate change crisis and changing geopolitical conditions, there is frenetic research activity to explore alternative PV sources. Additional criteria in the choice of PV technology is the energy payback period, i.e. recovery of the energy expended during manufacturing over a period of operation. A payback period of fewer than two years is a promising outlook towards Net-zero emissions. A complete analysis of the energy-assessment cycle is complex since the drastically varying supply chain costs for manufacturing, needs to be incorporated besides manufacturing costs. However, it is clear that solution-processed solar cells based on organic photovoltaics and hybrid perovskites fall under this class of manufacturing technology which ensures energy recovery within a fraction of its life cycle of usage.

In this chapter, we largely focus on the progress, status and directions of organic solar cells (OSCs). The looming presence of hybrid organic-inorganic perovskites (HOIP) in the research landscape with impressive performance parameters is briefly discussed in the initial sections. The HOIP solar cells also serve as a benchmark for OSCs. An obvious advantage the OSCs have is the absence of toxic metals in soluble form and flexible form factors. A critical test for both these solution-processed technology are issues related to stability and degradation. We address certain important issues prevailing in solution-processed PVs.

## 2. Hybrid Organic-Inorganic Perovskites

Hybrid perovskites have emerged as promising materials for solar cell applications. The salient features of HOIPs, such as high absorption coefficient ($10^5$ $cm^{-1}$), band-like charge transport, long carrier lifetime, and diffusion length (> 1μm), make them a prominent choice for solution-processed PVs. Since Miyasaka and co-workers in 2009 reported solar cells based on $MAPbI_3$ and $MAPbBr_3$, the last decade has witnessed a dramatic spurt of activity in these classes of materials.[1-2] The record efficiency of perovskite solar cells has exceeded the certified efficiency of

25.2% on a laboratory scale, and module-level outdoor cells have reached 21.4%.[3-5]

The chemical formula of hybrid perovskites is typically $ABX_3$, where A (organic or metallic) and B (metallic) represent cations, and X is an anion. The ideal perovskite crystal structure exists in a cubic form with the corner-sharing $BX_6$ octahedral network (e.g. $CH_3NH_3PbX_3$). The HOIPs can be synthesised by various scalable methods that involve combining a metal salt such as $PbI_2$ with an organic halide salt like MA iodide in a single step and spin-coating the solution of both salts.[6] The low-temperature processing route enables the choice of various substrates, thus opening up the realm of printed and flexible solar cells. The thickness of the active matter is an important parameter in this aspect of flexibility.[7]

Apart from efficient light harvesting, the large absorption coefficient of hybrid perovskites helps to achieve high open-circuit voltage ($V_{OC}$) in the device. As a result of the high absorption coefficient, the device requires only a thin active layer to almost completely absorb light with energy above the bandgap, which can reduce the charge recombination-induced reverse saturation dark current. This interdependence is described by the Shockley–Queisser model of the relationship between film thickness and $V_{OC}$:

$$V_{OC} = \frac{k_B T}{q} \ln\left(\frac{J_{SC}}{J_0}\right) = \frac{k_B T}{q} \ln\left(\frac{J_{SC} N_D \tau_{eff}}{q n_i^2 d}\right)$$

where $k_B$ is the Boltzmann constant, T is the temperature, $q$ is the elemental charge, $J_{SC}$ is the short-circuit current density, $J_0$ is the initial current density, $N_D$ is the doping concentration, $\tau_{eff}$ is the effective carrier recombination lifetime, $n_i$ is the intrinsic carrier concentration, and $d$ is the thickness of the active layer. The thickness of most reported high-efficiency HOIP solar cells is in the range of 0.3–0.6 μm. In contrast, crystalline silicon solar cells are usually about 300 μm thick, indicating the benefit of a large absorption coefficient in HOIPs.[8-10]

This advantage of HOIP results in a very low $V_{OC}$ deficit (= $E_g/q$ - $V_{OC}$) of ~ 0.37 eV in comparison to the silicon solar cells (~ 0.6 eV).[11] The performance of perovskite solar cells is dependent on the processing conditions and post-treatment, as it controls the crystallinity and morphology of the film. The optoelectronic properties and performance of

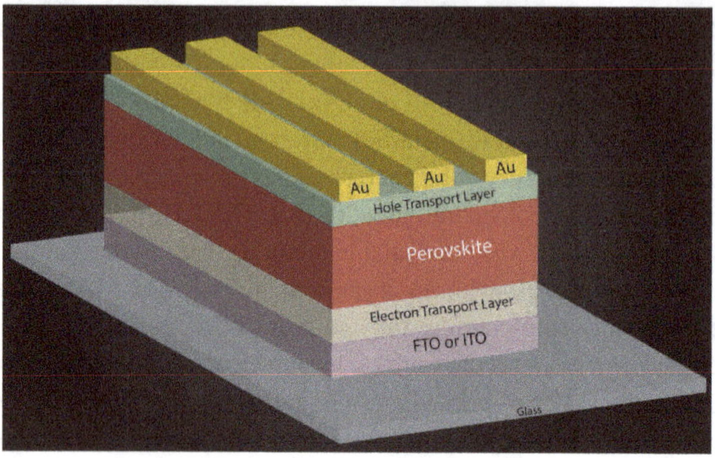


Figure 1. Schematic diagram depicting the device architecture of a typical planar perovskite solar cell in conventional (n-i-p) configuration.

the device can be readily tuned by varying the composition of the hybrid perovskite.[6,12-13]

In comparison with any other low-temperature solution-processed technology, the key feature of defect tolerance (significantly reduced sub-bandgap defect states) in HOIPs can be attributed to the reduced energy losses and high-performance metrics.[14] Several experimental studies on the estimation of sub-bandgap defect distribution (Urbach tail) in HOIPs indicate a low defect density (Urbach energy ~ 15 meV < $k_BT$) which is close to the established technologies such as c-Si and GaAs systems.[15] The presence of large sub-bandgap defects may increase the energy loss via SRH recombination leading to an overall $V_{OC}$ reduction in the device. Estimates based on internal photoluminescence quantum efficiency indicate a low energy loss in this system which is comparable with the benchmark GaAs, and the $V_{OC}$ can reach close to the radiative limit (~ 1.3 eV for $CH_3NH_3PbI_3$).[16]

The thin film aspect in these solar cells gets relevant for tandem cells, building-integrated structures and transparent solar cells. The versatile feature of bandgap tunability in HOIPs enables them to be integrated with Si in multi-junction solar cells. This approach has already been demonstrated as an efficient strategy with the PCE of monolithic perovskite-Si tandem exceeding 29%.[17]

Presently, the major challenge is the rapid degradation of the HOIP solar cells under real operating conditions (oxygen, moisture, light, heat and mechanical stress). HOIPs are highly sensitive to moisture, and the hydration reaction leads to the formation of $PbI_2$ crystals and an increase in the concentration of mobile ions. The moisture-induced degradation gets accelerated in the presence of heat and electric field and is identified to be irreversible. On the other hand, the reaction of perovskite ions with oxygen leads to the formation of electronic traps, and the photooxidation causes deprotonation of the methylammonium cation of the perovskite.[18] The degradation induced by the moisture and oxygen can be minimized by encapsulating the perovskite solar cell, and a long lifetime has been demonstrated with efficient encapsulation.[19-21] The chemical decomposition and formation of hydrogen iodide (HI) under illumination severely cause performance deterioration. The lattice distortion and phase transition (tetragonal to the cubic phase in the case of $CH_3NH_3PbI_3$) under thermal stress cause performance degradation. Apart from this, different transport layers, such as spiro-OMeTAD, degrade at elevated temperatures and also under illumination conditions. Further, light-induced degradation of commonly used electron transport layers such as PCBM and $TiO_2$ are known to increase carrier recombination and expedite the degradation of the device.

An important strategy to improve the stability is employing low-dimensional perovskites, such as 2D perovskite, in the device. The organic spacer in the 2D perovskite improves the lattice stability, and its hydrophobic nature inhibits moisture ingression.[22-23] Further, 2D perovskite has shown to have a reduced defect density and suppressed ion migration leading to improved device stability.[24] Utilizing additives while processing the perovskite layer has also been shown to improve the stability. Processing additives in the perovskite precursor improve the crystallinity of the film and reduce the trap density.[25-26] Other strategies to enhance the stability of perovskite solar cells include using mixed cation and mixed halide perovskite as absorbers.[27-29]

## 3. Organic Solar Cells

Organic solar cells rely on processes that have different photophysics from the hybrid perovskite and conventional p-n junction-based solar cells. A typical device architecture consists of a photo-responsive active layer sandwiched between suitable buffer layers and counter electrodes. The photo-responsive active layer can be a bilayer or a blend-mix consisting of donor polymer and acceptor molecules in a suitable proportion (see Figure 2). Initial reports on OSCs based on a donor-acceptor (D-A) bilayer heterojunction efficiency (PCE) of about 1%, essentially relied on the photocarriers generated by absorption within the vicinity of the interface. This planar junction was limited by the D-A interfacial area representing a bottleneck in charge collection. In order to overcome the limitations of bilayer (donor layer and acceptor layer) architecture, the concept of bulk heterojunction (BHJ) was introduced.[30] The BHJ concept utilized the inherent phase separation length scale of a two-component polymer mixture. In the bulk heterojunction strategy, a blend of donor polymer and acceptor molecule in a suitable stoichiometry forms the active layer of the OSC with the advantage of an intercalating network of donor and acceptor interface throughout the film (1995 by Heeger and co-workers).[30-32] The bulk heterojunction strategy of two components can be extended with one more donor (acceptor) element constituting the active layer and is known as the ternary blend OSC. The schematic in Figure 2 depicts the different types of OSC architecture with representative morphology.[33]

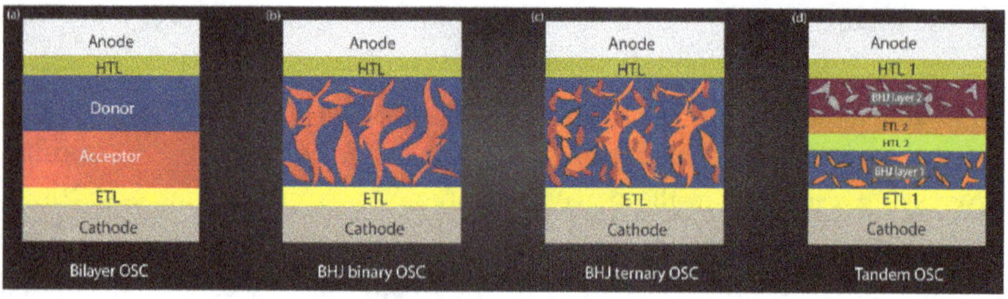


Figure 2. Schematic diagram depicting the device architecture of (a) bilayer, (b) binary bulk heterojunction, (c) ternary bulk heterojunction and (d) tandem organic solar cells, respectively.

It has been shown that in certain systems, the photoexcitation almost instantaneously generates vibrationally excited hot excitons that relax to the vibrational ground state in ~ 100 *fs* time scale.[34] If the exciton generation occurs in the vicinity of the D-A interface, then the hot exciton dissociates with extra free energy according to the Frank-Condon principle. The singlet excitons generated upon photoexcitation can also diffuse to the D-A interface in the picosecond (*ps*) time scale, and hence the active layer morphology with a sufficient number of D-A interfaces and optimum domain sizes is necessary for efficient device performance. The diffusion length of excitons in a typical organic solar cell is estimated to be nearly 10 nm. At this point, instead of diffusing to the D-A interface, the singlet excitons can also undergo radiative recombination. When a singlet exciton successfully reaches a D-A interface, an ultrafast quasi-adiabatic electron transfer process with a time scale of less than 100 *fs* occurs. The electron transfer leads to the formation of a charge-transfer (CT) state in which the electron is in the LUMO of the acceptor and the hole in the HOMO of the donor at the D-A interface. The CT exciton can directly dissociate into free electrons and holes through a hot electron transfer process typically in the timescale of fewer than 50 *fs*. Also, the CT exciton can thermally relax to the ground state CT (timescale ~ *fs*-*ps*) and eventually, either charge separation or geminate recombination occurs.[35-36] As mentioned earlier, the CT excitons have a Coulombic barrier due to the low dielectric constant ($\varepsilon_r$ ~ 3-4) of organic semiconductors given by,

$$V = \frac{e^2}{4\pi\varepsilon_0\varepsilon_r r}$$

where $e$ is the charge of an electron, $\varepsilon_r$ is the dielectric constant, $\varepsilon_0$ is the electric permittivity in vacuum, and $r$ is the electron-hole separation distance. This mutual interaction of CT excitons results in a part of excitons suffering geminate recombination in a time scale of ~ 100 *ps* – 100 *ns*.[37-38]

Recent breakthroughs in the design and development of novel non-fullerene molecular acceptors (NFAs), starting from the ITIC (Lin *et al.* 2015), ushered a new era in bulk-heterojunction photovoltaic research with power conversion efficiencies now approaching 20% for single-junction BHJ organic solar cells. These narrow band gap acceptors having strong complementary absorption (Visible and NIR region of the solar radiation spectrum) coupled with low energetic disorder and reduced recombination losses highlight the unprecedented improvements in the device performance. The added features of a thicker active layer and higher stability estimates of non-fullerene BHJs as compared to fullerene-based BHJs offer a prospect for scalability. Challenges including large non-radiative recombination losses (0.3–0.5 eV), synthetic complexity of the high-performing NFAs and rapid degradation of the device under standard operating conditions, still need to be addressed.

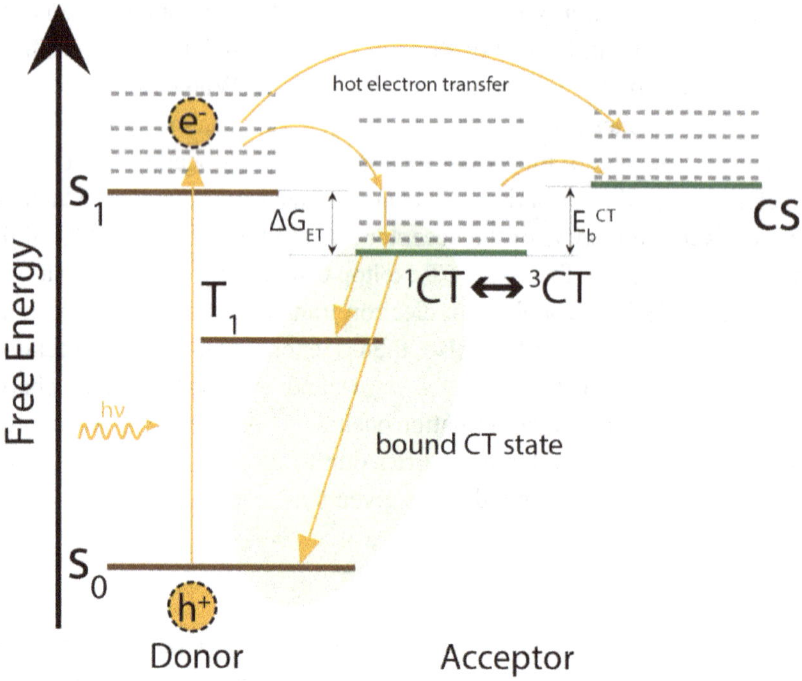


Figure 3. The schematic diagram depicts the possible ways for the formation of a charge-separated state (free charges) from the singlet exciton in the donor molecule (adapted from Shivanna and K.S. Narayan, 2018).

### 3.1. *Non-fullerene acceptors*

Fullerene derivatives such as $PC_{60}BM$ or $PC_{70}BM$ were the primary choice for the electron acceptors in OSCs till a decade ago.[39-40] Advancements in the device performance were achieved largely by the design and synthesis of polymer donors. The polymer: fullerene OSCs were able to achieve a maximum PCE of ≈ 11% during this period, and much of the device physics was understood using this model system.[41-42] The high electron mobility, the spherical structure of the molecule (3D character) and the LUMO level are suitable for the exciton dissociation with most of the donors, and making it favourable for balanced charge transport and optimal microstructure. Despite having these remarkable features intrinsically, there were many limitations to the fullerene derivative acceptors.[43] Importantly, these sets of acceptors have no significant absorption in the visible region of the solar spectrum. Apart from this, the chemical modification of the molecule is not straightforward, resulting in synthetic complexity and considerable limitation toward the bandgap tunability (tuning the LUMO level). The purification of fullerene derivatives, particularly $PC_{70}BM$, is time-consuming, resulting in a high cost of material, making it difficult to achieve commercial viability. Further, devices with fullerene derivatives as the acceptor show significantly low thermal stability and photostability.[44] Rapid aggregation of fullerene molecules in the active layer severely affects the charge transport and limits the device lifetime.[45-46] These issues defined the need for the design and development of acceptors with strong absorption in the visible region of the spectrum complementary to the donor and readily tunable electronic energy levels. The quest for alternatives to cross the efficiency barriers enabled the development of several sets of non-fullerene acceptors (NFAs).[47-48] This includes rylene diimides such as perylene diimides (PDI) and naphthalene diimides (NDI), fused ring electron acceptors (FREAs) and non-fused ring electron acceptors (NFREAs).[47,49-50] The FREAs such as ITIC and Y6 molecules brought in a paradigm shift with excellent improvements in PCE outperforming the fullerene-based devices and PCE over 19% has been reported to date.[51-56]

The major advantages of NFAs over fullerene derivatives are:

- Absorption in the visible and NIR region of the spectrum with a high absorption coefficient (~ $10^5$ $cm^{-1}$).[43,57]
- Synthetic flexibility to tune the electronic energy levels matching appropriately with the polymer donors.[58]
- Molecular design with strong intramolecular charge transfer effect.[59]

- Molecular design is suitable for fine-tuning crystallinity and miscibility.[60]
- High molecular weight is suitable for robust film formation and better thermal stability in the blend.[61]

The absorption spectra and electronic energy levels can be tuned readily by changing the push-pull strength in the NFA molecules. The ease of bandgap tunability in NFAs resulted in the synthesis of a large pool of acceptor molecules with significant enhancement in the PCE.[62] Along with some of the newly developed donors like PBDB-T and D18 having complementary absorption and excellent charge transport properties, it appears that the efficiency of over 20% is within reach using NFAs.[63]

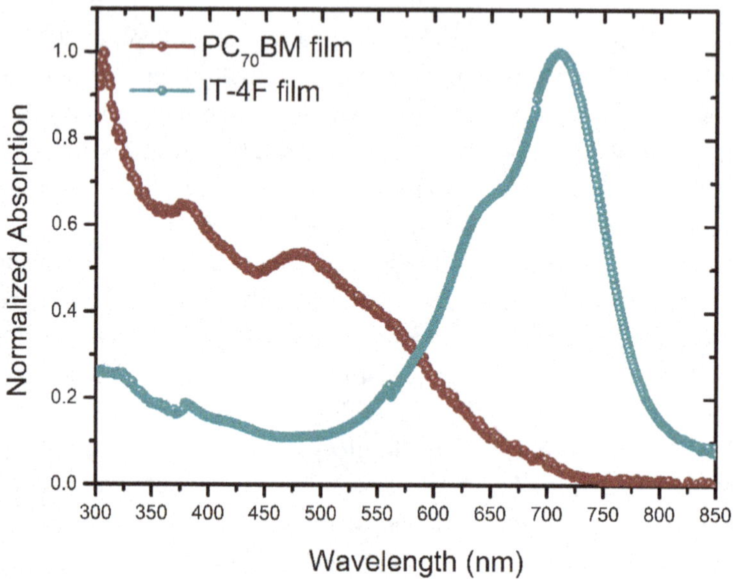


Figure 4. Typical normalized absorption spectra of fullerene derivative acceptor $PC_{70}BM$ and non-fullerene acceptor IT-4F thin films.

The strong absorption in the visible and NIR region (Figure 4) and high absorption coefficient enable superior photon harvesting in NFA-based OSCs. Further, an analysis of the absorption spectra of ITIC derivatives demonstrates excellent tunability and potential to blend with a large spectrum of donors and compatible acceptors in binary and ternary strategies, respectively.[64]

The initial stages of developing NFAs adopted basic design strategies using perylene diimide (PDI) based molecules such as that from JNCASR.[65-66] The crystallization of perylenes in BHJs is prevented by reducing the co-facial stacking without adversely impacting the electronic properties to form a twisted perylene molecule where the two PDI units are brought together using hydrazine as a linker. These materials indicated performance parameters comparable to that of PCBM and were viable options. Subsequently, as the molecular design rules expanded, different families of acceptors were reported that far exceeded the performance of fullerene systems (Figure 5e). Perylene diimide (PDI) small molecules, perylene diimide (PDI) polymers, and naphthalene diimide (NDI) polymers belong to the class of rylene diimides.[67] The history of rylene diimides can be traced back to the very first bilayer OSC in which a perylene diimide-based acceptor was used by Tang and co-workers in 1986.[68] The early developments in rylene diimides were limited by the PCE of the solar cell employing these acceptors, while fullerene derivatives were performing comparatively better in BHJ OSCs.[55] Despite having strong absorption in the visible region, one of the important reasons behind this efficiency lag for rylene diimides was rapid morphological evolution with strong aggregation behaviour and a tendency to form large crystals in the BHJ active layer. The 3D molecular structure of fullerene derivatives (Figure 5a) forms intercalating pathways for efficient charge transport with a large number of donors, while the 2D planar structure of rylene diimides with strong aggregation behaviour possessed a poor active layer morphology leading to weak charge separation and transport.[50,55,69] Fused ring electron acceptors such as IT4F and Y6 (Figures 5b and 5c) have multiple fused rings in the core of the molecule, and this central core has some of the key functionalities.[52,56] The presence of this donor unit in the core of the molecule primarily enables a strong intramolecular charge transfer effect. The intramolecular charge transfer effect can improve the absorption and extend to the NIR region. Further, the coplanar backbone of the molecule helps the delocalization of π electrons and enables strong intermolecular π-π interactions resulting in improved charge transport. In FREAs importantly, two types of molecular design are present: A-D-A and A-DA'D-A models.[70] The ITIC molecule and its derivatives (IT-4F, IT-M) belong to the A-D-A type, which constitutes a central

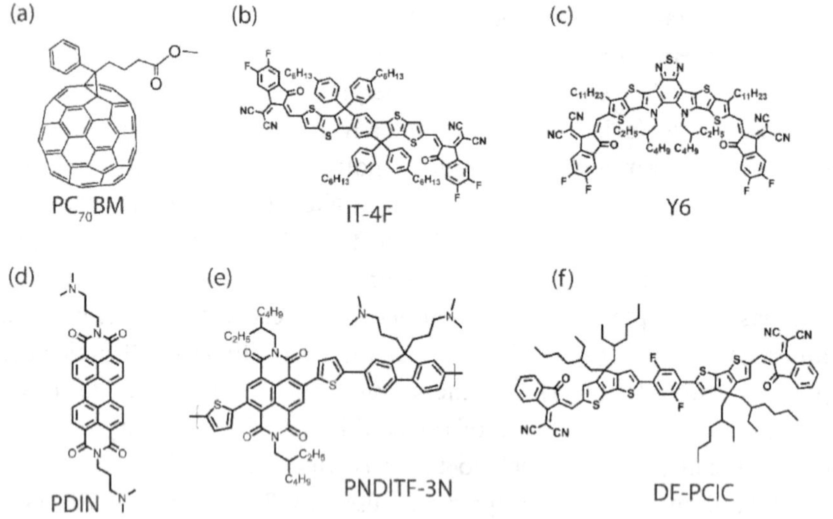


Figure 5. The chemical structure of (a) fullerene derivative acceptor molecule $PC_{70}BM$, (b) non-fullerene acceptor molecules belonging to the class of fused-ring electron acceptors IT-4F and (c) Y6, (d) non-fullerene acceptor molecules belonging to the class of rylene diimides PDIN and (e) PNDITF-3N and (f) non-fullerene acceptor molecules belonging to the class of non-fused ring electron acceptor DF-PCIC, respectively.

electron-donating core unit (D) and two electron-withdrawing acceptor units (A) as the end groups. The modification of the push-pull strength of these D and A units can readily tune the electronic energy level and the absorption spectra of the acceptor molecule. The fused core unit in this A-D-A type acceptor molecule is indacenodithiophene (IDT) class of derivatives, improving absorption to the NIR region. The electron-withdrawing end groups also play a crucial role in deciding the morphology and solubility of the acceptor and tuning the electronic properties. The molecular design of the Y6 acceptor belongs to an A-DA'D-A design in which an electron deficit benzothiadiaole (BT) unit is incorporated in the fused ring.[71] This modification can improve the D-A interaction and, consequently, the charge carrier mobility. The presence of a pyrrole bridging group with a strong electron-donating tendency helps to upshift the HOMO level, thereby reducing the bandgap. At the same time, the large alkyl chain attached to the N atom in the pyrrole group significantly reduces the aggregation behaviour due to steric hindrance.[59] This feature of narrow bandgap in the Y6 acceptor molecule enables a strong NIR absorption and enhanced $J_{SC}$ in the devices. These properties arising from the rational design resulted in all the highest PCE-reported OSCs based on FREAs as the acceptor component, surpassing both fullerene derivatives and rylene diimides. Figure 6 shows the J-V characteristics of a typical BHJ OSC based donor PM7 and fused ring electron acceptor Y6 with $J_{SC}$ exceeding 25 mA/cm$^2$.

The excellent improvement in PCE of OSCs employing FREAs exceeding 19% indicates commercialization capabilities in the near future. However, despite having high performance, the commercial viability of FREAs is limited due to high synthetic complexity and production cost. In order to overcome these challenges, concentrated efforts are in place to design and develop non-fused ring electron acceptors (NFREAs) with largely reduced synthetic complexity. The key design strategy in the development of NFREAs is replacing the fused ring core with a single bonded structure. Further, the electronic energy levels and molecular packing can also be

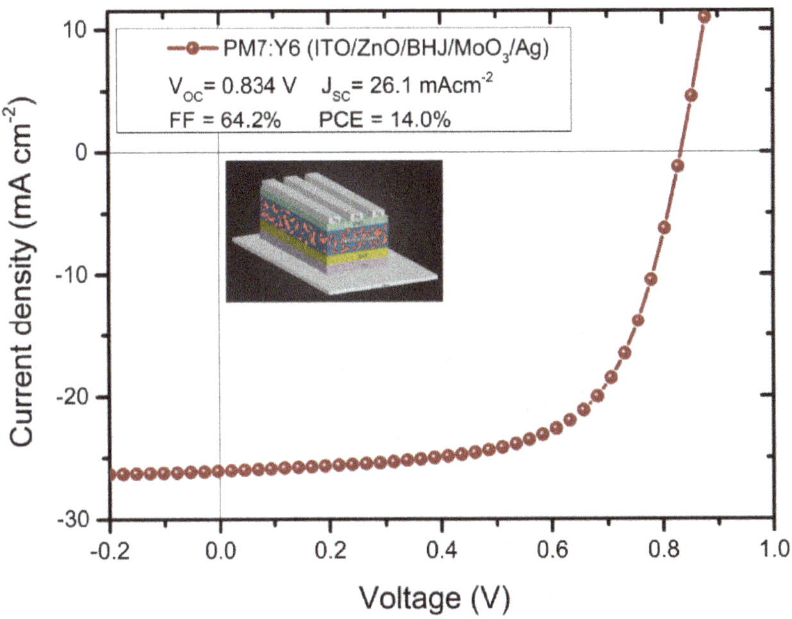


Figure 6. Typical current-voltage characteristic of a BHJ organic solar cell using donor PM7 and fused ring electron acceptor Y6 fabricated in our laboratory (inset shows device architecture).

readily modified with appropriate units. Recently, an impressive PCE of ~15.2% was able to achieve using NFREA (A14-16) based on a bithiophene-based building block.[72] This efficiency is almost comparable with that of high-performing FREAs with superior photostability, indicating the importance NFREAs in OSC technology development.[73] Figure 5f shows NFREA molecule DF-PCIC when blended with a suitable donor PBDB-T, exhibits PCE > 10%.[74] This type of NFREA molecule has an A-D-D'-D-A type structure. In DF-PCIC, the central non-fused ring core unit is connected to two cyclopentadithiophene units by single bonds, whereas later modifications with a strong electron-withdrawing end group to this molecule (acceptor HF-PCIC) are introduced to increase the intra-molecular charge transfer effect. The molecules exhibit similar absorption spectra to that of the ITIC molecule and performance parameters close to it with significantly reduced synthetic complexity.[73-75]

From the device point of view, the excellent performance of NFAs employed OSCs is characterized by the exciton dissociation with a small energy offset. Figure 7 shows the electronic energy levels of $PC_{70}BM$ molecule, NFAs ITIC and Y6, along with the most compatible donor PM6. The energy levels indicate that the LUMO offset is negligible when a blend of donor PM6 and acceptor ITIC is considered, and without significant energy offset, excellent exciton dissociation and PCE are achieved with OSCs employing ITIC and its derivatives. The low energetic disorder of NFAs can be attributed to the sharp absorption onset and reduced voltage loss. For a good OSC, the blend morphology is critical in which importantly, optimum domain size and purity are necessary. Apart from this, percolate and continuous charge transport pathways and a favourable molecular orientation are also important. The planar conformation and better molecular packing lead to a more stable morphology of NFA-based OSCs. On the other hand, the phase purity in the blend is better for fullerene derivatives due to their spherical geometry, but the phase purity can be improved in NFAs as well by improving the molecular crystallinity. Table 1 shows the J-V parameters of the highest efficiency reported systems to date using various types of NFAs along with the champion fullerene derivative $PC_{70}BM$ based system. The high short-circuit current density and PCE are obtained when FREAs are blended with suitable donors.[50]

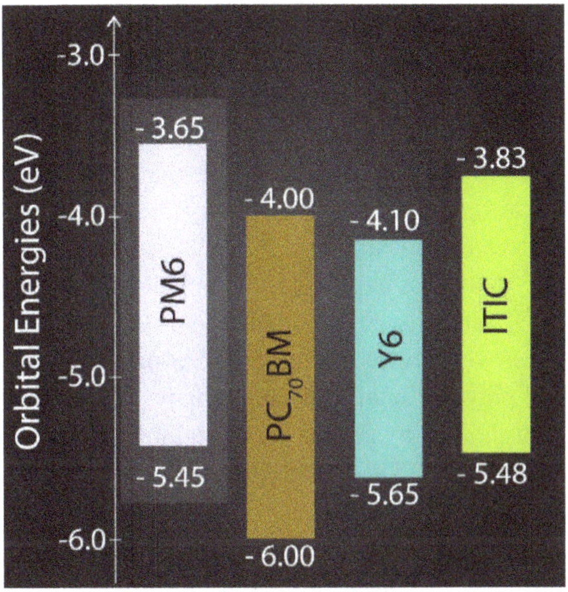


Figure 7. Orbital energy diagram of donor PM6 and acceptors $PC_{70}BM$, Y6 and IT-4F.

Table 1. The performance parameters of highest efficiency reported NFAs belonging to various types of molecular designs along with the champion fullerene derivative-based OSC.

| Donor(s) | Acceptor | $V_{OC}$ | $J_{SC}$ | FF | PCE | Reference |
|---|---|---|---|---|---|---|
| DR3TSBDT:PTB7-Th | $PC_{70}BM$ | 0.772 | 23.31 | 70.44 | 12.10 | 41 |
| PDBT-T1 | 6a[(a)] | 1.001 | 12.53 | 71.7 | 8.98 | 76 |
| PBDB-T-2Cl | IT-4F[(b)] | 0.890 | 20.58 | 78.0 | 14.29 | 54 |
| PBDB-T-2F | Y6[(b)] | 0.856 | 25.73 | 76.8 | 16.94 | 77 |
| PBDB-T-2F | HF-PCIC[(c)] | 0.910 | 17.81 | 70.77 | 11.49 | 75 |
| PBDB-T-2F | A14-16[(c)] | 0.876 | 21.8 | 79.8 | 15.2 | 72 |

(a) Rylene diimide, (b) fused-ring electron acceptor and (c) non-fused ring electron acceptor.

### 3.2. *Ternary blend organic solar cells*

Ternary strategy in which a blend of three photoactive materials constituting the active layer of OSC has been shown to improve the device performance.[78-80] Ternary and tandem strategies are generally adopted to improve binary OSCs device performance further.[81-82] Due to significantly reduced fabrication complexity, single-junction BHJ OSCs are preferred over tandem device architecture. In this regard, ternary blend organic solar cells (TBOSCs) are ideal due to improved photon harvesting and facile device fabrication processes.[83] TBOSCs have several advantages over binary devices, summarized in Figure 8. The highest PCE reported single-junction OSCs to date are based on ternary blend strategy whether its fullerene derivative or non-fullerene molecule as the acceptor component (see Table 2).[41,84] TBOSCs based on a combination of polymers and small molecules along with all-polymer and all-small molecule blends have significantly improved device performance. A small molecule as the third component in a polymer/small molecule binary BHJ has been shown to optimize the morphology by improving crystallinity along with improving the overall absorption.[78,85] The polymer-polymer-small molecule TBOSCs

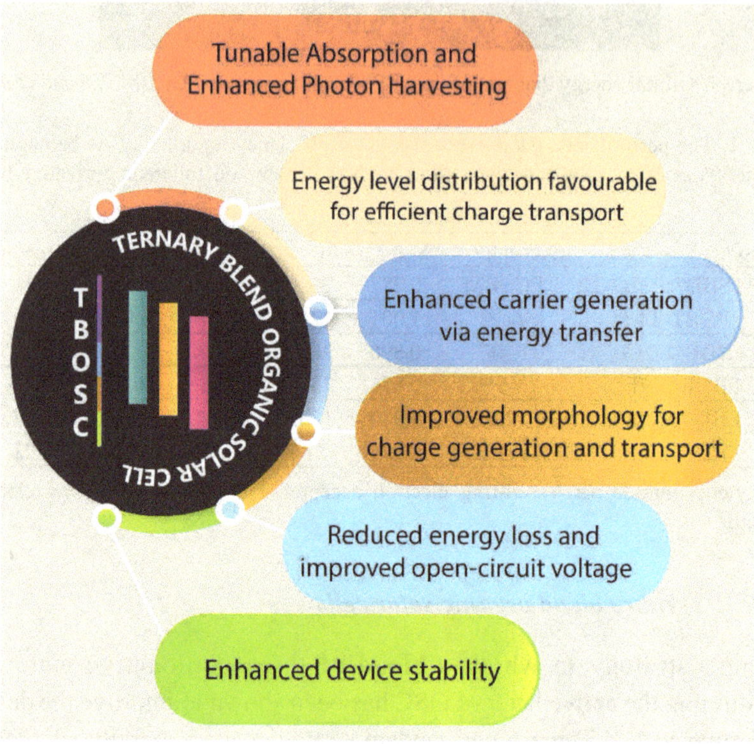


Figure 8. Infographic depicting the advantages of ternary blend organic solar cells over binary organic solar cells.

Table 2. The performance parameters of highest efficiency reported ternary blend organic solar cells and the corresponding binary devices in which the third component is introduced.

| Binary Components | Third Component | Voc (V) | Jsc (mA cm$^{-2}$) | FF (%) | PCE (%) | Reference |
|---|---|---|---|---|---|---|
| PTB7-Th:Y6 | | 0.67 | 20.16 | 56.77 | 7.71 | 92 |
| PTB7-Th:Y6 | $PC_{70}BM$ | 0.67 | 24.68 | 58.02 | 9.55 | 92 |
| PM6:Br-ITIC | | 0.937 | 18.53 | 72.7 | 12.63 | 93 |
| PM6:Br-ITIC | J71 | 0.930 | 19.39 | 78.4 | 14.13 | 93 |
| PM6:IT-4F | | 0.869 | 20.26 | 77.4 | 13.62 | 94 |
| PM6:IT-4F | N7IT | 0.890 | 22.54 | 74.8 | 15.02 | 94 |
| PM7:IT-4F | | 0.890 | 20.58 | 78.0 | 14.29 | 54 |
| PM7:IT-4F | $PC_{60}BM$ | 0.875 | 20.7 | 77.2 | 14.0 | 95 |
| PM7:IT-4F | $PC_{70}BM$ | 0.872 | 21.2 | 77.3 | 14.3 | 95 |
| PM6:Y6 | | 0.820 | 25.20 | 76.1 | 15.7 | 52 |
| PM6:Y6 | $PC_{70}BM$ | 0.861 | 25.10 | 77.2 | 16.7 | 95 |
| PM6:Y6 | D18 | 0.852 | 25.55 | 72.8 | 15.85 | 96 |
| PM6:Y6 | ITIC | 0.879 | 25.50 | 74.20 | 16.6 | 97 |
| PM6:Y6 | BTP-M | 0.875 | 26.56 | 73.46 | 17.03 | 98 |
| PM6:Y6 | IT-4F | 0.844 | 25.40 | 75.9 | 16.27 | 99 |
| PM6:Y6 | BTF | 0.853 | 26.11 | 74.22 | 16.53 | 100 |
| PM6:Y6 | J71 | 0.850 | 25.55 | 76.0 | 16.50 | 101 |
| PM6:Y6 | S3 | 0.856 | 25.86 | 79.17 | 17.53 | 102 |
| PM6:Y6 | AQx-3 | 0.870 | 26.82 | 77.2 | 18.01 | 77 |
| PM6:BTP-4F-12 | | 0.850 | 24.50 | 77.9 | 16.2 | 103 |
| PM6:BTP-4F-12 | MeIC | 0.863 | 25.40 | 79.2 | 17.4 | 103 |
| PM6:BTP-eC9 | | 0.845 | 26.81 | 77.5 | 17.58 | 104 |
| PM6:BTP-eC9 | BTP-F | 0.858 | 26.99 | 79.7 | 18.45 | 104 |
| PM6:N3 | | 0.837 | 25.81 | 73.9 | 15.98 | 105 |
| PM6:N3 | $PC_{70}BM$ | 0.840 | 26.85 | 78.0 | 17.60 | 106 |
| PM6:C9 | | 0.852 | 26.58 | 76.73 | 17.38 | 107 |
| PM6:C9 | PM6-Si30 | 0.870 | 26.90 | 78.04 | 18.27 | 107 |
| PM7:Y6 | | 0.868 | 24.98 | 71.42 | 15.49 | 108 |
| PM7:Y6 | $PC_{70}BM$ | 0.868 | 25.44 | 75.66 | 16.71 | 108 |
| D18-Cl:Y6 | | 0.881 | 25.53 | 75.88 | 17.07 | 109 |
| D18-Cl:Y6 | Y6-1O | 0.900 | 25.87 | 76.92 | 17.91 | 109 |
| D18-Cl:L8-BO | | 0.888 | 25.10 | 77.80 | 17.37 | 104 |
| D18-Cl: L8-BO | L8-BO-F | 0.902 | 25.30 | 78.30 | 17.83 | 104 |
| D18-Cl:N3 | | 0.859 | 27.85 | 75.70 | 18.13 | 110 |
| D18-Cl:N3 | $PC_{60}BM$ | 0.849 | 28.22 | 78.0 | 18.69 | 110 |

| Binary Components | Third Component | Voc (V) | Jsc (mA cm$^{-2}$) | FF (%) | PCE (%) | Reference |
|---|---|---|---|---|---|---|
| PBQx-TF:eC9-2Cl | | 0.868 | 25.9 | 78.6 | 17.7 | 111 |
| PBQx-TF:eC9-2Cl | F-BTA3 | 0.879 | 26.7 | 80.9 | 19.0 | 111 |
| D18:L8-BO | | 0.907 | 24.9 | 78.5 | 17.7 | 82 |
| D18:L8-BO | PM6 | 0.896 | 26.7 | 81.9 | 19.6 | 82 |

Table 2 *continued.*

are set forth to improve the absorption of the blend, as the polymer donors usually exhibit a narrow absorption window.[86] The third component, usually, a polymer donor, in this case, assists the charge transport and dissociation, along with broadening the absorption window. As a natural extension to the all-polymer binary BHJ OSCs, all-polymer TBOSCs are also introduced, which show superior device performance.[83] All small-molecule TBOSCs are interesting due to the superior reproducibility of high-performing devices owing to the characteristics of small molecules, such as well-defined structure, molecular weight and high purity.[87-88]
Rational choice of the third component in an optimal amount can enhance the photon harvesting and, subsequently, the device performance metrics of the binary device. Morphological modification brought in by the third component has shown to improve the device stability in specific combinations of donor-acceptor (D-A) systems (Table 2). Though ternary strategy is a well-established technique in BHJ OSCs, with the development of NFAs, this technique has consistently improved the device efficiency of the binary system on many champion D-A combinations.[89] Apart from the enhanced absorption window broadening and morphological modifications of the active layer, different device-engineering strategies for these high-performing ternary devices can also push the PCE further. In this regard, a comprehensive understanding of the role of the third component in charge transport physics is necessary. The presence of the third component (either a donor or an acceptor) in a D-A system can impact the device physics of the OSC in different ways depending on the electronic energy levels and various other molecular properties.[79-80]
Multiple charge transport models have been proposed depending on the alignment of electronic energy levels of the third component in the binary system. These are charge transfer, energy transfer, parallel-like and alloy

models.[90] The charge transfer mechanism refers to an ideal step-like alignment of energy levels, the overlap between the emission of one component and absorption of another can lead to energy transfer. Efficient charge transfer between the components equivalent to two D-A systems refers to the parallel-like model. In the case of the alloy model, either a donor alloy (or an acceptor alloy) establishes charge transport with the acceptor (or donor), similar to a single D-A system. On the other hand, optimization of active layer morphology is a deciding factor in enhancing the efficiency and stability aspects of the ternary device.[90] Though using this simplified charge transport model based on material energetics and morphology of the blend can explain the working mechanism of ternary devices, the validity of some of these models is still debatable.[91]

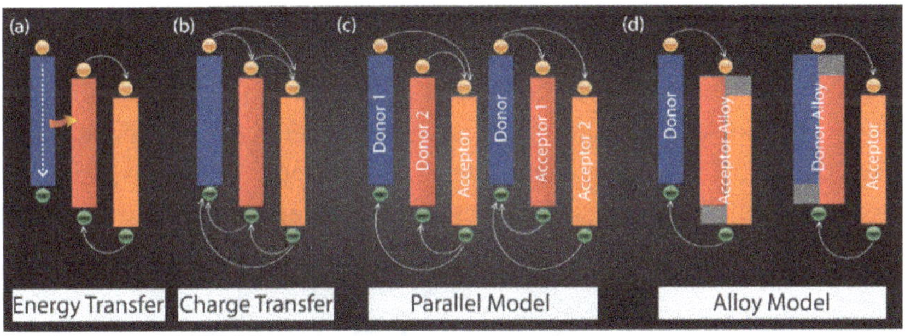


Figure 9. Graphical representation depicting the various charge transport models in a ternary blend organic solar cell using orbital energy levels of constituent elements.[90]

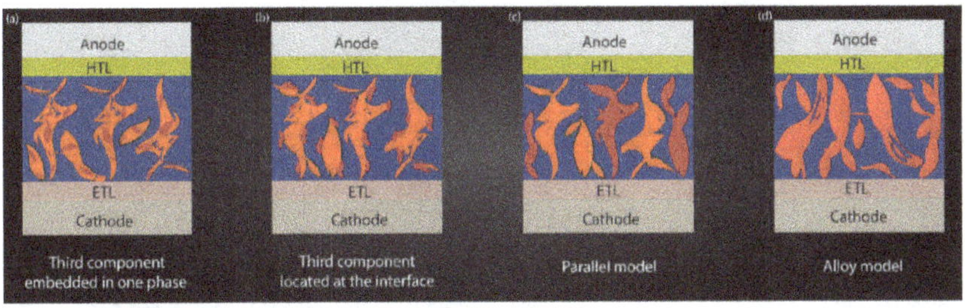


Figure 10. Graphical representation depicting the various charge transport models in a ternary blend organic solar cell using the morphology of the bulk heterojunction active layer in inverted device architecture.[90]

Efforts to correlate the distribution of the third component in a bi-continuous D-A phase to the device performance and morphological stability still need a neat framework. The thermal properties and miscibility of various constituting components are crucial in extracting the best performance of the device. Further, the weight ratio of individual components in the BHJ and its correlation to the device parameters, specifically the $V_{OC}$, needs to be understood.[112] The presence of multiple chromophores in the BHJ blend provides the advantage of the possibility of energy transfer between the constituent components in the blend. The fluorescence or Förster resonance energy transfer (FRET) mechanism is an effective strategy to enhance the performance of BHJ OSCs.[61] FRET is a process in which a donor transfers energy to an acceptor non-radiatively through dipole-dipole interaction. The necessary condition for FRET to occur in a system is that the emission spectrum of the FRET donor must overlap with the absorption spectra of one of the constituent elements in the system (Figure 11). Apart from this, for an efficient FRET, the donor and acceptor elements must be within a specific length scale called the Förster radius ($R_0$). The value of $R_0$ is system dependent and can vary from nearly 1 nm to 10 nm. The fluorescence lifetime and quantum yield of the donor and the quenching efficiency of the acceptor also influence the FRET occurrence in the blend.[61,112-113]

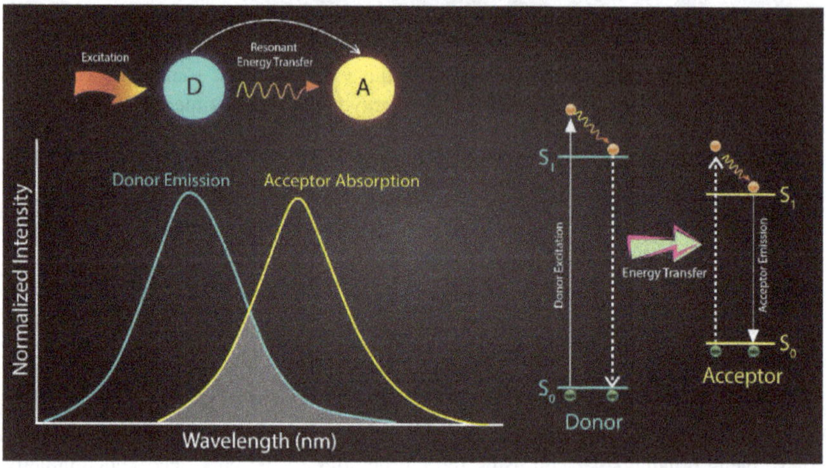


Figure 11. Schematic diagram depicting the FRET mechanism in a ternary blend organic solar cell.

In the FRET mechanism, the photo-excited donor ($D^*$) relaxes to the ground state under suitable conditions by exciting the acceptor ($A$) by the reaction,

$$D^* + A \rightarrow D + A^*$$

The rate constant of FRET is given by the expression,

$$\mathrm{K_{FRET}} = \frac{1}{\tau_{\mathrm{D}}}\left(\frac{\mathrm{R_0}}{\mathrm{R}}\right)^6$$

Where $\tau_D$ is the fluorescence lifetime of the FRET donor in pristine form, and R is the distance between the donor and acceptor.
The FRET efficiency can be calculated as,

$$E_{FRET} = \frac{1}{1 + \left({}^{R}/_{R_0}\right)^6} = 1 - \frac{\tau_{DA}}{\tau_D}$$

here $\tau_{DA}$ is the lifetime of the donor emission in presence of the acceptor. The steady-state photoluminescence (PL), time-resolved photo-luminescence (TRPL) and transient absorption spectroscopy (TAS) can be used as experimental tools to probe the FRET mechanism in a blend. The presence of FRET in the BHJ system can benefit in several ways, including enhanced exciton diffusion length. For instance, McGehee *et al.* showed that in a strongly Coulomb-coupled D-A system, exciton harvesting over a distance of more than 25 nm is possible with an order of magnitude increase in the effective exciton diffusion length.[114] Thus, the FRET mechanism in a BHJ system can improve the $J_{SC}$, whereas the effect of FRET on $V_{OC}$ and FF still needs to be uncovered in detail.

### 3.3. *Stability aspects of NFA-based organic solar cells*

The stability of the OSCs is highly dependent on the material properties of the acceptor molecule employed. But in general, the stability of NFA based devices is superior to their fullerene-based counterparts.[64,115-116] Recently, Forrest *et al.* reported an extrapolated device lifetime of ~ 30 years after accelerated aging studies in NFA BT-CIC based devices.[117] Earlier, in a similar analysis along with long-term J-V monitoring, Brabec and co-workers estimated ~ 10 years of a lifetime for IT-4F based inverted

OSCs.[118] The essential takeaway from these results is the potential of high performing NFA based OSCs' ability to perform intrinsically, similar to the inorganic counterpart. As promising efficiency is obtained using recently developed NFAs, understanding stability implications in these systems is extremely important. The thermal stability of NFA based OSCs is superior compared to the fullerene counterpart, partly due to the anisotropic molecular structure with a rigid backbone, lending stability to the blend.[64] The thermal transition temperature of the NFA molecule plays a critical role in morphological stability, and in general, a high thermal transition temperature is desired (> 100°C).[119-120] High thermal transition temperature can slow down the diffusion and crystallization of the BHJ components providing stable morphology over a long period.[119] Apart from this, the parameters such as miscibility, crystallinity, domain size and purity, and morphological evolution during the operation, such as mixing and de-mixing of the constituents, are also critical.[119] It has been shown that optimal miscibility, domain size, and domain purity, along with balanced nucleation, aggregation and crystallization, can be achieved in OSCs by fine tuning NFAs.[64]

The availability of a vast amount of D-A combinations combined with system-specific degradation mechanisms makes it challenging to propose a general set of frameworks for long term stability applicable to all OSCs.[119,121] A general framework to identify and eliminate rapidly degrading combinations may help to isolate and concentrate on promising combinations, and this requires a detailed understanding of degradation and its relation with material properties. The significant 'burn-in' degradation in most of the OSCs is another major challenge that needs to be addressed.[122] A set of design rules for 'burn-in'-free OSCs requires detailed investigations to understand the degradation mechanism induced by various contributing factors. The intrinsic chemical stability of NFAs needs to be improved by considering the effects of interaction with interlayers and photo-thermal stress.[116] The fragile C=C in NFAs tends to weaken in a short period of operation due to the interaction with interlayers in the presence of external stress factors such as light, heat, moisture and oxygen.[122-124] The morphological instabilities of NFA-based devices are largely dependent on the compatibility with the choice of donor molecule, and it is important to screen the potential of a particular NFA with the material properties. This requires an in-depth understanding of the morphology formation and evolution in BHJ OSCs. The planar conformations of NFAs combined with the insights on molecular packing and interaction from single crystals of NFAs can shed more light on desired film formation. The real operation condition of OSCs further involves thermal and light cycles, and the impact of this phenomenon is particularly important.[125] Effective encapsulation techniques with facile processing requirements need to be developed to minimize the effect of extrinsic factors.[20,126] Along with it, optimizing the device by simultaneously considering both efficiency and long-term stability is necessary for high-performing OSCs.[122,127]

## 4. Summary

It is foreseen that these new generations of solar cells will find their utility in niche applications. The remarkable improvements in the performance of NFA-based single junction and tandem organic solar cells present a promising way forward in solar energy harvesting in various applications involving flexible, semi-transparent and wearable devices. Further stability improvements could lead to viable indoor applications. Material design and device engineering strategies to reduce the high $V_{OC}$ loss by suppressing the non-radiative recombination loss are needed to improve the commercial viability. A combination of OSCs and hybrid perovskites in tandem structures also offers attractive options to get to high efficiencies.

## References

1. Kojima, A.; Teshima, K.; Shirai, Y.; Miyasaka, T., Organometal Halide Perovskites as Visible-Light Sensitizers for Photovoltaic Cells. *Journal of the American Chemical Society* **2009**, *131*, 6050-6051.
2. Gao, P.; Grätzel, M.; Nazeeruddin, M. K., Organohalide Lead Perovskites for Photovoltaic Applications. *Energy & Environmental Science* **2014**, *7*, 2448-2463.
3. NREL Golden, C., Best Research-Cell Effiency Chart | Photovoltaic Research | Nrel. https://www.nrel.gov/pv/cell-efficiency.html **accessed 2021**.
4. Liu, C.; Yang, Y.; Rakstys, K.; Mahata, A.; Franckevicius, M.; Mosconi, E.; Skackauskaite, R.; Ding, B.; Brooks, K. G.; Usiobo, O. J., Tuning Structural Isomers

of Phenylenediammonium to Afford Efficient and Stable Perovskite Solar Cells and Modules. *Nature Communications* **2021**, *12*, 1-9 2041-1723.

5. Yoo, J. J.; Seo, G.; Chua, M. R.; Park, T. G.; Lu, Y.; Rotermund, F.; Kim, Y.-K.; Moon, C. S.; Jeon, N. J.; Correa-Baena, J.-P., Efficient Perovskite Solar Cells Via Improved Carrier Management. *Nature* **2021**, *590*, 587-593 1476-4687.
6. Ansari, M. I. H.; Qurashi, A.; Nazeeruddin, M. K., Frontiers, Opportunities, and Challenges in Perovskite Solar Cells: A Critical Review. *Journal of Photochemistry and Photobiology C: Photochemistry Reviews* **2018**, *35*, 1-24 1389-5567.
7. Zhang, J.; Zhang, W.; Cheng, H.-M.; Silva, S. R. P., Critical Review of Recent Progress of Flexible Perovskite Solar Cells. *Materials Today* **2020**, *39*, 66-88 1369-7021.
8. Jiang, Q.; Zhao, Y.; Zhang, X.; Yang, X.; Chen, Y.; Chu, Z.; Ye, Q.; Li, X.; Yin, Z.; You, J., Surface Passivation of Perovskite Film for Efficient Solar Cells. *Nature Photonics* **2019**, *13*, 460-466.
9. Yang, W. S.; Park, B.-W.; Jung, E. H.; Jeon, N. J.; Kim, Y. C.; Lee, D. U.; Shin, S. S.; Seo, J.; Kim, E. K.; Noh, J. H., Iodide Management in Formamidinium-Lead-Halide–Based Perovskite Layers for Efficient Solar Cells. *Science* **2017**, *356*, 1376-1379.
10. Zhao, J.; Wang, A.; Green, M. A.; Ferrazza, F., 19.8% Efficient "Honeycomb" Textured Multicrystalline and 24.4% Monocrystalline Silicon Solar Cells. *Applied Physics Letters* **1998**, *73*, 1991-1993.
11. Huang, J.; Yuan, Y.; Shao, Y.; Yan, Y., Understanding the Physical Properties of Hybrid Perovskites for Photovoltaic Applications. *Nature Reviews Materials* **2017**, *2*, 1-19.
12. Hao, F.; Stoumpos, C. C.; Cao, D. H.; Chang, R. P. H.; Kanatzidis, M. G., Lead-Free Solid-State Organic–Inorganic Halide Perovskite Solar Cells. *Nature Photonics* **2014**, *8*, 489-494 1749-4893.
13. Pang, S.; Hu, H.; Zhang, J.; Lv, S.; Yu, Y.; Wei, F.; Qin, T.; Xu, H.; Liu, Z.; Cui, G., Nh2ch Nh2pbi3: An Alternative Organolead Iodide Perovskite Sensitizer for Mesoscopic Solar Cells. *Chemistry of Materials* **2014**, *26*, 1485-1491 0897-4756.
14. Brandt, R. E.; Stevanović, V.; Ginley, D. S.; Buonassisi, T., Identifying Defect-Tolerant Semiconductors with High Minority-Carrier Lifetimes: Beyond Hybrid Lead Halide Perovskites. *Mass Communications* **2015**, *5*, 265-275 2159-6859.
15. De Wolf, S.; Holovsky, J.; Moon, S.-J.; Loper, P.; Niesen, B.; Ledinsky, M.; Haug, F.-J.; Yum, J.-H.; Ballif, C., Organometallic Halide Perovskites: Sharp Optical Absorption Edge and Its Relation to Photovoltaic Performance. *Journal of Physical Chemistry Letters* **2014**, *5*, 1035-1039 1948-7185.
16. Braly, I. L.; DeQuilettes, D. W.; Pazos-Outón, L. M.; Burke, S.; Ziffer, M. E.; Ginger, D. S.; Hillhouse, H. W., Hybrid Perovskite Films Approaching the Radiative Limit with over 90% Photoluminescence Quantum Efficiency. *Nature Photonics* **2018**, *12*, 355-361 1749-4893.
17. Al-Ashouri, A.; Köhnen, E.; Li, B.; Magomedov, A.; Hempel, H.; Caprioglio, P.; Márquez, J. A.; Morales Vilches, A. B.; Kasparavicius, E.; Smith, J. A., Monolithic

Perovskite/Silicon Tandem Solar Cell with > 29% Efficiency by Enhanced Hole Extraction. *Science* **2020**, *370*, 1300-1309 0036-8075.

18. Aristidou, N.; Sanchez-Molina, I.; Chotchuangchutchaval, T.; Brown, M.; Martinez, L.; Rath, T.; Haque, S. A., The Role of Oxygen in the Degradation of Methylammonium Lead Trihalide Perovskite Photoactive Layers. *Angewandte Chemie* **2015**, *127*, 8326-8330 0044-8249.
19. Raman, R. K.; Thangavelu, S. A. G.; Venkataraj, S.; Krishnamoorthy, A., Materials, Methods and Strategies for Encapsulation of Perovskite Solar Cells: From Past to Present. *Renewable and Sustainable Energy Reviews* **2021**, *151*, 111608 1364-0321.
20. Uddin, A.; Upama, M. B.; Yi, H.; Duan, L., Encapsulation of Organic and Perovskite Solar Cells: A Review. *Coatings* **2019**, *9*, 65.
21. Bella, F.; Griffini, G.; Correa-Baena, J.-P.; Saracco, G.; Grätzel, M.; Hagfeldt, A.; Turri, S.; Gerbaldi, C., Improving Efficiency and Stability of Perovskite Solar Cells with Photocurable Fluoropolymers. *Science* **2016**, *354*, 203-206 0036-8075.
22. Wang, F.; Jiang, X.; Chen, H.; Shang, Y.; Liu, H.; Wei, J.; Zhou, W.; He, H.; Liu, W.; Ning, Z., 2d-Quasi-2d-3d Hierarchy Structure for Tin Perovskite Solar Cells with Enhanced Efficiency and Stability. *Joule* **2018**, *2*, 2732-2743 2542-4351.
23. Spanopoulos, I.; Hadar, I.; Ke, W.; Tu, Q.; Chen, M.; Tsai, H.; He, Y.; Shekhawat, G.; Dravid, V. P.; Wasielewski, M. R., Uniaxial Expansion of the 2d Ruddlesden–Popper Perovskite Family for Improved Environmental Stability. *Journal of the American Chemical Society* **2019**, *141*, 5518-5534 0002-7863.
24. Lin, Y.; Bai, Y.; Fang, Y.; Chen, Z.; Yang, S.; Zheng, X.; Tang, S.; Liu, Y.; Zhao, J.; Huang, J., Enhanced Thermal Stability in Perovskite Solar Cells by Assembling 2d/3d Stacking Structures. *Journal of Physical Chemistry Letters* **2018**, *9*, 654-658 1948-7185.
25. Pereyra, C.; Xie, H.; Lira-Cantu, M., Additive Engineering for Stable Halide Perovskite Solar Cells. *Journal of Energy Chemistry* **2021**, *60*, 599-634 2095-4956.
26. Bai, S.; Da, P.; Li, C.; Wang, Z.; Yuan, Z.; Fu, F.; Kawecki, M.; Liu, X.; Sakai, N.; Wang, J. T.-W., Planar Perovskite Solar Cells with Long-Term Stability Using Ionic Liquid Additives. *Nature* **2019**, *571*, 245-250 1476-4687.
27. Saliba, M.; Matsui, T.; Domanski, K.; Seo, J.-Y.; Ummadisingu, A.; Zakeeruddin, S. M.; Correa-Baena, J.-P.; Tress, W. R.; Abate, A.; Hagfeldt, A., Incorporation of Rubidium Cations into Perovskite Solar Cells Improves Photovoltaic Performance. *Science* **2016**, *354*, 206-209 0036-8075.
28. Noh, J. H.; Im, S. H.; Heo, J. H.; Mandal, T. N.; Seok, S. I., Chemical Management for Colorful, Efficient, and Stable Inorganic–Organic Hybrid Nanostructured Solar Cells. *Nano Letters* **2013**, *13*, 1764-1769 1530-6984.
29. Zarick, H. F.; Soetan, N.; Erwin, W. R.; Bardhan, R., Mixed Halide Hybrid Perovskites: A Paradigm Shift in Photovoltaics. *Journal of Materials Chemistry A* **2018**, *6*, 5507-5537.

30. Yu, G.; Gao, J.; Hummelen, J. C.; Wudl, F.; Heeger, A. J., Polymer Photovoltaic Cells: Enhanced Efficiencies Via a Network of Internal Donor-Acceptor Heterojunctions. *Science* **1995**, *270*, 1789-1791 0036-8075.
31. Zhao, F.; Wang, C.; Zhan, X., Morphology Control in Organic Solar Cells. *Advanced Energy Materials* **2018**, *8*, 1703147.
32. Maturová, K.; van Bavel, S. S.; Wienk, M. M.; Janssen, R. A. J.; Kemerink, M., Morphological Device Model for Organic Bulk Heterojunction Solar Cells. *Nano Letters* **2009**, *9*, 3032-3037 1530-6984.
33. Ameri, T.; Dennler, G.; Lungenschmied, C.; Brabec, C. J., Organic Tandem Solar Cells: A Review. *Energy & Environmental Science* **2009**, *2*, 347-363.
34. Grancini, G.; Maiuri, M.; Fazzi, D.; Petrozza, A.; Egelhaaf, H.-J.; Brida, D.; Cerullo, G.; Lanzani, G., Hot Exciton Dissociation in Polymer Solar Cells. *Nature Materials* **2013**, *12*, 29-33 1476-4660.
35. Shivanna, R.; Rajaram, S.; Narayan, K. S., Role of Charge-Transfer State in Perylene-Based Organic Solar Cells. *Chemistry Select* **2018**, *3*, 9204-9210 2365-6549.
36. Vithanage, D. A.; Devižis, A.; Abramavičius, V.; Infahsaeng, Y.; Abramavičius, D.; MacKenzie, R. C. I.; Keivanidis, P. E.; Yartsev, A.; Hertel, D.; Nelson, J., Visualizing Charge Separation in Bulk Heterojunction Organic Solar Cells. *Nature Communications* **2013**, *4*, 1-6 2041-1723.
37. Etzold, F.; Howard, I. A.; Mauer, R.; Meister, M.; Kim, T.-D.; Lee, K.-S.; Baek, N. S.; Laquai, F., Ultrafast Exciton Dissociation Followed by Nongeminate Charge Recombination in Pcdtbt: Pcbm Photovoltaic Blends. *Journal of the American Chemical Society* **2011**, *133*, 9469-9479 0002-7863.
38. Menke, S. M.; Ran, N. A.; Bazan, G. C.; Friend, R. H., Understanding Energy Loss in Organic Solar Cells: Toward a New Efficiency Regime. *Joule* **2018**, *2*, 25-35 2542-4351.
39. He, Y.; Li, Y., Fullerene Derivative Acceptors for High Performance Polymer Solar Cells. *Physical Chemistry Chemical Physics* **2011**, *13*, 1970-1983.
40. Li, C.-Z.; Yip, H.-L.; Jen, A. K. Y., Functional Fullerenes for Organic Photovoltaics. *Journal of Materials Chemistry* **2012**, *22*, 4161-4177.
41. Kumari, T.; Lee, S. M.; Kang, S.-H.; Chen, S.; Yang, C., Ternary Solar Cells with a Mixed Face-on and Edge-on Orientation Enable an Unprecedented Efficiency of 12.1%. *Energy & Environmental Science* **2017**, *10*, 258-265.
42. Huang, J.; Carpenter, J. H.; Li, C. Z.; Yu, J. S.; Ade, H.; Jen, A. K. Y., Highly Efficient Organic Solar Cells with Improved Vertical Donor–Acceptor Compositional Gradient Via an Inverted Off-Center Spinning Method. *Advanced Materials* **2016**, *28*, 967-974 0935-9648.
43. Li, S.; Liu, W.; Li, C. Z.; Shi, M.; Chen, H., Efficient Organic Solar Cells with Non-Fullerene Acceptors. *Small* **2017**, *13*, 1701120 1613-6810.

44. Zhao, W.; Qian, D.; Zhang, S.; Li, S.; Inganäs, O.; Gao, F.; Hou, J., Fullerene-Free Polymer Solar Cells with over 11% Efficiency and Excellent Thermal Stability. *Advanced Materials* **2016**, *28*, 4734-4739 0935-9648.
45. Pan, Q.-Q.; Li, S.-B.; Duan, Y.-C.; Wu, Y.; Zhang, J.; Geng, Y.; Zhao, L.; Su, Z.-M., Exploring What Prompts Itic to Become a Superior Acceptor in Organic Solar Cell by Combining Molecular Dynamics Simulation with Quantum Chemistry Calculation. *Physical Chemistry Chemical Physics* **2017**, *19*, 31227-31235.
46. Piersimoni, F.; Degutis, G.; Bertho, S.; Vandewal, K.; Spoltore, D.; Vangerven, T.; Drijkoningen, J.; Van Bael, M. K.; Hardy, A.; D'Haen, J., Influence of Fullerene Photodimerization on the Pcbm Crystallization in Polymer: Fullerene Bulk Heterojunctions under Thermal Stress. *Journal of Polymer Science Part B: Polymer Physics* **2013**, *51*, 1209-1214 0887-6266.
47. Li, H.; Wang, J.; Wang, Y.; Bu, F.; Shen, W.; Liu, J.; Huang, L.; Wang, W.; Belfiore, L. A.; Tang, J., The Progress of Non-Fullerene Small Molecular Acceptors for High Efficiency Polymer Solar Cells. *Solar Energy Materials and Solar Cells* **2019**, *190*, 83-97 0927-0248.
48. Cheng, P.; Li, G.; Zhan, X.; Yang, Y., Next-Generation Organic Photovoltaics Based on Non-Fullerene Acceptors. *Nature Photonics* **2018**, *12*, 131-142 1749-4893.
49. Yan, C.; Barlow, S.; Wang, Z.; Yan, H.; Jen, A. K. Y.; Marder, S. R.; Zhan, X., Non-Fullerene Acceptors for Organic Solar Cells. *Nature Reviews Materials* **2018**, *3*, 1-19 2058-8437.
50. Zhang, G.; Zhao, J.; Chow, P. C. Y.; Jiang, K.; Zhang, J.; Zhu, Z.; Zhang, J.; Huang, F.; Yan, H., Nonfullerene Acceptor Molecules for Bulk Heterojunction Organic Solar Cells. *Chemical Reviews* **2018**, *118*, 3447-3507 0009-2665.
51. Liu, Q.; Jiang, Y.; Jin, K.; Qin, J.; Xu, J.; Li, W.; Xiong, J.; Liu, J.; Xiao, Z.; Sun, K., 18% Efficiency Organic Solar Cells. *Science Bulletin* **2020**, *65*, 272-275.
52. Yuan, J.; Zhang, Y.; Zhou, L.; Zhang, G.; Yip, H.-L.; Lau, T.-K.; Lu, X.; Zhu, C.; Peng, H.; Johnson, P. A., Single-Junction Organic Solar Cell with over 15% Efficiency Using Fused-Ring Acceptor with Electron-Deficient Core. *Joule* **2019**, *3*, 1140-1151 2542-4351.
53. Zhang, S.; Qin, Y.; Zhu, J.; Hou, J., Over 14% Efficiency in Polymer Solar Cells Enabled by a Chlorinated Polymer Donor. *Advanced Materials* **2018**, *30*, 1800868 0935-9648.
54. Zhang, Y.; Yao, H.; Zhang, S.; Qin, Y.; Zhang, J.; Yang, L.; Li, W.; Wei, Z.; Gao, F.; Hou, J., Fluorination Vs. Chlorination: A Case Study on High Performance Organic Photovoltaic Materials. *Science China Chemistry* **2018**, *61*, 1328-1337 1869-1870.
55. Armin, A.; Li, W.; Sandberg, O. J.; Xiao, Z.; Ding, L.; Nelson, J.; Neher, D.; Vandewal, K.; Shoaee, S.; Wang, T., A History and Perspective of Non-Fullerene Electron Acceptors for Organic Solar Cells. *Advanced Energy Materials* **2021**, *11*, 2003570 1614-6832.

56. Lin, Y.; Wang, J.; Zhang, Z. G.; Bai, H.; Li, Y.; Zhu, D.; Zhan, X., An Electron Acceptor Challenging Fullerenes for Efficient Polymer Solar Cells. *Advanced Materials* **2015**, *27*, 1170-1174 0935-9648.
57. Lin, H.; Chen, S.; Li, Z.; Lai, J. Y. L.; Yang, G.; McAfee, T.; Jiang, K.; Li, Y.; Liu, Y.; Hu, H., High-Performance Non-Fullerene Polymer Solar Cells Based on a Pair of Donor–Acceptor Materials with Complementary Absorption Properties. *Advanced Materials* **2015**, *27*, 7299-7304 0935-9648.
58. Li, S.; Ye, L.; Zhao, W.; Zhang, S.; Mukherjee, S.; Ade, H.; Hou, J., Energy-Level Modulation of Small-Molecule Electron Acceptors to Achieve over 12% Efficiency in Polymer Solar Cells. *Advanced Materials* **2016**, *28*, 9423-9429 0935-9648.
59. Wei, Q.; Liu, W.; Leclerc, M.; Yuan, J.; Chen, H.; Zou, Y., A-Da′ Da Non-Fullerene Acceptors for High-Performance Organic Solar Cells. *Science China Chemistry* **2020**, *63*, 1352-1366 1869-1870.
60. Yang, Y.; Zhang, Z.-G.; Bin, H.; Chen, S.; Gao, L.; Xue, L.; Yang, C.; Li, Y., Side-Chain Isomerization on an N-Type Organic Semiconductor Itic Acceptor Makes 11.77% High Efficiency Polymer Solar Cells. *Journal of the American Chemical Society* **2016**, *138*, 15011-15018 0002-7863.
61. Wang, T.; Wang, X.; Yang, R.; Li, C., Recent Advances in Ternary Organic Solar Cells Based on Förster Resonance Energy Transfer. *Solar RRL* **2021**, *5*, 2100496 2367-198X.
62. Wadsworth, A.; Moser, M.; Marks, A.; Little, M. S.; Gasparini, N.; Brabec, C. J.; Baran, D.; McCulloch, I., Critical Review of the Molecular Design Progress in Non-Fullerene Electron Acceptors Towards Commercially Viable Organic Solar Cells. *Chemical Society Reviews* **2019**, *48*, 1596-1625.
63. Firdaus, Y.; Le Corre, V. M.; Khan, J. I.; Kan, Z.; Laquai, F.; Beaujuge, P. M.; Anthopoulos, T. D., Key Parameters Requirements for Non-Fullerene-Based Organic Solar Cells with Power Conversion Efficiency > 20%. *Advanced Science* **2019**, *6*, 1802028 2198-3844.
64. Gurney, R. S.; Lidzey, D. G.; Wang, T., A Review of Non-Fullerene Polymer Solar Cells: From Device Physics to Morphology Control. *Reports on Progress in Physics* **2019**, *82*, 036601 0034-4885.
65. Rajaram, S.; Shivanna, R.; Kandappa, S. K.; Narayan, K. S., Nonplanar Perylene Diimides as Potential Alternatives to Fullerenes in Organic Solar Cells. *Journal of Physical Chemistry Letters* **2012**, *3*, 2405-2408 1948-7185.
66. Shivanna, R.; Shoaee, S.; Dimitrov, S.; Kandappa, S. K.; Rajaram, S.; Durrant, J. R.; Narayan, K. S., Charge Generation and Transport in Efficient Organic Bulk Heterojunction Solar Cells with a Perylene Acceptor. *Energy & Environmental Science* **2014**, *7*, 435-441.
67. Li, C.; Wonneberger, H., Perylene Imides for Organic Photovoltaics: Yesterday, Today, and Tomorrow. *Advanced Materials* **2012**, *24*, 613-636 0935-9648.
68. Tang, C. W., Two-Layer Organic Photovoltaic Cell. *Applied Physics Letters* **1986**, *48*, 183-185 0003-6951.

69. Sun, H.; Chen, F.; Chen, Z.-K., Recent Progress on Non-Fullerene Acceptors for Organic Photovoltaics. *Materials Today* **2019**, *24*, 94-118 1369-7021.
70. Hou, J.; Inganäs, O.; Friend, R. H.; Gao, F., Organic Solar Cells Based on Non-Fullerene Acceptors. *Nature Materials* **2018**, *17*, 119-128 1476-4660.
71. Zou, X.; Wen, G.; Hu, R.; Dong, G.; Zhang, C.; Zhang, W.; Huang, H.; Dang, W., An Insight into the Excitation States of Small Molecular Semiconductor Y6. *Molecules* **2020**, *25*, 4118 1420-3049.
72. Ma, L.; Zhang, S.; Zhu, J.; Wang, J.; Ren, J.; Zhang, J.; Hou, J., Completely Non-Fused Electron Acceptor with 3d-Interpenetrated Crystalline Structure Enables Efficient and Stable Organic Solar Cell. *Nature Communications* **2021**, *12*, 1-12 2041-1723.
73. Yang, M.; Wei, W.; Zhou, X.; Wang, Z.; Duan, C., Non-Fused Ring Acceptors for Organic Solar Cells. *Energy Materials* **2021**, *1*, 100008.
74. Li, S.; Zhan, L.; Liu, F.; Ren, J.; Shi, M.; Li, C. Z.; Russell, T. P.; Chen, H., An Unfused-Core-Based Nonfullerene Acceptor Enables High-Efficiency Organic Solar Cells with Excellent Morphological Stability at High Temperatures. *Advanced Materials* **2018**, *30*, 1705208 0935-9648.
75. Li, S.; Zhan, L.; Zhao, W.; Zhang, S.; Ali, B.; Fu, Z.; Lau, T.-K.; Lu, X.; Shi, M.; Li, C.-Z., Revealing the Effects of Molecular Packing on the Performances of Polymer Solar Cells Based on a–D–C–D–a Type Non-Fullerene Acceptors. *Journal of Materials Chemistry A* **2018**, *6*, 12132-12141.
76. Meng, D.; Fu, H.; Xiao, C.; Meng, X.; Winands, T.; Ma, W.; Wei, W.; Fan, B.; Huo, L.; Doltsinis, N. L., Three-Bladed Rylene Propellers with Three-Dimensional Network Assembly for Organic Electronics. *Journal of the American Chemical Society* **2016**, *138*, 10184-10190 0002-7863.
77. Liu, F.; Zhou, L.; Liu, W.; Zhou, Z.; Yue, Q.; Zheng, W.; Sun, R.; Liu, W.; Xu, S.; Fan, H., Organic Solar Cells with 18% Efficiency Enabled by an Alloy Acceptor: A Two-in-One Strategy. *Advanced Materials* **2021**, *33*, 2100830 0935-9648.
78. Yu, R.; Yao, H.; Hou, J., Recent Progress in Ternary Organic Solar Cells Based on Nonfullerene Acceptors. *Advanced Energy Materials* **2018**, *8*, 1702814 1614-6832.
79. Zhao, C.; Wang, J.; Zhao, X.; Du, Z.; Yang, R.; Tang, J., Recent Advances, Challenges and Prospects in Ternary Organic Solar Cells. *Nanoscale* **2021**, *13*, 2181-2208.
80. Gasparini, N.; Salleo, A.; McCulloch, I.; Baran, D., The Role of the Third Component in Ternary Organic Solar Cells. *Nature Reviews Materials* **2019**, *4*, 229-242 2058-8437.
81. Zheng, Z.; Wang, J.; Bi, P.; Ren, J.; Wang, Y.; Yang, Y.; Liu, X.; Zhang, S.; Hou, J., Tandem Organic Solar Cell with 20.2% Efficiency. *Joule* **2022**, *6*, 171-184 2542-4351.
82. Zhu, L.; Zhang, M.; Xu, J.; Li, C.; Yan, J.; Zhou, G.; Zhong, W.; Hao, T.; Song, J.; Xue, X., Single-Junction Organic Solar Cells with over 19% Efficiency Enabled by a Refined Double-Fibril Network Morphology. *Nature Materials* **2022**, 1-8 1476-4660.

83. Benten, H.; Nishida, T.; Mori, D.; Xu, H.; Ohkita, H.; Ito, S., High-Performance Ternary Blend All-Polymer Solar Cells with Complementary Absorption Bands from Visible to near-Infrared Wavelengths. *Energy & Environmental Science* **2016**, *9*, 135-140.
84. Zhan, L.; Li, S.; Li, Y.; Sun, R.; Min, J.; Bi, Z.; Ma, W.; Chen, Z.; Zhou, G.; Zhu, H., Desired Open-Circuit Voltage Increase Enables Efficiencies Approaching 19% in Symmetric-Asymmetric Molecule Ternary Organic Photovoltaics. *Joule* **2022**, *6*, 662-675 2542-4351.
85. Baran, D.; Kirchartz, T.; Wheeler, S.; Dimitrov, S.; Abdelsamie, M.; Gorman, J.; Ashraf, R. S.; Holliday, S.; Wadsworth, A.; Gasparini, N., Reduced Voltage Losses Yield 10% Efficient Fullerene Free Organic Solar Cells with > 1 V Open Circuit Voltages. *Energy & Environmental Science* **2016**, *9*, 3783-3793.
86. Liu, X.; Yan, Y.; Yao, Y.; Liang, Z., Ternary Blend Strategy for Achieving High-Efficiency Organic Solar Cells with Nonfullerene Acceptors Involved. *Advanced Functional Materials* **2018**, *28*, 1802004 1616-301X.
87. An, Q.; Zhang, F.; Yin, X.; Sun, Q.; Zhang, M.; Zhang, J.; Tang, W.; Deng, Z., High-Performance Alloy Model-Based Ternary Small Molecule Solar Cells. *Nano Energy* **2016**, *30*, 276-282 2211-2855.
88. Chang, Y.; Chang, Y.; Zhu, X.; Zhou, X.; Yang, C.; Zhang, J.; Lu, K.; Sun, X.; Wei, Z., Constructing High-Performance All-Small-Molecule Ternary Solar Cells with the Same Third Component but Different Mechanisms for Fullerene and Non-Fullerene Systems. *Advanced Energy Materials* **2019**, *9*, 1900190 1614-6832.
89. Bi, P.; Hao, X., Versatile Ternary Approach for Novel Organic Solar Cells: A Review. *Solar RRL* **2019**, *3*, 1800263 2367-198X.
90. Xu, X.; Li, Y.; Peng, Q., Recent Advances in Morphology Optimizations Towards Highly Efficient Ternary Organic Solar Cells. *Nano Select* **2020**, *1*, 30-58 2688-4011.
91. Huang, X.; Liu, X.; Ding, K.; Forrest, S. R., Is There Such a Thing as a Molecular Organic Alloy? *Materials Horizons* **2020**, *7*, 244-251.
92. Yin, Z.; Mei, S.; Chen, L.; Gu, P.; Huang, J.; Li, X.; Wang, H.-Q.; Song, W., Efficient Ptb7-Th: Y6: Pc71bm Ternary Organic Solar Cell with Superior Stability Processed by Chloroform. *Organic Electronics* **2021**, *99*, 106308 1566-1199.
93. An, Q.; Wang, J.; Zhang, F., Ternary Polymer Solar Cells with Alloyed Donor Achieving 14.13% Efficiency and 78.4% Fill Factor. *Nano Energy* **2019**, *60*, 768-774 2211-2855.
94. Liu, T.; Ma, R.; Luo, Z.; Guo, Y.; Zhang, G.; Xiao, Y.; Yang, T.; Chen, Y.; Li, G.; Yi, Y., Concurrent Improvement in J Sc and V Oc in High-Efficiency Ternary Organic Solar Cells Enabled by a Red-Absorbing Small-Molecule Acceptor with a High Lumo Level. *Energy & Environmental Science* **2020**, *13*, 2115-2123.
95. Pan, M.-A.; Lau, T.-K.; Tang, Y.; Wu, Y.-C.; Liu, T.; Li, K.; Chen, M.-C.; Lu, X.; Ma, W.; Zhan, C., 16.7%-Efficiency Ternary Blended Organic Photovoltaic Cells

with Pcbm as the Acceptor Additive to Increase the Open-Circuit Voltage and Phase Purity. *Journal of Materials Chemistry A* **2019**, *7*, 20713-20722.

96. Liu, Z., Enhancing the Photovoltaic Performance with Two Similar Structure Polymers as Donors by Broadening the Absorption Spectrum and Optimizing the Molecular Arrangement. *Organic Electronics* **2021**, *93*, 106153 1566-1199.
97. Su, Z.; Zhang, Z.; Xie, G.; Zhang, Y.; Zhang, X.; Zhang, W.; Zhang, J., Over 16.5% Efficiency in Ternary Organic Solar Cells by Adding an Alloyed Acceptor with Energy Transfer Process. *Dyes and Pigments* **2021**, *192*, 109434 0143-7208.
98. Zhan, L.; Li, S.; Lau, T.-K.; Cui, Y.; Lu, X.; Shi, M.; Li, C.-Z.; Li, H.; Hou, J.; Chen, H., Over 17% Efficiency Ternary Organic Solar Cells Enabled by Two Non-Fullerene Acceptors Working in an Alloy-Like Model. *Energy & Environmental Science* **2020**, *13*, 635-645.
99. An, Q.; Ma, X.; Gao, J.; Zhang, F., Solvent Additive-Free Ternary Polymer Solar Cells with 16.27% Efficiency. *Sci. Bull* **2019**, *64*, 504-506.
100. Ma, Y.; Zhou, X.; Cai, D.; Tu, Q.; Ma, W.; Zheng, Q., A Minimal Benzo [C][1, 2, 5] Thiadiazole-Based Electron Acceptor as a Third Component Material for Ternary Polymer Solar Cells with Efficiencies Exceeding 16.0%. *Materials Horizons* **2020**, *7*, 117-124.
101. Xie, G.; Zhang, Z.; Su, Z.; Zhang, X.; Zhang, J., 16.5% Efficiency Ternary Organic Photovoltaics with Two Polymer Donors by Optimizing Molecular Arrangement and Phase Separation. *Nano Energy* **2020**, *69*, 104447 2211-2855.
102. An, Q.; Wang, J.; Ma, X.; Gao, J.; Hu, Z.; Liu, B.; Sun, H.; Guo, X.; Zhang, X.; Zhang, F., Two Compatible Polymer Donors Contribute Synergistically for Ternary Organic Solar Cells with 17.53% Efficiency. *Energy & Environmental Science* **2020**, *13*, 5039-5047.
103. Ma, X.; Wang, J.; Gao, J.; Hu, Z.; Xu, C.; Zhang, X.; Zhang, F., Achieving 17.4% Efficiency of Ternary Organic Photovoltaics with Two Well-Compatible Nonfullerene Acceptors for Minimizing Energy Loss. *Advanced Energy Materials* **2020**, *10*, 2001404 1614-6832.
104. Li, Y.; Cai, Y.; Xie, Y.; Song, J.; Wu, H.; Tang, Z.; Zhang, J.; Huang, F.; Sun, Y., A Facile Strategy for Third-Component Selection in Non-Fullerene Acceptor-Based Ternary Organic Solar Cells. *Energy & Environmental Science* **2021**, *14*, 5009-5016.
105. Jiang, K.; Wei, Q.; Lai, J. Y. L.; Peng, Z.; Kim, H. K.; Yuan, J.; Ye, L.; Ade, H.; Zou, Y.; Yan, H., Alkyl Chain Tuning of Small Molecule Acceptors for Efficient Organic Solar Cells. *Joule* **2019**, *3*, 3020-3033 2542-4351.
106. Qin, Y.; Xu, Y.; Peng, Z.; Hou, J.; Ade, H., Low Temperature Aggregation Transitions in N3 and Y6 Acceptors Enable Double-Annealing Method That Yields Hierarchical Morphology and Superior Efficiency in Nonfullerene Organic Solar Cells. *Advanced Functional Materials* **2020**, *30*, 2005011 1616-301X.
107. Peng, W.; Lin, Y.; Jeong, S. Y.; Genene, Z.; Magomedov, A.; Woo, H. Y.; Chen, C.; Wahyudi, W.; Tao, Q.; Deng, J., Over 18% Ternary Polymer Solar Cells Enabled by a Terpolymer as the Third Component. *Nano Energy* **2022**, *92*, 106681 2211-2855.

108. Gao, J.; Wang, J.; An, Q.; Ma, X.; Hu, Z.; Xu, C.; Zhang, X.; Zhang, F., Over 16.7% Efficiency of Ternary Organic Photovoltaics by Employing Extra Pc71bm as Morphology Regulator. *Science China Chemistry* **2020**, *63*, 83-91 1869-1870.
109. Ma, X.; Zeng, A.; Gao, J.; Hu, Z.; Xu, C.; Son, J. H.; Jeong, S. Y.; Zhang, C.; Li, M.; Wang, K., Approaching 18% Efficiency of Ternary Organic Photovoltaics with Wide Bandgap Polymer Donor and Well Compatible Y6: Y6-1o as Acceptor. *National Science Review* **2021**, *8*, nwaa305 2095-5138.
110. Jin, K.; Xiao, Z.; Ding, L., 18.69% Pce from Organic Solar Cells. *J. Semicond* **2021**, *42*, 060502.
111. Cui, Y.; Xu, Y.; Yao, H.; Bi, P.; Hong, L.; Zhang, J.; Zu, Y.; Zhang, T.; Qin, J.; Ren, J., Single-Junction Organic Photovoltaic Cell with 19% Efficiency. *Advanced Materials* **2021**, *33*, 2102420 0935-9648.
112. Doumon, N. Y.; Yang, L.; Rosei, F., Ternary Organic Solar Cells: A Review of the Role of the Third Element☆. *Nano Energy* **2022**, 106915 2211-2855.
113. Mohapatra, A. A.; Tiwari, V.; Patil, S., Energy Transfer in Ternary Blend Organic Solar Cells: Recent Insights and Future Directions. *Energy & Environmental Science* **2021**, *14*, 302-319.
114. Scully, S. R.; Armstrong, P. B.; Edder, C.; Fréchet, J. M. J.; McGehee, M. D., Long-Range Resonant Energy Transfer for Enhanced Exciton Harvesting for Organic Solar Cells. *Advanced Materials* **2007**, *19*, 2961-2966 0935-9648.
115. Li, W.; Liu, D.; Wang, T., Stability of Non-Fullerene Electron Acceptors and Their Photovoltaic Devices. *Advanced Functional Materials* **2021**, *31*, 2104552 1616-301X.
116. Speller, E. M.; Clarke, A. J.; Luke, J.; Lee, H. K. H.; Durrant, J. R.; Li, N.; Wang, T.; Wong, H. C.; Kim, J.-S.; Tsoi, W. C., From Fullerene Acceptors to Non-Fullerene Acceptors: Prospects and Challenges in the Stability of Organic Solar Cells. *Journal of Materials Chemistry A* **2019**, *7*, 23361-23377.
117. Li, Y.; Huang, X.; Ding, K.; Sheriff, H. K. M.; Ye, L.; Liu, H.; Li, C.-Z.; Ade, H.; Forrest, S. R., Non-Fullerene Acceptor Organic Photovoltaics with Intrinsic Operational Lifetimes over 30 Years. *Nature Communications* **2021**, *12*, 1-9 2041-1723.
118. Du, X.; Heumueller, T.; Gruber, W.; Classen, A.; Unruh, T.; Li, N.; Brabec, C. J., Efficient Polymer Solar Cells Based on Non-Fullerene Acceptors with Potential Device Lifetime Approaching 10 Years. *Joule* **2019**, *3*, 215-226.
119. Qin, Y.; Balar, N.; Peng, Z.; Gadisa, A.; Angunawela, I.; Bagui, A.; Kashani, S.; Hou, J.; Ade, H., The Performance-Stability Conundrum of Btp-Based Organic Solar Cells. *Joule* **2021**, *5*, 2129-2147 2542-4351.
120. Mateker, W. R.; McGehee, M. D., Progress in Understanding Degradation Mechanisms and Improving Stability in Organic Photovoltaics. *Advanced Materials* **2017**, *29*, 1603940.
121. Guo, J.; Wu, Y.; Sun, R.; Wang, W.; Li, J.; Zhou, E.; Guo, J.; Wang, T.; Wu, Q.; Luo, Z., Photooxidation Analysis of Two Isomeric Nonfullerene Acceptors: A

Systematic Study of Conformational, Morphological, and Environmental Factors. *Solar RRL* **2021**, *5*, 2000704 2367-198X.

122. Azeez, A.; Narayan, K. S., Dominant Effect of Uv-Light-Induced "Burn-in" Degradation in Non-Fullerene Acceptor Based Organic Solar Cells. *The Journal of Physical Chemistry C* **2021**, *125*, 12531-12540 1932-7447.
123. Tournebize, A.; Mattana, G.; Gorisse, T. r. s.; Bousquet, A.; Wantz, G.; Hirsch, L.; Chambon, S., Crucial Role of the Electron Transport Layer and Uv Light on the Open-Circuit Voltage Loss in Inverted Organic Solar Cells. *ACS Applied Materials & Interfaces* **2017**, *9*, 34131-34138.
124. Classen, A.; Heumueller, T.; Wabra, I.; Gerner, J.; He, Y.; Einsiedler, L.; Li, N.; Matt, G. J.; Osvet, A.; Du, X., Revealing Hidden Uv Instabilities in Organic Solar Cells by Correlating Device and Material Stability. *Advanced Energy Materials* **2019**, *9*, 1902124.
125. Lee, H. K. H.; Durrant, J. R.; Li, Z.; Tsoi, W. C., Stability Study of Thermal Cycling on Organic Solar Cells. *Journal of Materials Research* **2018**, *33*, 1902-1908 0884-2914.
126. Corsini, F.; Griffini, G., Recent Progress in Encapsulation Strategies to Enhance the Stability of Organometal Halide Perovskite Solar Cells. *Journal of Physics: Energy* **2020**, *2*, 031002.
127. Vijayan, R.; Azeez, A.; Narayan, K., Toward Reliable High Performing Organic Solar Cells: Molecules, Processing, and Monitoring. *APL Materials* **2020**, *8*, 040908.

https://doi.org/10.1142/9789811270956_0002

# Chapter 2

# Towards the Post-lithium Technologies: Advances in Electrode Materials for Na-ion and K-ion Batteries

Vinita Ahuja, Arindam Ghosh and Premkumar Senguttuvan*

*New Chemistry Unit, International Centre for Materials Science and School of Advanced Materials, Jawaharlal Nehru Centre for Advanced Scientific Research, Jakkur, Bangalore 560064*
**prem@jncasr.ac.in*

In the present state of art, the post-lithium batteries such as Na-ion (NIBs) and K-ion batteries (KIBs) are considered as attractive alternatives because of the cheap and earth-abundant precursors. To continue this quest, the identification of suitable electrode materials and electrolyte systems is essential and crucial for the successful implementation of these storage systems in the market. In this account, we have reviewed developments and recent advances in electrode materials such as layered oxides cathodes, carbon and alloy-based anodes, in Na- and K-ion batteries.

## 1. Introduction

The modern society is largely reliant on the non-renewable resources such as coal, fossil fuels, natural gas resources for its basic energy needs such as power generation and ground transportation. The ever-increasing population further worsens the depletion of finite availability of such non-renewable resources on earth's crust and demands for viable and sustainable energy sources like sun, wind, water and geothermal energy.[1,2] The intelligent and practical approach to create a sustainable future is to gather and store this energy in all of its vast potential for later use.[3,4] Nevertheless, the intermittent nature of renewable energy sources, makes it essential to manage the renewable energy systems both in terms of energy harvesting and storage.[5,6] For instance, the solar energy is dilute and irregularly dispersed, and wind and tides deliver much lower energy than the required consumption. This huge mismatch between the rate of renewable energy conversion and energy consumption makes energy storage research one of the most important challenges for modern technology. At present, batteries and supercapacitors are the most widely used and recognized systems, for their broad range of applications, including those in the mobile industry. Due to their adaptability, low cost, easy maintenance, and high energy conversion efficiency, secondary rechargeable alkali-ion batteries have emerged as one of the storage systems that can efficiently integrate renewable energy into the grid.

## 2. History and Challenges

The history of battery can be associated with the Baghdad battery that was dated back to 250 B.C. Since then, batteries have experienced a series of evolution which include lead-acid, lithium-ion, nickel-cadmium, nickel metal hydride, sodium-sulfur, zinc-bromine, nickel hydrogen, nickel-zinc, molten salt, and metal-air batteries, and continued as the modern-day portable alkali-ion rechargeable batteries i.e., Li-ion batteries.[6] Eventually, the shortcomings of these systems were improved, and new materials were explored over the years to produce recyclable and safer batteries. In recent years, there has also been a lot of research done on other chemistries beyond alkali-ion chemistry.

Lithium-ion batteries (LIBs), which was first commercialized in 1991 by Sony Corp., have many advantages, including increased energy density, long performance life, and low self-discharge, and till date, have a dominant share in the global market of portable electronics and electric vehicle market.[7–9] Lithium having the most negative reduction potential (−3.045 V vs. standard hydrogen electrode), offers high gravimetric (100–265 Wh $kg^{-1}$) and volumetric capacity (250–670 Wh $L^{-1}$) making this class of rechargeable batteries more successful and dominant. In addition, the applications of LIBs also have been expanded to grid-level applications, for instance, in 2017, Tesla installed the largest grid-scale LIBs in South Australia with a storage capacity of 129 MWh and a power of 100 MW. The limited reserves of lithium and cobalt on earth's crust, high cost and safety are one of the major concerns that have hampered the widespread implementation of LIBs. Therefore, for the development of large-scale grid storage, cost-effective alternative energy storage systems are the need of the hour.[9,10] The research in alternative alkali-ion systems, such as Na-ion batteries (NIBs) and K-ion batteries (KIBs), have gained a great momentum in the last few decades. The studies on LIBs and NIBs commenced parallelly in early 1980s,[11–13] whilst the K-ion batteries (KIBs) technology have been studied since 2004[14,15] and is still in its early stage. These technologies work on the common working principle of "rocking-chair" mechanism, which consist of back-and-forth movement of alkali-ions between two host electrodes through an ion-conducting medium known as electrolyte. Moreover, sodium and potassium are evenly spread in earth's crust unlike lithium, that constitutes merely 0.0017% of earth's crust. As a result, these two technologies certainly are the most promising alternatives for the post-lithium age. Sodium being a cost-efficient source, offers low energy density and high standard reduction potential (−2.71 V vs. $E^0$). Therefore, it can be explored and utilized in the grid-scale storage systems. Even potassium possesses a similar redox potential to that of lithium, with only difference of 0.12 V (Table 1). This suggests that KIBs can also serve as potential candidates for high voltage batteries similar to LIBs in terms of practical voltages. In this chapter we will discuss about the development of electrode materials for these alternative technologies, their challenges, limitations, and future outlook.[16,17]

Table 1. Comparison between the physical properties of lithium, sodium and potassium.[17]

| Property | Li | Na | K |
|---|---|---|---|
| **Atomic mass (g mol$^{-1}$)** | 6.94 | 23.00 | 39.10 |
| **Shannon's ionic radii (Å)** | 0.76 | 1.02 | 1.38 |
| **$E^0$ vs. SHE (V)** | −3.04 | −2.71 | −2.93 |
| **Melting point (°C)** | 180.5 | 97.7 | 63.4 |
| **Density (g cm$^{-3}$)** | 0.534 | 0.968 | 0.859 |
| **Theoretical gravimetric capacity (mAh g$^{-1}$)** | 3861 | 1166 | 685 |
| **Abundance (% by weight)** | Negligible | 2.8 | 2.6 |
| **Carbonate price ($ per metric ton)** | 5800 | 250–300 | 600–1200 |
| **Stokes radius in PC (Å)** | 4.8 | 4.6 | 3.6 |

## 3. State-of-the-Art: Electrode materials in NIBs and KIBs

Although a myriad of electrode materials has been explored in NIBs and KIBs, layered oxides, polyanionic frameworks as cathodes and carbonaceous based materials, alloys such as Sb, Bi, Sn based materials as anode are promising for practical applications. In this chapter, we will focus on some particular and most explored candidates of electrode materials such as layered oxide-based cathodes, carbonaceous materials and alloy-based anodes in NIBs and KIBs.

### 3.1. *Positive electrode materials in NIBs and KIBs*

An ideal cathode material should be cheap, safe, possess high voltage and storage capacity, high electronic and ion conductivity, and have great structural and chemical stability. Various classes of positive electrodes like metal oxides, polyanionic frameworks, hexacyanometalates, and organic compounds, have been widely explored till date. Among these, layered oxide cathode materials are very attractive due to their easy synthetic routes, huge capacities, and wide operating voltages. The following section will discuss recent progress in layered oxide cathode materials for NIBs and KIBs.

#### 3.1.1. *Layered oxides in NIBs*

Pioneering work on layered oxide cathode for NIBs was initiated by C. Delmas in the 1980s.[18] Their general composition can be written as $Na_xMeO_2$ ($0 < x \leq 1$; Me = Ti-Cu; Nb and Mo or a combination of elements). These are the most studied families of positive electrode materials because of their increased capacities, wide operating voltages, and easy synthesis routes. It is worth noting that in the Na-ion layered structures all the 3d transition metal ions (except Zn) are electrochemically active. Their structures are made of alternating metal-oxygen layers consisting of edge-shared $MeO_6$ octahedra which are separated by sodium ions. Layered oxides are classified into O2, O3, P2, P3, etc. groups. Where "P" or "O" indicates whether the coordination shell of $Na^+$ is prismatic or the octahedral and 2 or 3, describes the number of transition metal-oxygen

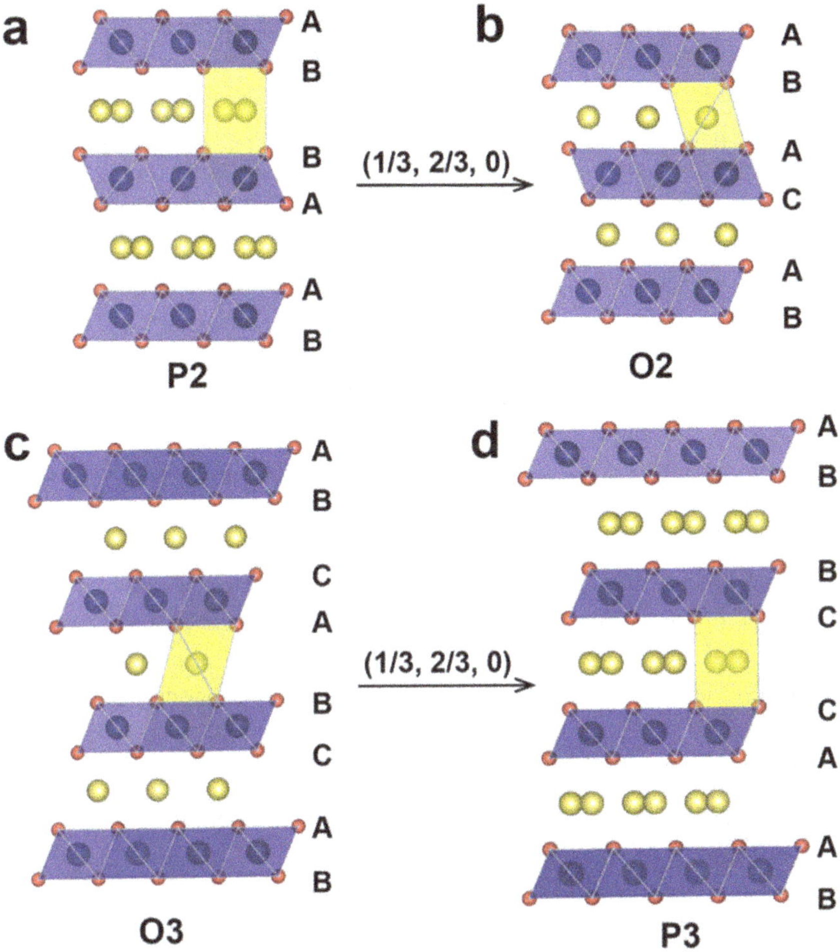


Figure 1. Representations of the crystal structures of (a) P2, (b) O2, (c) O3 and (d) P3 type layered $Na_xMeO_2$ oxides. The blue, yellow and red balls indicate the transition metal, sodium and oxygen ions respectively. Reproduced with permission.[19] © 2018, Wiley-VCH.

layers in a unit cell. The detailed crystal structures of these phases are described in Figure 1.[19] There are three possible ways to have different stacking positions of oxygen atoms in a transition metal-oxygen layer called A, B and C. In P2-type $Na_xMeO_2$ ($x \leq 0.7$, space group of $P6_3/mmc$) the oxide layer stacking follows the "ABBA......" pattern (Figure 1a), which signifies the presence of two kinds of transition metal-oxygen layers

(AB and BA) and the $Na^+$ ions in two distinct sites: $Na_1$ and $Na_2$ (Figure 1a). In O3-$Na_xMeO_2$ ($0.7 \leq x \leq 1.0$, space group: $R\bar{3}m$), the stacking order follows the "ABCABC......" pattern (Figure 1c), and all the $Na^+$ ions share one edge and one face with transition metal-oxygen layers. The P3-phase, mainly known to occur when $x \leq 0.5$, the stacking follows the "ABBCCA...." pattern (Figure 1d) and all Na share one face with one $MeO_6$ octahedra and three edges with three $MeO_6$ octahedra. O2 is another type of phase where the coordination shell of the $Na^+$ ion is also octahedral, and the oxygen-atoms stacking follows the "ABCB......" pattern (Figure 1b).

Electrochemical properties of a material are greatly governed by the structure of the layered oxide phase. Generally, O3 type cathodes show superior capacities due to large amounts of $Na^+$ ions, whereas the P2 type cathodes show higher-rate capabilities because of the fast diffusion of $Na^+$ ions. O3 type oxide cathodes undergo several phase transformations whilst P2 phases experience limited number of phase transitions during charge-discharge process, thus display better cyclability. However, recent studies show P2-O2 phase transition at high voltage due to the gliding of some transition metal-oxygen layers which leads to faster capacity decay.

#### 3.1.1.1. Single TM systems

Generally, this class of materials contains only one transition metal element, such as Cr, Fe, Mn, Co, Ni, etc. along with sodium and oxygen. After the discovery of the intercalation/de-intercalation mechanism of $Li^+$ in $LiCoO_2$ in 1980 by J. B. Goodenough, many research groups accelerated their attempts to apply the concept in Na-based systems. In 1981, the electrochemical properties of $NaCoO_2$ were first investigated by C. Delmas.[20] In the following section we will focus on some important findings in such systems.

**$Na_xCoO_2$**: O3-phase of $Na_xCoO_2$ can be electrochemically de-intercalated up to 0.5 moles of Na while experiencing a phase transition to P3. And multiple plateaus are observed in the electrochemical profile from $x = 0.67$ to $x = 0.8$ range. This series of phase transitions (O3–O′3–P′3) causes poor cycling performance of the material. The P2-type $Na_xCoO_2$ can retain its overall structure throughout an electrochemical process with multiple plateaus[21] (Figure 2a) and exhibit 126 mAh $g^{-1}$ of discharge capacity between 4.2–2.0 V vs. $Na^+/Na^0$ at 0.1 C with 90% capacity retention after 10 cycles.[22] Eventually, the higher price of cobalt becomes a prime factor to look for cheaper element-based layered oxides.

**$Na_xMnO_2$**: It is a very attractive choice for positive electrode material as it is low-cost and possesses a large specific capacity. With a high sodium content (x = 1), it exists in two phases: α-$NaMnO_2$ and β-$NaMnO_2$. α-$NaMnO_2$ displayed a discharge capacity of 197 mAh $g^{-1}$ with multiple plateaus in the range of 3.8–2.0 V vs. $Na^+/Na^0$ at C/30 rate.[23] However, in successive cycles due to Jahn–Teller (J. T.) distortion of $Mn^{3+}$, the capacity decays faster making it difficult to use it as an electrode material in the NIBs. Interestingly, β-$NaMnO_2$ possesses a different layered structure from that of conventional $NaMeO_2$ type structures. It is composed of zigzag layers of $MnO_6$ octahedra and in between the layers, sodium resides.[24] It exhibits a greater capacity of 190 mAh $g^{-1}$ in 4.25–2.0 V vs. $Na^+/Na^0$ at C/20 rate (Figure 2b) and can sustain decent capacity at faster C-rate (142 mAh $g^{-1}$ at 2 C). Although the cathode becomes disordered while charging, it shows remarkable cyclability.[25]

**$NaNiO_2$**: Like $NaMnO_2$, $NaNiO_2$ is also an attractive choice as a cathode material due to the cost-efficacy of Na and Mn and high redox potential of $Ni^{2+}/Ni^{3+}/Ni^{4+}$. It shows a capacity of 112 mAh $g^{-1}$ in 4.0–2.0 V vs. $Na^+/Na^0$ 0.1 C (Figure 2c) but undergoes several phase transformations during $Na^+$ intercalation/de-intercalation processes when cycled between 4.0 and 1.5 V vs. $Na^+/Na^0$ at 0.1 C rate (Figure 2d).[26] G. Ceder *et al.* showed a better capacity retention (94% over 20 cycles) by fixing the upper cutoff voltage to 3.75 V vs. $Na^+/Na^0$.[27] However, J. T. effect of $Ni^{3+}$ leads to structural instability which causes capacity fading in long term cycling. Recently, it was found out that when a high voltage is applied, the structure gets irreversibly transformed and electrolyte oxidation also occurs. And also during cycling, an irreversible $Na_{0.17}NiO_2$ phase evolves which greatly affects material's $Na^+$ storage performances.[28]

**$NaFeO_2$**: Due to its inexpensive and easily available raw materials, α-$NaFeO_2$, an O3-type layered oxide cathode, has been greatly explored. However, its practical capacity is limited to 85 mAh $g^{-1}$, which is about 1/3 of its theoretical capacity (240 mAh $g^{-1}$) (Figure 2e).[29] The cation-migration phenomenon of $Fe^{3+}$ causes complex phase transition processes during

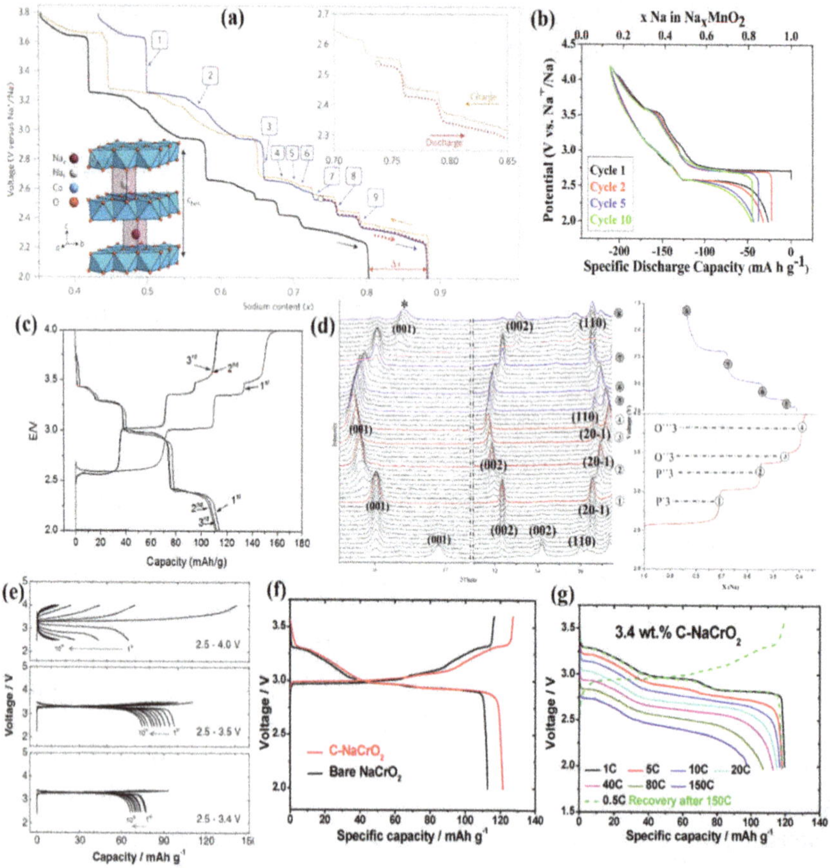


Figure 2. (a) Variation in potential during $Na^+$ ion insertion in P2-$Na_xCoO_2$. Reproduced with permission.[21] ©s 2011 Nature Publishing Group. (b) Load curves for β-$NaMnO_2$ at a rate of C/20. Reproduced with permission.[24] © 2014, American Chemical Society. (c) Voltage vs. specific capacity at C/10 rate. (d) *Operando* XRD patterns during the $1^{st}$ charge-discharge process and voltage vs. x plot with corresponding phase transitions of $NaNiO_2$. Reproduced with permission.[26] © 2014, Elsevier. (e) Voltage vs. specific capacity plot varying the potential for α-$NaFeO_2$.[29] (f) Electrochemical performances of bare and carbon coated $NaCrO_2$ (C-$NaCrO_2$) from 3.6–2.0 V vs. $Na^+/Na^0$ at 20 mA $g^{-1}$ and (g) Rate performances of C-$NaCrO_2$. Reproduced with permission.[33] © 2015, Royal Society of Chemistry.

cycling of the material. β-$NaFeO_2$, a P2-type layered oxide cathode, exhibits a high $1^{st}$ discharge capacity of 335 mAh $g^{-1}$ at 0.1 C and 136 mAh $g^{-1}$ after $150^{th}$ cycle.[30]

**$NaCrO_2$**: Though having a capacity of only 26 mAh $g^{-1}$, $NaCrO_2$ received a lot of attention for its excellent thermal stability.[31] When a positive potential is applied to the material, it transforms into O′3 and P′3 phases. However, upon extracting more than 0.5 moles of sodium *pfu,* the $Cr^{3+}$ ions migrate into the interslab spacing and an irreversible structural modification happens.[32] Recently, Yu *et al.* prepared carbon-coated $NaCrO_2$ particles *via* carbonization of pitch, which displayed an initial discharge capacity of 100–110 mAh $g^{-1}$ (Figure 2f) with a superb charge-discharge repetition and endurance of an ultrafast 150 C rate (Figure 2g).[33]
**$NaTiO_2$ & $NaVO_2$**: In search of low-voltage sodium insertion compounds, Ti and V based layered oxide have also been developed. O3-$NaVO_2$ delivers a highly reversible capacity of 120 mAh $g^{-1}$ in 2.4–1.2 V vs. $Na^+/Na^0$ at C/20 rate.[34] Like $NaCrO_2$, it also shows monoclinic distortion with multiple phase changes in the electrochemical profile. When the $Na^+$ extraction is > 0.5, its Na-storage performance deteriorates indicating the irreversible migration of $V^{3+}$ into interslab spacing. Similar phenomenon was also observed in the case of $NaTiO_2$ at higher charge potentials.[35] But, in the potential window of 1.6–0.6 V vs. $Na^+/Na^0$, it shows a specific capacity of 150 mAh $g^{-1}$ with $V_{avg}$ = ~1.0 V. Due to low energy density of Ti/V based layered oxides and toxic nature of vanadium precursors, their practical applications are limited.

#### 3.1.1.2. Binary TM systems

The main concern with single-TM based O3-$Na_xMeO_2$ materials is that they undergo several structural changes during charge-discharge process in the wide potential range. To address this dispute, binary metals were added in transition metal oxide layers, where the synergistic effect of two metals can reduce the multiple phase transitions, increasing the average potential, thereby improving the electrochemical performances overall.
**Ni/Mn based oxides**: To begin with, nickel and manganese-based systems are the one of the most explored electrode materials, primarily for two reasons: (1) a reversible redox process from $Ni^{2+}$ to $Ni^{4+}$ can supply $2e^-$ at a high working voltage; (2) the presence of J. T. inactive $Ni^{2+}$ and $Mn^{4+}$ ions, ensures better structural integrity during cycling. However, the electrochemical profile shows multiple plateaus indicating complicated

phase transitions. For instance, The $NaNi_{0.5}Mn_{0.5}O_2$ electrode demonstrated the capacity of 125 & 185 mAh $g^{-1}$ in the voltage range of 3.8–2.2 V & 4.5–2.2 V vs. $Na^+/Na^0$ (Figure 3a) with 75% capacity of retention after 50 cycles. It undergoes a complex phase change of O3-P3-P′3-P″3 during the sodium ion extraction process, which eventually results in capacity fade (Figure 3f).[36] In 2001, J. Dahn *et al.* reported P2-$Na_{0.67}Ni_{0.33}Mn_{0.67}O_2$, in which all $Na^+$ ions can be extracted till 4.5V utilizing the $Ni^{2+}/Ni^{4+}$ redox couple, leading to an impressive capacity of 160 mAh $g^{-1}$ with a $V_{avg}$ = 3.7 V vs. $Na^+/Na^0$.[37] In this case, an extended-voltage plateau at 4.2 V implies the P2-O2 transition with a 20% volume change in the crystal structure, which was further confirmed from the *in-situ* XRD experiments. Instead of showing a higher voltage among the reported P2 and O3 type materials, poor cyclability inhibits its large-scale practical application.

**Fe/Mn based oxides**: Fe/Mn based layered oxides are very appealing due to cheap, earth-abundant precursors, high working potential of $Fe^{3+}/Fe^{4+}$ and huge specific capacity of $Mn^{3+}/Mn^{4+}$. In 2012, S. Komaba *et al.* first demonstrated the interaction process of O3-$Na[Fe_{0.5}Mn_{0.5}]O_2$ cathode delivering the capacity of 110 mAh $g^{-1}$ in between 4.2–1.5 V vs. $Na^+/Na^0$ with a large polarization.[38] However, its P2 counterpart shows better performances as it requires much lesser energy to remove $Na^+$ from interlayer of P2 oxygen stacking. Although the phase transition for O3 phase at higher voltage (> 4.0 V) remains unclear, a clear P2–OP4 transition is reported for P2-$Na_{0.67}[Fe_{0.5}Mn_{0.5}]O_2$. This P2–OP4 phase change can be avoided, if Mn content is increased such as in case of P2-$Na_{0.67}[Fe_{0.4}Mn_{0.6}]O_2$. Here the P2 phase is retained, even in discharged state (Figure 3d).[39]

**Ni/Fe based oxides**: The J.-T. effect of $Ni^{3+}$ leads to structural imbalances in $NaNiO_2$, thereby substituting a small amount of Ni with Fe can effectively suppress this effect. In one such report, Wang *et al.* prepared a solid solution of O3-$NaFe_{1-x}Ni_xO_2$ ($0 \leq x \leq 1$)[40] with $Fe^{3+}/Fe^{4+}$ and $Ni^{3+}/Ni^{4+}$ as prominent redox couples. O3-$NaFe_{0.5}Ni_{0.5}O_2$ and $NaFe_{0.3}Ni_{0.7}O_2$ show a capacity of 112 and 135 mAh $g^{-1}$ respectively in between 3.8–2.0 V vs. $Na^+/Na^0$. Amongst both electrodes, O3-$NaFe_{0.3}Ni_{0.7}O_2$ c retained 74% of its initial capacity after 30 cycles with the improved stability mainly due to the to a lesser extent of unstable $Fe^{4+}$ in the charge state and full major redox contribution from $Ni^{3+}/Ni^{4+}$ redox.

#### 3.1.1.3. Ternary TM systems

Cathode materials of single and binary-metal-based TM layered oxides exhibit different unique electrochemical characteristics, such as excellent thermal stability for $NaCrO_2$, high capacity for $NaMnO_2$, and enhanced voltage and energy density for $NaNi_{0.5}Mn_{0.5}O_2$. However, they suffer from poor cycling stability when operated at broad potential range. So, utilizing synergistic contribution from multi-metal oxides is a promising approach to achieve better electrochemical performances. For example, $NaNi_{0.5}Mn_{0.5}O_2$ can act as a good electrode material below 4.0 V vs. $Na^+/Na^0$ but beyond this range it shows quick cycling fade. To address this concern, X. Yang *et al.* showed that Fe substitution in $NaNi_{0.5}Mn_{0.5}O_2$ leads to a smoother electrochemistry with the reversible OP2 phase formation.[41] Y.-K. Sun *et al.* probed the effect of Ni and Fe contents on the crystal structure and electrochemistry of O3-$Na[Ni_{0.75-x}Fe_xMn_{0.25}]O_2$ ($x = 0.4, 0.45, 0.5, 0.55$).[45] They proposed that increasing Ni content increases the capacity but compromises the capacity retention, while increasing Fe content improves the capacity retention but decreases the capacity. This study demonstrates that by appropriately balancing the Fe and Ni ratio, the electrochemical performances can be improved. In another case study, Z. Ma *et al.* examined the air-stable O3-$NaNi_{0.33}Fe_{0.33}Mn_{0.33}O_2$ material which shows a capacity of 136 mAh $g^{-1}$ with $V_{avg} = 3.15$ V vs. $Na^+/Na^0$ between 4.0–2.0 V at 0.1 C with 80% capacity retention for 100 cycles.[46] When the upper limit of the charging potential is fixed to 4.3 V vs. $Na^+/Na^0$, an O3-O3′-P3′ phase transformation occurs while charging, and discharge occurs *via* O3′-P3′-O3 phase transition until it regains its original structure. To further improve its performance, various cations such as $Li^+$, $Ti^{4+}$, $Mg^{2+}$ were doped in the course of time. In 2018, Z. Ma *et al.* achieved 92% retention in capacity after 200 cycles at 1 C *via* Ca doping.[47] The material $Na_{0.9}Ca_{0.05}Ni_{0.33}Fe_{0.33}Mn_{0.33}O_2$ demonstrated the 126.9 mAh $g^{-1}$ with ~400 Wh $kg^{-1}$ energy density. In 2014, the first report was published demonstrating the enhanced structural stability of Ni-Fe-Mn based system using $Li^+$ doping.[48] They found $Na[Li_{0.05}(Ni_{0.25}Fe_{0.25}Mn_{0.5})_{0.95}]O_2$ to show a reversible capacity of 180.1 mAh $g^{-1}$ at 0.1 C and retain 92.1% capacity after 40 cycles at 0.5 C. A full cell with hard carbon anode showed excellent long-term cycling with holding 76.2% of its initial capacity over 200 cycles at 0.5 C. X. Liu *et al.* synthesized $Mg^{2+}$ doped $NaMn_{0.48}Ni_{0.2}Fe_{0.3}Mg_{0.02}O_2$[42] and the system delivered an impressive capacity of 160 mAh $g^{-1}$ with a $V_{avg}$ = 3.0 V vs. $Na^+/Na^0$ at 0.05 C between 4.2–1.5 V. This material withstands a wide range of C-rates from 0.05 C to 50 C (Figure 3b) and retained ~99% of its initial capacity after 100 cycles at 0.1 C (Figure 3c). In case of P2-type layered oxides, charging voltages above 4.0 V vs. $Na^+/Na^0$ causes P2 to O2 transition.[37] In the same line of work, similar doping strategies can alleviate this transformation and thus can improve the overall electrochemical performance. For instance, Cu-doped P2-type $Na_{0.67}Ni_{0.1}Cu_{0.2}Mn_{0.7}O_2$ showed the complete P2 structure throughout the charge-discharge process. From *in-situ* XRD patterns it is evident that for the undoped sample O2, phase appears at higher voltage but in case of Cu-doped $Na_{0.67}Ni_{0.1}Cu_{0.2}Mn_{0.7}O_2$, a complete retention of P2 phase in the overall charge-discharge process can be seen (Figure 3e).[43] The presence of $Cu^{2+}$ greatly improves the structural integrity as well as the high voltage $Cu^{2+}/Cu^{3+}$ redox enhances the specific capacity greatly. Besides, multiple cation doping proved to be an interesting and useful approach. O3-$NaLi_{0.05}Mn_{0.50}Ni_{0.30}Cu_{0.10}Mg_{0.05}O_2$ can show 172 mAh $g^{-1}$ with $V_{avg}$ = 2.93 V vs. $Na^+/Na^0$ in 4.0–2.0 V potential window at 0.5 C rate.[44] The synergistic effect of Li, Cu and Mg markedly improves the air stability and rate capability. It was observed that no significant structural change occurred despite immersing the material in water for 12 hr and the capacity retention achieved was 70.4% at 20 C after 1000 cycles (Figure 3g). Similarly, a quinary layered oxide $NaNi_{0.25}Co_{0.25}Fe_{0.25}Mn_{0.125}Ti_{0.125}O_2$ showed 91% capacity retention at 4 C after 300 cycles with the decent capacity of 62.3 mAh $g^{-1}$ at 30 C.[49]

### 3.1.2. *Layered oxides in KIBs*

Though the same "rocking-chair" principle-based NIBs and KIBs have emerged as an alternative for LIBs, KIBs have some advantages over NIBs. Firstly, they possess higher operating voltages. Secondly, commercially available graphite, as an anode, can be used to accommodate

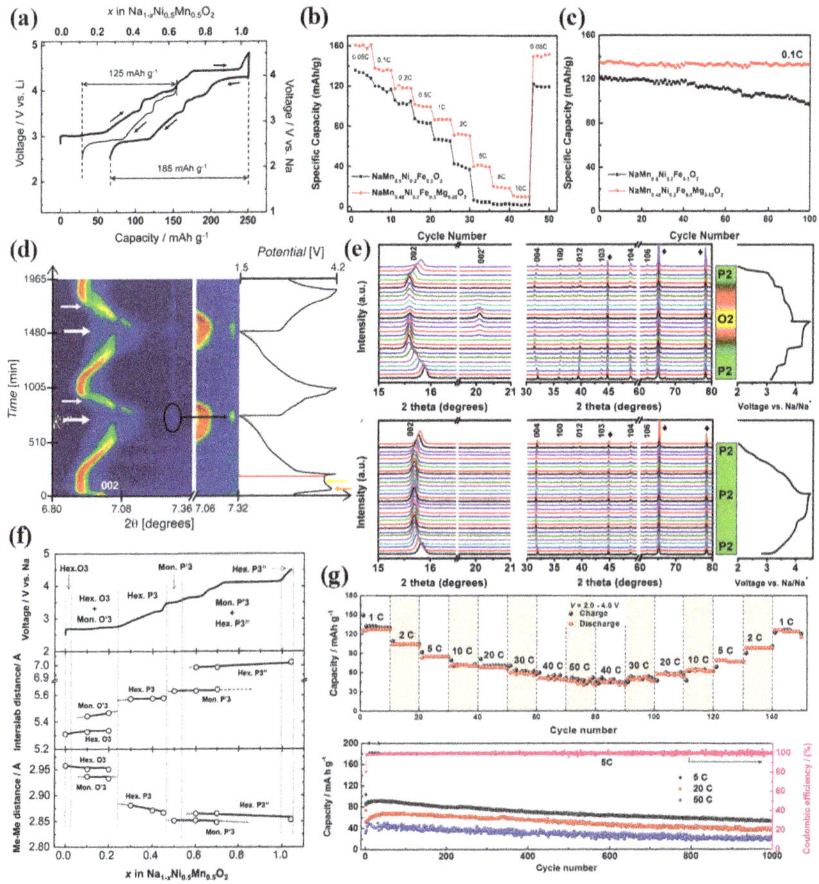


Figure 3. (a) First charge-discharge profile of $NaNi_{0.5}Mn_{0.5}O_2$ cathode at 1/50 C cycled between 2.2–3.8 and 2.2–4.5 V vs. Na and Li-metal. Reproduced with permission.[36] © 2012, American Chemical Society. (b) Specific capacity at different C-rates and (c) Cycling performance at 0.1C between 4.2–1.5 V vs. $Na^+/Na^0$ of $NaMn_{0.48}Ni_{0.2}Fe_{0.3}Mg_{0.02}O_2$. Reproduced with permission.[42] © 2012, American Chemical Society. (d) Contour diagram of synchrotron XRD data showing the evolution of the (002) peak in $Na_{0.67}Fe_{0.4}Mn_{0.6}O_2$. Reproduced with permission.[39] © 2016, American Chemical Society. (e) *Operando* XRD patterns for the first charge-discharge process of $Na_{0.67}Ni_{0.3}Mn_{0.7}O_2$ and $Na_{0.67}Ni_{0.1}Cu_{0.2}Mn_{0.7}O_2$ cathodes between 4.5–2.0 V vs. $Na^+/Na^0$ at 0.1 C. Reproduced with permission.[43] © 2017, Royal Society of Chemistry (f) Change in the crystallographic parameters and phase evolution of $Na_{1-x}Ni_{0.5}Mn_{0.5}O_2$ as a function of the x value. Reproduced with permission.[36] © 2012, American Chemical Society. (g) Rate performance from 1 C to 50 C, and cycling performance at 5, 20, and 50 C for $Na[Li_{0.05}Mn_{0.50}Ni_{0.30}Cu_{0.10}Mg_{0.05}]O_2$ electrode between 2.0 and 4.0 V vs. $Na^+/Na^0$. Reproduced with permission.[44] © 2017, Wiley-VCH.

$K^+$ ions during intercalation/de-intercalation. But large radius and higher atomic weight of potassium results in slow kinetics and low specific capacity for KIBs.[50] On the cathode side, the research on layered oxides further flourished in the recent decades and the following section will highlight the important materials studied in KIBs systems.

#### 3.1.2.1. Single TM systems

The earliest report on potassium based layered oxide can be dated back to 1965 for the synthesis and structural characterization of $KScO_2$.[51] Almost after 50 years, S. Passerini *et al.* rejuvenated the field by reporting the first layered oxide cathode for KIBs. The P2-type $K_{0.3}MnO_2$ cathode delivers 65 mAh $g^{-1}$ and 136 mAh $g^{-1}$ between 3.5–1.5 V and 4.0–1.5 V vs. $K^+/K^0$ respectively.[52] The P3-$K_{0.5}MnO_2$ shows a specific capacity of 100 mAh $g^{-1}$ with a reversible P3–O3 transition, when cycled between 3.9–1.5 V.[53] However, excessive $K^+$ extraction at higher potential leads to rapid capacity fading in both the cases. Moreover, $K_xMnO_2$ cathode suffers from two main problems: first, the disproportionation reaction of $Mn^{3+}$ to $Mn^{2+}$ and $Mn^{4+}$, and second, the Jahn–Teller distortion. To deal with such issues, in 2019, J. Chen *et al.*[54] reported a dual interphase strategy on $K_{0.67}MnO_2$, with improved $K^+$ diffusion rate and an inert spinel layer is formed on bulk $K_{0.67}MnO_2$ that alleviate Jahn-Teller distortion. Mn dissolution enables a reversible phase transition during charging-discharging. This strategy greatly improves the retention of capacity to 90.5% after 300 cycles at 0.1 C.

Like in LIBs and NIBs, cobalt can be used in KIBs. G. Ceder *et al.* reported P2-$K_{0.6}CoO_2$ cathode material showing a reversible capacity of 80 mAh $g^{-1}$ with $V_{avg}$ = 2.7 V vs. $K^+/K^0$ between 4.0–1.7 V.[55] In the same year, Hironaka *et al.* reported P3-$K_{0.67}CoO_2$ and P2-$K_{0.41}CoO_2$ cathodes for their K-ion storage performances between 2.0–3.9 V vs. $K^+/K^0$.[56] In both the structures, the phase transitions at higher voltages are not so prominent thereby retaining their P2 & P3 structures throughout the charge-discharge process, unlike its Na-counterpart, i.e. $Na_{0.67}CoO_2$. These reversible phase transformations are likely to be originated from $K^+$/vacancy ordering rather than gliding of $CoO_2$ slabs.

Among various O3-$K_xTMO_2$, $KCrO_2$ has the lowest energy and so it is thermodynamically very stable.[57] Ceder and co-workers showed that

$KCrO_2$ exhibits a discharge capacity of 92 mAh $g^{-1}$ between 4.0–1.5 V vs. $K^+/K^0$ due to higher K-content. The peaks of oxidation and reduction process in dQ/dV plot match well in terms of position and intensity which explains the electrochemical reversibility of $K^+$ intercalation/de-intercalation reactions. *In-situ* XRD shows the complex and reversible phase transition of O3–O′3–P′3–P3–P′3–P3–O3 during the $K^+$ extraction process. The reverse path is taken during K-insertion however at the end of discharge it is converted into a different O′3 structure. Y-K. Sun *et al.* reported P3-$K_{0.69}CrO_2$, synthesized *via* ion-exchange, showing the capacity of 80 mAh $g^{-1}$ with an extremely good retention of capacity i.e. 65% after 1000 cycles at 1 C rate, which is one of the best cycling performances for KIB layered oxide cathodes till date.[58]

#### 3.1.2.2. Binary and ternary TM systems

Fe-Mn based systems are also a good choice for KIBs like NIBs and LIBs. In 2016, L. Mai *et al.* first reported the P3-$K_{0.7}Fe_{0.5}Mn_{0.5}O_2$ nanowires with a high discharge capacity of 178 mAh $g^{-1}$ with 87% capacity retention after 450 cycles at a current density of 1 A $g^{-1}$.[59] Here, the special three-dimensional network structure provides an easy pathway for $K^+$ diffusion. Also, when a full cell was assembled with soft carbon as an anode, it could maintain 76% of the original capacity even after 250 cycles. To find out the optimal Fe substitution amount, a series of P3-$K_{0.45}Mn_{1-x}Fe_xO_2$ ($x = 0$–$0.5$) were synthesized by solid-state method.[60] Among them, $K_0Mn_{0.8}Fe_{0.2}O_2$ demonstrates the best electrochemical performance with a discharge capacity of 106.2 mAh $g^{-1}$ with a retention of 77.3% after 100th cycle. In recent years, metal oxides based on nickel and manganese, have become very popular in KIBs. Substitution with a little portion of $Ni^{2+}$ in Mn-based systems has proven to be an effective way to alleviate J.T. distortion. The Ni-substituted P3-$K_{0.5}Ni_{0.1}Mn_{0.9}O_2$ displays an increased capacity of 121 mAh $g^{-1}$ and retains 82% of initial capacity after 100 cycles while P3-$K_{0.5}MnO_2$ shows only 57% capacity retention.[61] Recently, P2-$K_{0.44}Ni_{0.22}Mn_{0.78}O_2$ was reported to maintain a single phase over the charge-discharge process with a minimal 1.5% volume change which is implicated in good cycling stability and retention in capacity (67% over 500 cycles).[62] Sometimes difficulties in a direct synthesis of these potassium based systems lead to an alternative way of synthesis i.e. electrochemical ion exchange from sodium based compounds. This concept was applied to synthesize a ternary Mn-rich layered oxide cathode, $K_{0.5}Mn_{0.8}Fe_{0.1}Ni_{0.1}O_2$, that displays a discharge capacity of 120 mAh $g^{-1}$ at 50 mA $g^{-1}$ with a retention of 74% in capacity after 300 cycles.[63] The shared oxygen atoms in $MnO_6$, $FeO_6$, and $NiO_6$ octahedra prevent the J. T. distortion and reversible phase transitions in charging-discharging process leads to its excellent structural stability. The full cell with hard carbon also shows excellent performances, including 113 mAh $g^{-1}$ of reversible capacity and holding of 89% capacity after 150 cycles. It signifies the practical applicability in KIBs. To summarize, we have briefly described the present development on the layered transition metal oxide cathode materials for Na-ion and K-ion batteries with the special emphasis on their electrochemical properties, alkali-ion storage mechanism, while discussing the key examples, their problems and solutions.

### 3.2. *Negative electrode materials*

To construct high-performing batteries and to reduce the gap between research and commercialization, enormous efforts have been concentrated towards the development of new negative electrodes and electrolytes in the past few decades. The negative electrode of the rechargeable alkali-ion battery shuffles the alkali ion back and forth during the cycling processes. In the case of alkali-ion batteries, metallic anode is not appropriate to use practically due to its high reactivity leading to safety hazards. In general, the wide class of anode materials can be broadly categorized into three main classes: insertion, alloying and conversion type materials.[64,65] Insertion-based anodes, primarily consist of titanium-based oxides and carbonaceous materials, that allow the intercalation or migration of alkali ions into the sites within the material. They are often desirable and attractive due to their high-rate capability, small volume change, good cycling stability and therefore they are closest to commercialization. The elements belonging to the groups 14 and 15 such as Si, Ge, Sn, P Sb, Bi, or binary alloys of these, e.g., BiSb, *etc.* define the most common alloying-based and promising anode materials for NIBs and KIBs.[66,67] Then there are conversion based anodes that mainly consists of transition metal oxides, sulfides, selenides, and phosphides-based materials. This involves breaking and formation of chemical bonds during the cycling, that leads to potentially huge capacities. This section will discuss the recent advances in the negative electrode materials, specifically carbonaceous based materials and alloy-based anodes for NIBs and KIBs.

### 3.2.1. *Carbonaceous materials*

The carbonaceous materials such as graphite, carbon nanotubes (CNTs), amorphous carbons such as soft carbon, (SC) and hard carbon (HC), graphene, etc., are the most explored anodes for alkali metal-ion batteries.[68] Post commercialization in 1991, graphite anodes have been ruling the LIBs market as the most successful practical anode owing to their specific capacity of 372 mAh $g^{-1}$ and low operational voltage (~0.1 V vs. $Li^{+}$/Li). However, it fails to exhibit satisfactory results in NIBs (35 mAh $g^{-1}$).[69,70] The poor rate capacity, and disoriented intercalation of ions along the graphite layers, make this candidate inappropriate for high energy density applications in NIBs. While in case of KIBs, graphite anodes are known to exhibit the decent capacity of 279 mA h $g^{-1}$ by transforming into the ionic graphite intercalation compounds (GICs). In one of the recent reports, Komaba *et al.*[71] assessed graphite anodes and disclosed that the potassium plating arises at lesser voltage than that of lithium, which makes them promising candidates for high voltage KIBs. Although CNTs exhibit high capacities as anodes in LIBs, their performances in NIBs and KIBs are fairly poor.[72] Alternatively, when soft carbons lose their disordered structure and become graphitized at higher temperatures (> 2500°C), NIBs exhibit very limited capacity of 100 mAh $g^{-1}$.[73] Mainly because of decreasing Na insertion capacity, due to increase in graphitization at annealing temperatures greater than 1200°C, a smaller number of sodium is inserted into a given site.[74] However, in KIBs, the increase in heat treatment has a positive influence of $K^{+}$ storage mechanism. An-min Cao and workers studied soft carbon anodes, treated at 1200°C, and exhibited impressive reversible capacity of ~300 mAh $g^{-1}$ with better cycling stability when compared to that of graphite anodes in KIBs.[75] The other non-graphitic forms of carbons such as hard carbons and carbon black are exceptional intercalation

contenders in alkali-ion batteries due to their additional storage mechanism other than graphite-type intercalation–deintercalation.[76–78] Particularly, hard carbon has caught tremendous attention in alkali-ion batteries due to its randomly distributed graphitized micro-domains, environmentally friendly organic precursors and easy to process synthesis techniques. Moreover, presence of voids and distorted graphene nanosheets further restrain the development of graphitic structure even at temperature above 3000°C.[79]
The 'house of cards' model was developed by Dahn and co-workers to depict the structure of hard carbon and is considered as one of the most acceptable models.[76,80] This model suggests the presence amorphous regions with defects, and nanopores in the hard carbon structures, formed by the sequence of graphene layers.[76,81] This graphitic domain with uneven expanded layers accommodates the alkali ions depending on the interlayer spacing of hard carbon, radius of ions and stability of the GICs formed after the intercalation.[69] Therefore the insertion mechanism of alkali ions can happen via following process: chemisorption, intercalation between graphitic layers, adsorption, and pore filling. In case of NIBs, Dahn and Steven proposed the "intercalation-pore filling" mechanism that suggests the intercalation of sodium ions in the graphitic layers originates in the sloping region, followed by the pore filling occurring at the low-voltage plateau region.[76,80] While $Li^+$ ions form stable Li-GIC with the formation of $LiC_6$, on the other hand, the Na-rich GICs of $NaC_6$ and $NaC_8$ exhibit the positive formation enthalpies, which implies they are thermodynamically unstable.[70] In that case, the intercalation pore filling model is widely accepted model, that suggest intercalation of sodium ions occurs in high voltage region (1.2 V to 0.2 V vs. $Na^+/Na^0$) with the pore filling process in low voltage region (~ 0.1 V vs. $Na^+/Na^0$).[72,82] These results were further supported experimentally by various *in-operando* techniques such as *in-situ* SAXS, XRD, Raman and *operando* $^{23}$Na solid state NMR by various groups.[83–86] Interestingly, other mechanisms such as "adsorption–pore filling" and "adsorption–intercalation" were equally explored, proposed and supported by various studies. For instance, Ding and co-workers concluded that the insertion kinetics is not entirely due to pore filling process by presenting a direct relationship between the porosity of hard carbon and the capacity achieved.[87] In one such study, Cao *et al.* explained the "adsorption-intercalation" mechanism from the observation that with

higher turbostatic nanodomains, the insertion capacity is increased in the plateau region than the sloping region.[85] Some reports proposed an alternate Na-ion storage mechanism which suggests insertion of $Na^+$ in the slope region, and pore filling in the low voltage region at the end of the process (Figure 4a).[88] Overall, based on the experimental reports, all three kinds of $Na^+$ storage behaviors may exist simultaneously, as the structure of hard carbon is comprised of pore defects and distorted graphene sheets. In one comparative study it was stated that Li ions are primarily adsorbed at defects, edges, and functional groups, that results in a sloping voltage profile, with very little contribution from the pore filling to the overall Li storage capacity. In case of NIBs, a large fraction of storage capacity arises from pore filling process at the low voltage region.[72]

The typical voltage profile for hard carbon in KIBs distinctly consist of two regions: slope region and quasi-plateau region. The K-ion storage in sloping region is mainly originates by the surface adsorption process while the quasi-plateau region is dominated by intercalation mechanism.[89] When compared with Na cells, Jian and co-workers demonstrated the average K-ion insertion plateau region in hard carbon spheres is situated at 0.2 V versus $K^+/K$, compared to that of Na-ion insertion region at 0.05 V versus $Na^+/Na$ (Figures 4b and 4c).[89] Therefore, in this aspect, the problems like metal plating and dendrite formation can be circumvented more effectively in KIBs than in NIBs. Hence, different from Li and Na storage behavior (Figure 4d), K ions are adsorbed at defects results in reduced voltage hysteresis and facilitated the insertion of ions into the micropores at a high voltage region.[72] It is also evident that weaker Lewis acidity and smaller desolvation energy of $K^+$ in aprotic solvents, enables its higher ion diffusivity in electrolytes.[71,90–92] Hence, soft carbon and hard carbon show improved rate performance in KIBs than in NIBs.

Komaba and Palacin co-workers demonstrated that the overall cycling performance of hard carbon anodes in NIBs can be further improved with the interplay between electrolyte systems using relevant salts and solvents.[93–95] In the most recent reports, many studies combined the computational methods and experimental results and summarized the $NaPF_6$ PC/FEC-based electrolyte as the conventional and widely accepted electrolyte system to effectively drive the hard-carbon electrode for higher cycling performance of NIBs.[72,96,97] The formation and evolution of the

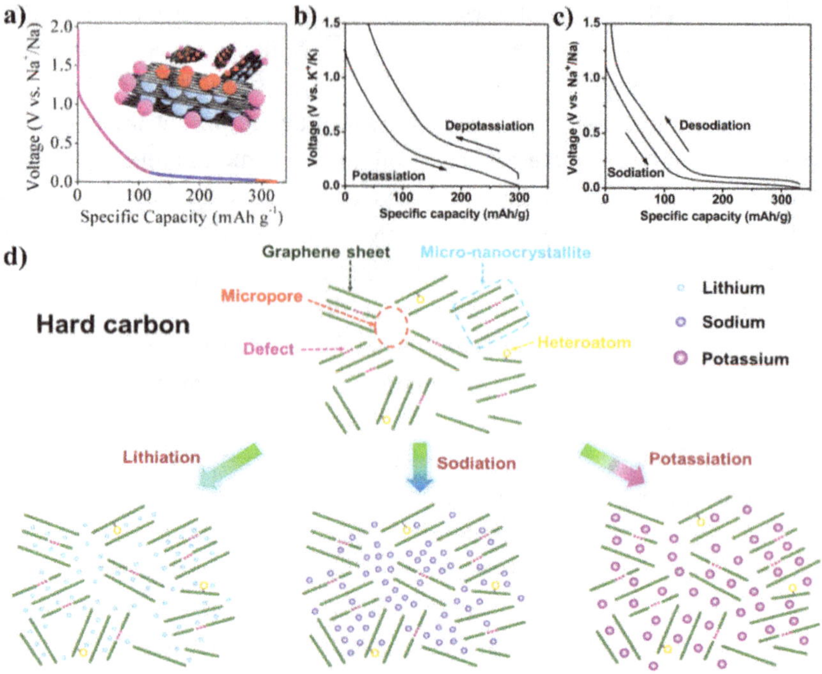


Figure 4. (a) The representation of "adsorption–intercalation" mechanism. Reproduced with permission.[88] © 2015, American Chemical Society. Discharging/charging profiles at C/10 in the second cycle of hard carbon microspheres in (b) KIBs and (b) NIBs. Reproduced with permission.[89] © 2015, Wiley-VCH. (d) Schematic structure illustration of (top), lithiated (bottom left), sodiated (bottom middle), and potassiated (bottom right) hard carbons. Reproduced with permission.[72] © 2021 American Society of Chemistry.

SEI and low initial coulombic efficiency (ICE) also determines the cycling performance of hard carbon anodes in alkali-ion batteries, thus limiting the viability of the full cells. In recent times, efforts are devoted to grasp a better understanding of these challenges. Based on various spectroscopic and surface techniques such as TOF-SIMS, XPS, TEM, SEM and Raman, it is observed that a comparatively thinner layer of SEI is formed on anodes in LIBs than in NIBs.[98,99] Also, the SEI layer in LIBs is primarily comprised of organic compounds whereas in NIBs it is majorly composed of inorganic species. The addition of FEC as an additive can further help to reduce the initial irreversible capacity, thereby stabilizing the SEI formation, resulting into the to higher cycling performance.[94] Although the

SEI is studied extensively in LIBs and NIBs, it is poorly understood in KIBs. Mainly because of the highly reactive nature of potassium metal, which results in un-stable formation of SEI at the interface and thereby adds to the lower ICE in graphite anodes. In one of the recent reports, it was found that the DEC solvent in the standard EC/DEC carbonate-based electrolyte is both chemically and electrochemically unstable against the common anode materials (i.e., graphite and metal K) in KIBs.[100] Henceforth, the stable and highly $K^+$-conductive SEI film is desirable. Soto and co-workers reported the in-depth first-principles investigation studies on SEI and explained the role of pre-cycling of an anode to form the stable SEI layer in alkali-ion based electrolyte.[101] The ionic transport of $Li^+$, $Na^+$ and $K^+$ can be kinetically controlled in the specific ether-based electrolyte systems due to significant difference in their activation energy barriers. Several studies suggest that SEI in ether-based electrolytes is thinner, compact and more ionically conductive than that of ester-based systems, which enable faster alkali-ion diffusion and capacitive storage. Certainly, the improved ICE, up to 75% was realized in diglyme system, for the microporous carbon anodes which is way more higher than in carbonate-based electrolytes (~40%) in general.[102] Also, the pre-treatment of carbon-based electrode materials can further reduce the initial cycle capacity loss and thus can enhance the ICE.

### 3.2.2. *Alloys*

Alloying anodes, i.e., Group 14-16 elements-based materials, are the second most explored and promising anodes for the alkali-ion batteries due to their huge specific capacities, low cost, and safe working potentials. Many elements including Bi, Sb, Sn, P, Si, Pb, Ge etc. are reactive and form alloys towards alkali-ion metals (Figure 5a). Such materials tend to take the larger number of alkali ions, by forming intermediate phases at multiple steps, thereby delivering much higher specific capacities than insertion-based carbon materials.[66,103–105] For example, as shown in Figure 5b, the P-, Si- and Sb-based alloy anodes have theoretical capacities of 2590, 4400 and 660 mAh $g^{-1}$, respectively, which is higher than that of capacity achieved by the graphite anode used in the present LIBs. The

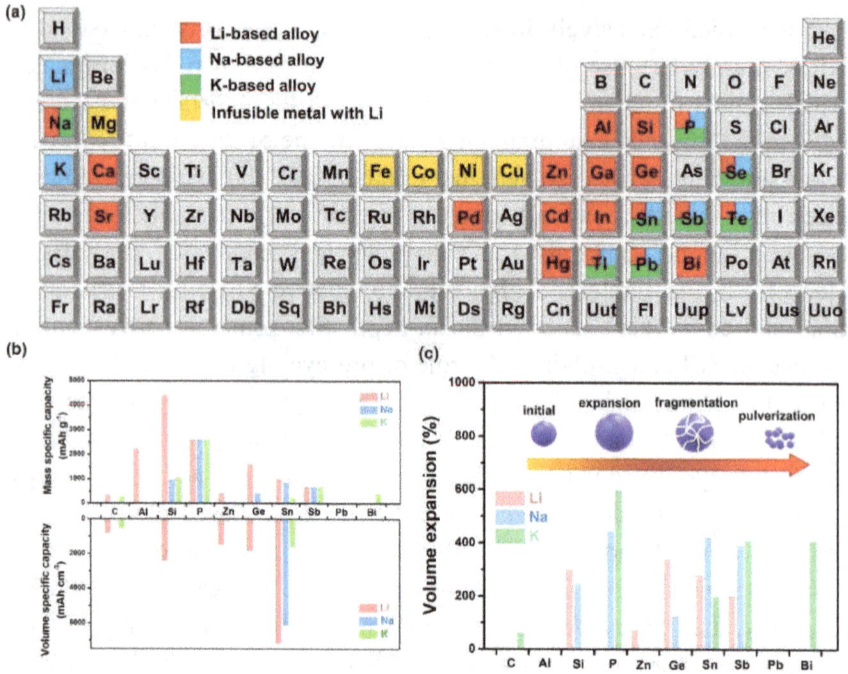


Figure 5. (a) Graphic illustration of conventional alloying-based anodes of alkali-ion batteries. (b) Theoretical gravimetric and volumetric specific capacities of prominent alloy anodes. (c) Percentage of volumetric variations during the charge-discharge cycles in ally anodes. Reproduced with permission.[103] © 2021 Elsevier.

reaction generally happens reversibly as per the general scheme presented in the following equation:

$$xA + yM^{+} + ye^{-} \leftrightarrow A_xM_y$$

(M = Li, Na, or K and A = an element that can alloy with Li, Na, or K)

Nevertheless, the practical application of such alloy anodes is hampered and is still limited by unsatisfactory cycle life as well as low Coulomb efficiency due to various reasons. Foremost, as per the alloying mechanism, this type of anode materials suffers from huge volumetric expansion and structural changes due to multi-electron transfer during the charge–discharge process. For instance, the volume expansion in Sb anode increases from 135% to ~300–400% in LIBs to NIBs and KIBs (Figure 5c), resulting from increase in the number of multistep intermediates from lithium to potassium cells. In addition, the continuously exposed surface arising from such dramatic volumetric changes leads to consumption of electrolyte, further leading to the pulverization and falling off active material from the current collector, resulting in the poor rate capability and capacity decay. However, to combat such issues, the reduction of the particle in nanoscale regime, or the addition of conductive carbon matrix to the alloys is one of the universal methods followed to establish the buffering effect to mitigate the volumetric strain in alloys.[66,103,106]

#### 3.2.2.1. Antimony based anodes

Antimony, which possesses good electrical conductivity, can also be described as a smart choice for anode in the alkali-ion batteries. Sb also undergoes multiple steps to form $Li_3Sb$ and $Na_3Sb$ in Li and Na-ion systems corresponding to the high theoretical capacity of 660 mAh $g^{-1}$.[107,108] However, it suffers from huge volumetric variations during electrochemical cycling. To tackle such critical issues, various approaches such as nanosizing, intermetallic alloying, carbon coating and developing carbon-based Sb materials have been employed. Akin to metallic antimony, other Sb-based such as oxides and chalcogenides are also explored as effective anodes in alkali-ion batteries and will be discussed in the subsequent section. Darwiche *et al.*[107] compared the electrochemical performances of metallic Sb anodes in LIBs and NIBs and suggested that alloying process of Sb is not same in both the systems. In case of Na, the intermediate stages are mainly amorphous and cannot be defined accurately, whereas for Li, these phases are crystalline.[107] In case of Na, the reaction processes do not proceed via alloying mechanism only, while in case of Li, the structural evolution remains unclear. However, employing the relevant electrode formation using carboxymethyl cellulose, carbon black, and carbon fibers and suitable electrolyte additive such as FEC, bulk Sb anodes performed better in NIBs than in LIBs.[66,105] In the recent times, intermetallic alloys such as Sb-Fe/Co/Ni/Cu, that are comprised of Sb and an inactive element, have also been explored and have demonstrated better performance compared to elemental Sb.[103] In such materials, the inactive element and alkali-ion will inhabit the sites of the alloy alternatively and thereby the volumetric alteration will be much smaller than that of the pure metal. At the same time, this will provide the structural reinforcement and electrical conductivity, thereby enhancing the overall rate performance of the electrodes. In the recent times, $Cu_2Sb$ is one of the most explored and popular Sb-based intermetallic anodes owing to the high conductivity of Cu. Fransson *et al.*[109] synthesized $Cu_2Sb$ Li-ion anode *via* ball milling method and systematically evaluated the phase transitions of the working anode using XRD. Combined with active metals such as Sb-In/Si/Sn/Bi based material that form Sb alloys, have also been examined as efficient anodes.[103,110] The enhanced execution in such dual systems can be ascribed to the synergistic effect of alloying between the two active metals which further smoothens the mechanical stress. In one such study by Manthiram's group,[111] they fabricated homogeneous alloys consisting of Bi and Sb with carbon using high energy ball milling approach and evaluated as potential anodes in both LIBs and NIBs. Specifically, $Bi_{0.57}Sb_{0.43}$-C and $Bi_{0.36}Sb_{0.64}$-C anodes exhibited electrochemical performances superior to those of their individual counterparts Bi-C and Sb-C. For example, the $Bi_{0.57}Sb_{0.43}$-C delivered good cycling stability for 300 cycles at the current density of 200 mA $g^{-1}$ in LIBs. While in Na cells, it delivered capacity of 326 mA h $g^{-1}$ at a high current rate of 2.6 C (1000 mA h $g^{-1}$). Xiao *et al.*[112] first reported SnSb alloy as the probable anode for NIBs, which underwent sequential (de)alloying of Sb and Sn with sodium during electrochemical cycling process. This mechanism was further studied and supported by Buriak *et al.*[113] using *in situ* TEM studies. In KIBs, McCulloch *et al.* first demonstrated Sb as anode for KIBs in 2015 in K-$O_2$ battery, with the huge reversible capacity of 650 mAh $g^{-1}$ corresponding to the presence of cubic $K_3Sb$ alloy phase.[114,115] The pure Sb electrode is known for its huge capacity fade owing to the large volumetric expansion during the cycling process. In another study by Wang's group, using DFT calculations, they proposed and demonstrated the sequential phase transition of Sb to $KSb_2$, KSb, $K_5Sb_4$, and $K_3Sb$, with the values of relative volume expansions of 73.7%, 147.6%, 191.7%, and 456.6%, respectively.[115] In spite of numerous reports on Sb anodes, the problem of fast capacity fading still remains unsolved. Future research can be directed towards the embedding of nano-Sb particles into carbon matrix. Introducing Sb with the conductive carbon on nanoscale can work two folds: first, it will increase the conductivity of electrodes and second, the carbon matrix will elevate their volume expansions eventually.

Ramireddy *et al.*[116] prepared Sb and carbon composites in one of the early works and demonstrated their electrochemical performances in NIBs and LIBs. Along this line, various reports discussed the role of different forms of carbon matrixes such as nanospheres, nanotubes to achieve the attractive cycling performances of Sb-C based anodes in alkali-ion systems.[66,105]

Apart from these, various oxides such as $Sb_2O_3$, chalcogenides such as $Sb_2S_3$ and $Sb_2Se_3$, are also studied and assessed as anodes in alkali-ion batteries.[105,110,117] $Sb_2O_3$ has been considered as one such suitable negative electrode material for NIBs due to its high theoretical capacity of 1102 mAh $g^{-1}$. In addition, the Na-ion intercalation advances via the formation of amorphous $Na_2O$ that accommodates the dramatic volumetric changes during the cycling.[117] Antimony sulfide ($Sb_2S_3$), which possesses high theoretical capacity of 947 $mAhg^{-1}$, has also gained considerable interest as a possible anode in alkali-ion batteries.[118] One mole of $Sb_2S_3$ can get 12 moles of $Na^+$ ions in its composition.[118] The formation of $Na_2S$ retains higher reversibility in case of sulfide anode than $Na_2S$ formed during sodiation process of $Sb_2O_3$. Lui *et al.*[119] designed the few-layered antimony sulfide/carbon sheet (SBS/C) anode using solution-triggered exfoliation. Based on the post cycling XRD analysis and *in-situ* TEM studies, they proposed that K-ion storage mechanism for the $Sb_2S_3$ involves uptake of eight electrons per unit, thereby leading to high-capacity values in $Sb_2S_3$-based anodes. The proposed reaction for this charge storage reaction can be proposed as $Sb_2S_3 + xK^+ + xe^- - M_yK_3Sb + zK_2S_3$, which was like the other reports.[119] Nevertheless, although Sb-based anodes are recognized as high-capacity anodes, but complete understanding of their reaction processes needs to be appropriately studied in order to harvest their complete utilization as successful advance anodes in alkali-ion batteries.

#### 3.2.2.2. Bismuth based anodes

Metallic bismuth and its intermetallic alloys are regarded as attractive anode materials in the alkali-ion batteries. Bismuth is an economical, benign, and environmentally friendly element forming distinct layered structures with interlayer spacing. Metallic bismuth closely resembles graphite in terms of structure and appearance. It shows a huge decent specific capacity of 385 mAh $g^{-1}$ operating at a low sodiation potential of 0.5 V. In one of the studies, Yang and co-workers also examined Bi@C anode for sodium/lithium-ion batteries, prepared *via* aerosol spray pyrolysis technique.[120] When evaluated as an anode in NIBs and LIBs, the Bi@C microspheres delivered an impressive specific capacity of 123.5 mAh $g^{-1}$ and 280 mAh $g^{-1}$ even after 100 cycles at 100 mA $g^{-1}$, respectively. It is realized that the sodiation of Bi proceeds via sequential formation of NaBi and $Na_3Bi$ (Bi → NaBi → $Na_3Bi$) according to the Na/Bi Phase diagram. However, depending upon the size of bismuth particle size, different $Na_3Bi$ phase can form as demonstrated by Sottman *et al.*[121] They suggested that during sodiation, micron sized Bi had a hexagonal crystallographic phase whereas the nanoparticle size Bi had a cubic phase (Figure 6a). Further applicability of bismuth was tested and demonstrated by Want *et al.*,[122] where they developed an innovative Na-ion full cell with bulk Bi as anode, and $Na_3V_2(PO_4)_3$/CNT composite as cathode in the $NaPF_6$-diglyme electrolyte. The full cell performed exceptionally, exhibiting both the high-power density of 2354.6 $Wkg^{-1}$ and energy density of 150 Wh $Kg^{-1}$ in addition to the superior cycling stability in the temperature range from −15 to 45°C.

Moving onto KIBs, Huang *et al.*[123] in one of their reports presented in-depth studies on alloying of Bi anodes in KIBs with the aid of *in-situ* XRD and density functional theory (DFT) analysis. They proposed that (de)-potassiation happens in three reversible two-phase reactions of $Bi \leftrightarrow KBi_2 \leftrightarrow K_3Bi_2 \leftrightarrow K_3Bi$ (Figure 6b).[123] The volume expansion increases up to 409% for the formation of $K_3Bi$ at the end of the discharge cycles. This problem of mechanical stress can be tackled by the introduction of conducting matrix, which could be of carbon or from other element, to alleviate the volumetric strain, thus, to improve the transport of ions. It was also realized that the overall anode performance can be further enhanced when binary or ternary alloy systems are utilized. In such cases, the volume expansion is alleviated synergistically when compared to their individual counterparts. For instance, systems with similar physio-chemical properties, Bi-Sb alloys. These are the most commonly studied anodes in alkali-ion batteries due to their capability to form solid solutions ($Bi_xSb_{1-x}$) at any molar ratios. Xiang and co-workers fabricated

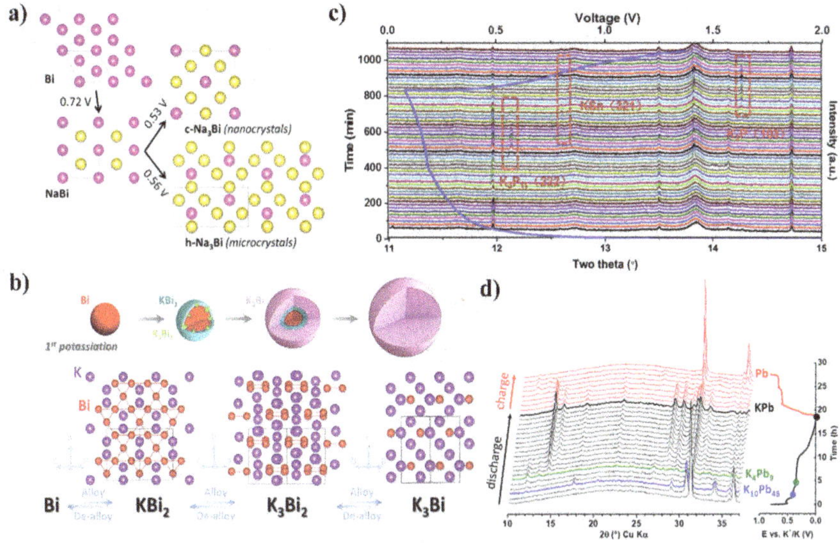


Figure 6. (a) Illustration of the different structural and electrochemical pathways in the Na–Bi system. Reproduced with permission.[121] © 2016 American Chemical Society. (b) Alloying and dealloying for Bi anode in KIBs for the subsequent cycles. Reproduced with permission.[123] © 2018 WILEY-VCH. (c) In operando synchrotron XRD patterns of $Sn_4P_3$ anode. Reproduced with permission.[139] © 2018 Elsevier Inc. (d) Operando XRD measurement of the Pb electrode during alloying (left) with the corresponding voltage profile plotted as a function of time (right). Black and red patterns represent the discharge and the charge, respectively. Reproduced with permission.[140] © 2016 American Chemical Society.

bismuth–antimony alloy nanoparticles embedded in a porous carbon matrix (BiSb@C) using facile freeze-drying and pyrolysis method. Operando X-ray diffraction studies proposed the reversible potassiation/depotassiation reaction mechanism of (Bi,Sb) ↔ K(Bi,Sb) ↔ $K_3$(Bi,Sb) for the composite.[124] Bi based materials such as oxides ($Bi_2O_3$), chalcogenides ($Bi_2S_3$, $Bi_2Se_3$, etc.) or bismuth-based oxyhalides (BiOCl) can be considered as potential candidates for exploring this anode to the fullest.[110,125] Bismuth oxide ($Bi_2O_3$) is known for its superior Na-ion storage capacity of 690 mAh $g^{-1}$.[126] Bismuth sulfide ($Bi_2S_3$) has also been explored as potential NIB anodes due to its notable gravimetric capacities up to 625 mAh $g^{-1}$. In most of the cases, such anodes suffer from huge irreversible capacity loss due to the immense surface area and continuous rupture of solid electrolyte interphase (SEI) occurring as a result of constant phase transitions. Another interesting strategy would be coupling of such derived anodes with electrolytes, where the electrochemical performance of Bi can be influenced along with its structural evolution.[123,127] In one such study reported by Wang *et al.*,[128] they utilized $NaPF_6$-diglyme electrolyte instead of the conventional carbonate-based PC solvent.[128] In this case, the bulk Bi preserves its porous integrity upon cycling in diglyme system. This further promotes faster charge kinetics with excellent capacity retention of 94.4% exhibiting the capacity of 400 mAh $g^{-1}$ even after 2000 cycles.

#### 3.2.2.3. Tin based anodes

Tin (Sn) has been extensively studied in LIBs and NIBs systems in the past few decades. In fact, Sn as an anode has been used in commercial rechargeable batteries "Nexelion" by the Sony Corp. In the above studied cases, the final product in all three alkali-ion systems is same, i.e., $M_3Sb$ or $M_3Bi$. In case of Sn, the alloying reaction of same material with different alkali metals may not be same. For instance, Tin (Sn) electrodes delivers reversible capacities of 640 and 655 mAh $g^{-1}$, corresponding to the growth of $Li_{17}Sn_4$ and $Na_{15}Sn_4$ phases in Li and Na cells.[129,130] Till date the alloying mechanism of Sn with Na remains indefinite and ambiguous mainly due to the formation of a series of metastable crystalline and amorphous phases upon cycling. According to the Sn–Na binary phase diagram, the thermodynamically stable and most identified are $NaSn_6$, $NaSn_4$, $NaSn_3$, $NaSn_2$, $Na_9Sn_4$, $Na_3Sn$, and $Na_{15}Sn_4$. Ellis *et al.*[131] confirmed the formation of $Na_{15}Sn_4$ as the terminal intermetallic phase which is also in line with other reports. Interestingly, with the increasing number of reports, it is realized that the phase sequences of Na–Sn alloys can be varied and are influenced by a variety of testing parameters such as electrolyte configuration, current rate, cutoff voltage, and cell assembly.[132,133] Baggetto *et al.*[134] discovered that $Na_7Sn_3$ (space group: $R\bar{3}m$) as one of the intermediate phase formed upon cycling instead of $Na_5Sn_2$. On one hand where Sn exhibits high theoretical capacity for Li and Na systems, the same anode delivers a very marginal reversible capacity of 110–150 mAh $g^{-1}$ ascribed to the presence of KSn on full potassiation.[135,136]

It has been demonstrated that Sn undergoes a series of phases in both Na and Li cells during the cycling, corresponding to the large volume expansion of 420% and ~300% respectively. Whereas in K cells, this alloying proceeds via two stages, first step corresponds to formation of $K_4Sn_9$ amorphous phase with the volume expansion of 113% followed by a final potassiated product of KSn phase with 197% volume expansion.[135,137] This implies that the electrochemical uptake of K ions is governed by both, thermodynamic and kinetic process. As mentioned earlier, it is also meaningful to include a synergistic element to buffer the volumetric change and thereby help to enhance the electrochemical performance.[124,138] For instance, the $Sn_4P_3$ system proposes synergistic reaction mechanisms for sodium and potassium storage. When P is electrochemically active and takes up the alkaline ions, the Sn will act like a conductive matrix to buffer the excess volumetric alteration and will increase the electrical conductivity. And likewise, when Sn is active, the P and the earlier formed $A_xP$ (A = Na, K) will offer mutual mitigation to alleviate the stress due to the volume variation. For instance, Zhang *et al.* investigated $Sn_4P_3$@N-doped carbon fibres as anode in KIBs and proposed synergistic K-storage mechanisms with the consecutive conversion of P to $K_3P_{11}$, $K_3P$ followed by the alloying of Sn to KSn, as analyzed by *operando* XRD studies (Figure 6c).[139] The anode exhibited reversible capacity of 160.7 mAh $g^{-1}$ after 1000 cycles at the rate of 500 mA $g^{-1}$.

#### 3.2.2.4. Lead based anodes

Lead, regarded as the neglected anode materials in alkali-ion batteries, can exhibit a theoretical capacity of 485 mAh $g^{-1}$ corresponding to $Na_{15}Pb_4$ and 569 mAh $g^{-1}$ corresponding to $Li_{22}Pb_5$ in Na- and Li-ion batteries, respectively. Nevertheless, because of its toxicity and weight, this material has always been put aside in alkali-ion batteries. Until last decade, it was considered that the sodiation of lead happens as per the Na-Pb phase diagram, i.e., alloying commences via sequential formation of $NaPb_3$, NaPb, $Na_5Pb_2$, and $Na_{15}Pb_4$ phases.[131] But in 2014, Obrovac's group reported an *in-situ* XRD study on lead electrode thin film in Na-cell and showed that the electrochemical processes do not precisely follow the Na–Pb phase diagram during sodiation, also no single-phase regions were evidenced and $Na_5Pb_2$ was not formed.[131] Lead (Pb) forms alloy with potassium giving a theoretical capacity of 129 mAh $g^{-1}$. Berthelot *et al.* studied the electrochemical properties of lead and through the *opernado* XRD analyses. They revealed that the potassiation of Pb undergoes the phase transformation by forming $K_{10}Pb_{48}$, $K_4Pb_9$, and final KPb alloys, whereas only $K_4Pb_9$ phase is formed in the de-potassiation process (Figure 6d).[140] Although, the reports on alloying of Pb in alkali-ion batteries are extremely limited, further studies on fundamental understanding of this anode is important. Moreover, with proper handling, the possibility of potential research and development of lead as the alloying type anode material for NIBs and KIBs can be achieved. In brief, the current state-of-the-art and preliminary studies on these Sb, Bi, Pb and Sn based anodes have shown that these anodes may have a satisfactory electrochemical performance in alkali-ion batteries, specifically in Na-ion and K-ion batteries.

## 4. Summary and Outlook

In summary, the present report briefly discusses the recent progress on the layered transition metal oxide cathode materials as well as carbonaceous and alloy-based anode materials for Na-ion and K-ion batteries. Regardless of their large reversible capacities, layered oxides materials still have a long way to go. They are not stable in air and undergo multiple phase transformations at higher voltages, that result in poor performance eventually. To overcome these issues, strategies such as multiple-cations doping in transition metal layer, morphology optimization, and protective layer coating are presented here.[19,50] As evident, some full cells performances are eye-catching but there is still room for the improvement to advance them towards the large-scale practical application. The present research requires to focus on cell components optimization and engineering for the commercialization of such material in full cell configurations. For the negative electrode, although a considerable amount of research has been concentrated on the carbon- and alloy-based anode materials, so far there is loads of room for supplementary chemical analysis and experimentations to understand their functioning to bridge the gap at commercial level. For instance, in the case of carbon-based anodes, pre-treatment of carbon materials can considerably reduce the initial capacity loss. However, further research efforts should be directed towards the understanding of SEI formation as well as devising of stable and robust SEI to utilise the full potential of such anodes. In case of alloy anodes, the fundamental knowledge of alkali-ion storage mechanism using advance operando techniques coupled with theoretical studies is required for the improvement of the electrochemical properties. In addition to the material itself, optimization of other key components is also essential to enhance the overall performance of such electrode materials.[65,66] For example, the present major issues in alloy anodes such as the uncontrollable interfacial reactions with the electrolyte, huge volumetric changes leading to the continuous SEI fracture and reformation which eventually leads to capacity failure, is urgently required. Therefore, the interplay between the chemistry of binder and electrolyte configurations can be employed and studied to harness the full potential of alloy anodes.[98,99] In the coming years, we believe that post-lithium systems such as NIBs and KIBs can serve as viable and appropriate solutions to the large-scale electrochemical grid storage systems.

## References

1. K. C. Divya and J. Østergaard, Battery Energy Storage Technology for Power Systems—An Overview, *Electr. Power Syst. Res.* **79**(4), 511–520, (2009).
2. S. Teleke, M. E. Baran, S. Bhattacharya and A. Q. Huang, Optimal Control of Battery Energy Storage for Wind Farm Dispatching, *IEEE Trans. Energy Convers.* **25**(3), 787–794, (2010).
3. J. B. Goodenough, Energy Storage Materials: A Perspective, *Energy Storage Mater.* **1**, 158–161, (2015).
4. J. B. Goodenough and Y. Kim, Challenges for Rechargeable Li Batteries, *Chem. Mater.* **22**(3), 587–603, (2010).
5. B. Dunn, H. Kamath and J.-M. Tarascon, Electrical Energy Storage for the Grid: A Battery of Choices, *Science (80).* **334**(6058), 928–935, (2011).
6. R. M. Dell and D. A. J. Rand, Understanding Batteries, (Royal Society of Chemistry, Cambridge, 2007).
7. M. Li, J. Lu, Z. Chen and K. Amine, 30 Years of Lithium-Ion Batteries, *Adv. Mater.* **30**(33), 1800561, (2018).

8. J. M. Tarascon, The Li-Ion Battery: 25 Years of Exciting and Enriching Experiences, *Interface Mag.* **25**(3), 79–83, (2016).
9. B. Scrosati and J. Garche, Lithium Batteries: Status, Prospects and Future, *J. Power Sources* **195**(9), 2419–2430, (2010).
10. P. K. Nayak, L. Yang, W. Brehm and P. Adelhelm, From Lithium-Ion to Sodium-Ion Batteries: Advantages, Challenges, and Surprises, *Angew. Chemie - Int. Ed.* **57**(1), 102–120, (2018).
11. V. Palomares, P. Serras, I. Villaluenga, K. B. Hueso, J. Carretero-González and T. Rojo, Na-ion Batteries, Recent Advances and Present Challenges to Become Low Cost Energy Storage Systems, *Energy Environ. Sci.* **5**(3), 5884–5901, (2012).
12. S. W. Kim, D. H. Seo, X. Ma, G. Ceder and K. Kang, Electrode Materials for Rechargeable Sodium-ion Batteries: Potential Alternatives to Current Lithium-ion Batteries, *Adv. Energy Mater.* **2**(7), 710–721, (2012).
13. S. Roberts and E. Kendrick, The Re-emergence of Sodium Ion Batteries: Testing, Processing, and Manufacturability, *Nanotechnol. Sci. Appl.* **11**, 23–33, (2018).
14. A. Eftekhari, Potassium Secondary Cell Based on Prussian Blue Cathode, *J. Power Sources* **126**(1–2), 221–228, (2004).
15. A. Eftekhari, Z. Jian and X. Ji, Potassium Secondary Batteries, *ACS Appl. Mater. Interfaces* **9**(5), 4404–4419, (2017).
16. R. Rajagopalan, Y. Tang, X. Ji, C. Jia and H. Wang, Advancements and Challenges in Potassium Ion Batteries: A Comprehensive Review, *Adv. Funct. Mater.* **30**(12), 1909486, (2020).
17. K. Kubota, M. Dahbi, T. Hosaka, S. Kumakura and S. Komaba, Towards K-Ion and Na-Ion Batteries as "Beyond Li-Ion", *Chem. Rec.* **18**(4), 459–479, (2018).
18. C. Delmas, C. Fouassier and Hagenmuller, Structural Classification and Properties of the Layered Oxides, *Phys. B+C* **99**(1–4), 81–85, (1980).
19. P.-F. Wang, Y. You, Y.-X. Yin and Y.-G. Guo, Layered Oxide Cathodes for Sodium-Ion Batteries: Phase Transition, Air Stability, and Performance, *Adv. Energy Mater.* 8(8), 1701912, (2018).
20. P. Ge and M. Fouletier, Electrochemical Intercalation of Sodium in Graphite, *Solid State Ionics* **28–30**(2), 1172–1175, (1988).
21. R. Berthelot, D. Carlier and C. Delmas, Electrochemical Investigation of the P2–$Na_xCoO_2$ Phase Diagram, *Nat. Mater.* **10**(1), 74–80, (2011).
22. B. V. Rami Reddy, R. Ravikumar, C. Nithya and S. Gopukumar, High Performance $Na_xCoO_2$ as a Cathode Material for Rechargeable Sodium Batteries, *J. Mater. Chem. A* **3**(35), 18059–18063, (2015).
23. X. Ma, H. Chen and G. Ceder, Electrochemical Properties of Monoclinic $NaMnO_2$, *J. Electrochem. Soc.* **158**(12), A1307, (2011).
24. J.-P. Parant, R. Olazcuaga, M. Devalette, C. Fouassier and P. Hagenmuller, Sur quelques nouvelles phases de formule $Na_xMnO_2$ ($x \leqslant 1$), *J. Solid State Chem.* **3**(1), 1–11, (1971).

25. J. Billaud, R. J. Clement, A. R. Armstrong, J. Canales-Vazquez, P. Rozier, C. P. Grey and P. G. Bruce, β-$NaMnO_2$: A High-Performance Cathode for Sodium-Ion Batteries., *J. Am. Chem. Soc.* **136**(14), 17243–17248, (2014).
26. M. H. Han, E. Gonzalo, M. Casas-Cabanas and T. Rojo, Structural Evolution and Electrochemistry of Monoclinic $NaNiO_2$ upon the First Cycling Process, *J. Power Sources* **258**, 266–271, (2014).
27. P. Vassilaras, X. Ma, X. Li and G. Ceder, Electrochemical Properties of Monoclinic $NaNiO_2$, *J. Electrochem. Soc.* **160**(2), A207–A211, (2013).
28. L. Wang, J. Wang, X. Zhang, Y. Ren, P. Zuo, G. Yin and J. Wang, Unravelling the Origin of Irreversible Capacity Loss in $NaNiO_2$ for High Voltage Sodium Ion Batteries, *Nano Energy* **34**, 215–223, (2017).
29. N. Yabuuchi, H. Yoshida and S. Komaba, Crystal Structures and Electrode Performance of Alpha-$NaFeO_2$ for Rechargeable Sodium Batteries, *Electrochemistry* **80**(10), 716–719, (2012).
30. L. Durai, A. Gopalakrishnan and S. Badhulika, One-step Solid-state Reaction Synthesis of β-$NaFeO_2$ Nanopebble as High Capacity Cathode Material for Sodium Ion Batteries, *Mater. Lett.* **270**, 127739, (2020).
31. S. Komaba, C. Takei, T. Nakayama, A. Ogata and N. Yabuuchi, Electrochemical Intercalation Activity of Layered $NaCrO_2$ vs. $LiCrO_2$, *Electrochem. Commun.* **12**(3), 355–358, (2010).
32. S. H. Bo, X. Li, A. J. Toumar and G. Ceder, Layered-to-Rock-Salt Transformation in Desodiated $Na_xCrO_2$ (x0.4), *Chem. Mater.* **28**(5), 1419–1429, (2016).
33. C. Y. Yu, J. S. Park, H. G. Jung, K. Y. Chung, D. Aurbach, Y. K. Sun and S. T. Myung, $NaCrO_2$ Cathode for High-rate Sodium-ion Batteries, *Energy Environ. Sci.* **8**(7), 2019–2026, (2015).
34. D. Hamani, M. Ati, J. M. Tarascon and P. Rozier, $Na_xVO_2$ as Possible Electrode for Na-ion Batteries, *Electrochem. Commun.* **13**(9), 938–941, (2011).
35. D. Wu, X. Li, B. Xu, N. Twu, L. Liu and G. Ceder, $NaTiO_2$: A Layered Anode Material for Sodium-ion Batteries, *Energy Environ. Sci.* **8**(1), 195–202, (2015).
36. S. Komaba, N. Yabuuchi, T. Nakayama, A. Ogata, T. Ishikawa and I. Nakai, Study on the Reversible Electrode Reaction of $Na_{1-x}Ni_{0.5}Mn_{0.5}O_2$ for a Rechargeable Sodium-ion Battery, *Inorg. Chem.* **51**(11), 6211–6220, (2012).
37. Z. Lu and J. R. Dahn, In Situ X-Ray Diffraction Study of P2-$Na_{2/3}Ni_{1/3}Mn_{2/3}O_2$, *J. Electrochem. Soc.* **148**(11), A1225, (2001).
38. N. Yabuuchi, M. Kajiyama, J. Iwatate, H. Nishikawa, S. Hitomi, R. Okuyama, R. Usui, Y. Yamada and S. Komaba, P2-type $Na_x[Fe_{1/2}Mn_{1/2}[O_2$ Made from Earth-abundant Elements for Rechargeable NaÂ Batteries, *Nat. Mater.* **11**(6), 512–517, (2012).
39. W. M. Dose, N. Sharma, J. C. Pramudita, J. A. Kimpton, E. Gonzalo, M. H. Han and T. Rojo, Cystallographic Evolution of P2 $Na_{2/3}Fe_{0.4}Mn_{0.6}O_2$ Electrodes During Electrochemical Cycling, *Chem. Mater.* **28**(17), 6342–6354, (2016).

40. X. Wang, G. Liu, T. Iwao, M. Okubo and A. Yamada, Role of Ligand-to-metal Charge Transfer in O3-type $NaFeO_2$-$NaNiO_2$ Solid Solution for Enhanced Electrochemical Properties, *J. Phys. Chem. C* **118**(6), 2970–2976, (2014).
41. D. D. Yuan, Y. X. Wang, Y. L. Cao, X. P. Ai and H. X. Yang, Improved Electrochemical Performance of Fe-substituted $NaNi_{0.5}Mn_{0.5}O_2$ Cathode Materials for Sodium-ion Batteries, *ACS Appl. Mater. Interfaces* **7**(16), 8585–8591, (2015).
42. C. Zhang, R. Gao, L. Zheng, Y. Hao and X. Liu, New Insights into the Roles of Mg in Improving the Rate Capability and Cycling Stability of O3-$NaMn_{0.48}Ni_{0.2}Fe_{0.3}Mg_{0.02}O_2$ for Sodium-Ion Batteries, *ACS Appl. Mater. Interfaces* **10**(13), 10819–10827, (2018).
43. L. Wang, Y. G. Sun, L. L. Hu, J. Y. Piao, J. Guo, A. Manthiram, J. Ma and A. M. Cao, Copper-substituted $Na_{0.67}Ni_{0.3-x}Cu_xMn_{0.7}O_2$ Cathode Materials for Sodium-ion Batteries with Suppressed P2-O2 Phase Transition, *J. Mater. Chem. A* **5**(18), 8752–8761, (2017).
44. J. Deng, W. Bin Luo, X. Lu, Q. Yao, Z. Wang, H. K. Liu, H. Zhou and S. X. Dou, High Energy Density Sodium-Ion Battery with Industrially Feasible and Air-Stable O3-Type Layered Oxide Cathode, *Adv. Energy Mater.* **8**(5),1701610 (2018).
45. J. Y. Hwang, S. T. Myung, D. Aurbach and Y. K. Sun, Effect of Nickel and Iron on Structural and Electrochemical Properties of O3 type Layer Cathode Materials for Sodium-ion Batteries, *J. Power Sources* **324**, 106–112, (2016).
46. Y. Xie, H. Wang, G. Xu, J. Wang, H. Sheng, Z. Chen, Y. Ren, C. J. Sun, J. Wen, J. Wang, D. J. Miller, J. Lu, K. Amine and Z. F. Ma, In Operando XRD and TXM Study on the Metastable Structure Change of $NaNi_{1/3}Fe_{1/3}Mn_{1/3}O_2$ under Electrochemical Sodium-Ion Intercalation, *Adv. Energy Mater.* **6**(24), 1601306, (2016).
47. L. Sun, Y. Xie, X. Z. Liao, H. Wang, G. Tan, Z. Chen, Y. Ren, J. Gim, W. Tang, Y. S. He, K. Amine and Z. F. Ma, Insight into Ca-Substitution Effects on O3-Type $NaNi_{1/3}Fe_{1/3}Mn_{1/3}O_2$ Cathode Materials for Sodium-Ion Batteries Application, *Small* **14**(21), 1704523, (2018).
48. S.-M. Oh, S.-T. Myung, J.-Y. Hwang, B. Scrosati, K. Amine and Y.-K. Sun, High Capacity O3-Type $Na[Li_{0.05}(Ni_{0.25}Fe_{0.25}Mn_{0.5})_{0.95}]O_2$ Cathode for Sodium Ion Batteries, *Chem. Mater.* **26**(21), 6165–6171, (2014).
49. J. L. Yue, W. W. Yin, M. H. Cao, S. Zulipiya, Y. N. Zhou and Z. W. Fu, A Quinary Layer Transition Metal Oxide of $NaNi_{1/4}Co_{1/4}Fe_{1/4}Mn_{1/8}Ti_{1/8}O_2$ as a High-rate-capability and Long-cycle-life Cathode Material for Rechargeable Sodium Ion Batteries, *Chem. Commun.* **51**(86), 15712–15715, (2015).
50. X. Zhang, Z. Wei, K. N. Dinh, N. Chen, G. Chen, F. Du and Q. Yan, Layered Oxide Cathode for Potassium-Ion Battery: Recent Progress and Prospective, *Small* **16**(38), 2002700, (2020).
51. R. Hoppe and H. Sabrowsky, Oxoscandate der Alkalimetalle: $KScO_2$ und $RbScO_2$, *ZAAC - J. Inorg. Gen. Chem.* **339**(3–4), 144–154, (1965).

52. C. Vaalma, G. A. Giffin, D. Buchholz and S. Passerini, Non-Aqueous K-Ion Battery Based on Layered $K_{0.3}MnO_2$ and Hard Carbon/Carbon Black, *J. Electrochem. Soc.* **163**(7), A1295–A1299, (2016).
53. Peng, Y. Li, J. Gao, F. Zhang, J. Li and G. Zhang, High energy K-ion Batteries Based on P3-Type $K_{0.5}MnO_2$ Hollow Submicrosphere Cathode, *J. Power Sources.* 437, 226913, (2019).
54. K. Lei, Z. Zhu, Z. Yin, P. Yan, F. Li and J. Chen, Dual Interphase Layers In Situ Formed on a Manganese-Based Oxide Cathode Enable Stable Potassium Storage, *Chem* **5**(12), 3220–3231, (2019).
55. H. Kim, J. C. Kim, S. Bo, T. Shi, D. Kwon and G. Ceder, K-Ion Batteries Based on a P2-Type $K_{0.6}CoO_2$ Cathode, *Adv. Energy Mater.* **7**(17), 1700098, (2017).
56. Y. Hironaka, K. Kubota and S. Komaba, P2- and P3-$K_xCoO_2$ as an Electrochemical Potassium Intercalation Host, *Chem. Commun.* **53**(26), 3693–3696, (2017).
57. H. Kim, D. H. Seo, A. Urban, J. Lee, D. H. Kwon, S. H. Bo, T. Shi, J. K. Papp, B. D. McCloskey and G. Ceder, Stoichiometric Layered Potassium Transition Metal Oxide for Rechargeable Potassium Batteries, *Chem. Mater.* **30**(18), 6532–6539, (2018).
58. J. Y. Hwang, J. Kim, T. Y. Yu, S. T. Myung, and Y. K. Sun, Development of P3-$K_{0.69}CrO_2$ as an Ultra-high-performance Cathode Material for K-ion Batteries, *Energy Environ. Sci.* **11**(10), 2821–2827, (2018).
59. X. Wang, X. Xu, C. Niu, J. Meng, M. Huang, X. Liu, Z. Liu and L. Mai, Earth Abundant Fe/Mn-Based Layered Oxide Interconnected Nanowires for Advanced K-Ion Full Batteries, *Nano Lett.* **17**(1), 544–550, (2017).
60. C. L. Liu, S. H. Luo, H. B. Huang, X. Liu, Y. C. Zhai and Z. W. Wang, Fe-doped Layered P3-type $K_{0.45}Mn_{1-x}Fe_xO_2$ ($x \leq 0.5$) as Cathode Materials for Low-cost Potassium-ion Batteries, *Chem. Eng. J.* **378**, 122167, (2019).
61. M. K. Cho, J. H. Jo, J. U. Choi and S. T. Myung, Cycling Stability of Layered Potassium Manganese Oxide in Nonaqueous Potassium Cells, *ACS Appl. Mater. Interfaces* **11**(31), 27770–27779, (2019).
62. X. Zhang, Y. Yang, X. Qu, Z. Wei, G. Sun, K. Zheng, H. Yu and F. Du, Layered P2-Type $K_{0.44}Ni_{0.22}Mn_{0.78}O_2$ as a High-Performance Cathode for Potassium-Ion Batteries, *Adv. Funct. Mater.* **29**(49), 1905679, (2019).
63. J. U. Choi, J. Kim, J. H. Jo, H. J. Kim, Y. H. Jung, D. C. Ahn, Y. K. Sun and S. T. Myung, Facile Migration of Potassium Ions in a Ternary P3-type $K_{0.5}[Mn_{0.8}Fe_{0.1}Ni_{0.1}]O_2$ Cathode in Rechargeable Potassium Batteries, *Energy Storage Mater.* **25**, 714–723, (2020).
64. M. R. Palacín, Recent Advances in Rechargeable Battery Materials: A Chemist's Perspective, *Chem. Soc. Rev.* **38**(9), 2565–2575, (2009).
65. J. Y. Hwang, S. T. Myung and Y. K. Sun, Sodium-ion batteries: Present and Future, *Chem. Soc. Rev.* **46**(12), 3529–3614, (2017).

66. H. Tan, D. Chen, X. Rui and Y. Yu, Peering into Alloy Anodes for Sodium-Ion Batteries: Current Trends, Challenges, and Opportunities, *Adv. Funct. Mater.* **29**(14), 1808745, (2019).
67. K.-X. Lei, J. Wang, C. Chen, S.-Y. Li, S.-W. Wang, S.-J. Zheng and F.-J. Li, Recent Progresses on Alloy-based Anodes for Potassium-ion Batteries, *Rare Met.* **39**(9), 989–1004, (2020).
68. B. Wang, Y. Peng, F. Yuan, Q. Liu, L. Sun, P. Zhang, Q. Wang, Z. Li and Y. A. Wu, A Comprehensive Review of Carbons Anode for Potassium-ion Battery: Fast Kinetic, Structure Stability and Electrochemical, *J. Power Sources* **484**, 229244, (2021).
69. D. P. Divincenzo and E. J. Mele, Cohesion and Structure in Stage-1 Graphite Intercalation Compounds, *Phys. Rev. B* **32**(4), 2538–2553, (1985).
70. P. Ge and M. Fouletier, Electrochemical Intercalation of Sodium in Graphite, *Solid State Ionics* **28–30**, 1172–1175, (1988).
71. S. Komaba, T. Hasegawa, M. Dahbi and K. Kubota, Potassium Intercalation into Graphite to Realize High-voltage/high-power Potassium-ion Batteries and Potassium-ion Capacitors, *Electrochem. commun.* **60**, 172–175, (2015).
72. Y. Huang, Y. Wang, P. Bai and Y. Xu, Storage Mechanism of Alkali Metal Ions in the Hard Carbon Anode: an Electrochemical Viewpoint, *ACS Appl. Mater. Interfaces* **13**(32), 38441–38449, (2021).
73. M. M. Doeff, Y. Ma, S. J. Visco and L. C. De Jonghe, Electrochemical Insertion of Sodium into Carbon, *J. Electrochem. Soc.* **140**(12), L169--L170, (1993).
74. R. Alcántara, J. M. Jiménez Mateos and J. L. Tirado, Negative Electrodes for Lithium- and Sodium-Ion Batteries Obtained by Heat-Treatment of Petroleum Cokes below 1000°C, *J. Electrochem. Soc.* **149**(2), A201, (2002).
75. Y. Liu, Y.-X. Lu, Y.-S. Xu, Q.-S. Meng, J.-C. Gao, Y.-G. Sun, Y.-S. Hu, B.-B. Chang, C.-T. Liu and A.-M. Cao, Pitch-Derived Soft Carbon as Stable Anode Material for Potassium Ion Batteries, *Adv. Mater.* **32**(17), 2000505, (2020).
76. D. A. Stevens and J. R. Dahn, High Capacity Anode Materials for Rechargeable Sodium-Ion Batteries, *J. Electrochem. Soc.* **147**(4), 1271, (2000).
77. H. Zhang, S. Zhao and F. Huang, A Comparative Overview of Carbon Anodes for Nonaqueous Alkali Metal-ion Batteries, *J. Mater. Chem. A* **9**(48), 27140–27169, (2021).
78. L. Fan, S. Chen, R. Ma, J. Wang, L. Wang, Q. Zhang, E. Zhang, Z. Liu and B. Lu, Ultrastable Potassium Storage Performance Realized by Highly Effective Solid Electrolyte Interphase Layer, *Small* **14**(30), 1801806, (2018).
79. B. Xiao, T. Rojo and X. Li, Hard Carbon as Sodium-Ion Battery Anodes: Progress and Challenges, *ChemSusChem* **12**(1), 133–144, (2019).
80. D. A. Stevens and J. R. Dahn, The Mechanisms of Lithium and Sodium Insertion in Carbon Materials, *J. Electrochem. Soc.* **148**(8), A803, (2001).
81. J. R. Dahn, T. Zheng, Y. Liu and J. S. Xue, Mechanisms for Lithium Insertion in Carbonaceous Materials, *Science (80).* **270**(5236), 590–593, (1995).

82. J. R. Dahn, W. Xing and Y. Gao, The "Falling Cards Model" for the Structure of Microporous Carbons, *Carbon N. Y.* **35**(6), 825–830, (1997).
83. B. Zhang, C. M. Ghimbeu, C. Laberty, C. Vix-Guterl and J.-M. Tarascon, Correlation Between Microstructure and Na Storage Behavior in Hard Carbon, *Adv. Energy Mater.* **6**(1), 1501588, (2016).
84. J. M. Stratford, P. K. Allan, O. Pecher, P. A. Chater and C. P. Grey, Mechanistic Insights into Sodium Storage in Hard Carbon Anodes using Local Structure Probes, *Chem. Commun.* **52**(84), 12430–12433, (2016).
85. D.-S. Bin, Y. Li, Y.-G. Sun, S.-Y. Duan, Y. Lu, J. Ma, A.-M. Cao, Y.-S. Hu and L.-J. Wan, Structural Engineering of Multishelled Hollow Carbon Nanostructures for High-Performance Na-Ion Battery Anode, *Adv. Energy Mater.* **8**(26), 1800855, (2018).
86. P. Bai, Y. He, X. Zou, X. Zhao, P. Xiong and Y. Xu, Elucidation of the Sodium-Storage Mechanism in Hard Carbons, *Adv. Energy Mater.* **8**(15), 1703217, (2018).
87. J. Ding, H. Wang, Z. Li, A. Kohandehghan, K. Cui, Z. Xu, B. Zahiri, X. Tan, E. M. Lotfabad, B. C. Olsen and D. Mitlin, Carbon Nanosheet Frameworks Derived from Peat Moss as High Performance Sodium Ion Battery Anodes, *ACS Nano* **7**(12), 11004–11015, (2013).
88. C. Bommier, T. W. Surta, M. Dolgos and X. Ji, New Mechanistic Insights on Na-Ion Storage in Nongraphitizable Carbon, *Nano Lett.* **15**(9), 5888–5892, (2015).
89. Z. Jian, Z. Xing, C. Bommier, Z. Li and X. Ji, Hard Carbon Microspheres: Potassium-Ion Anode Versus Sodium-Ion Anode, *Adv. Energy Mater.* **6**(3), 1501874, (2016).
90. Q. Zhao, Y. Hu, K. Zhang and J. Chen, Potassium–Sulfur Batteries: A New Member of Room-Temperature Rechargeable Metal–Sulfur Batteries, *Inorg. Chem.* **53**(17), 9000–9005, (2014).
91. J. C. Pramudita, D. Sehrawat, D. Goonetilleke and N. Sharma, An Initial Review of the Status of Electrode Materials for Potassium-Ion Batteries, *Adv. Energy Mater.* **7**(24), 1602911, (2017).
92. T. Hosaka, K. Kubota, H. Kojima and S. Komaba, Highly Concentrated Electrolyte Solutions for 4 V Class Potassium-ion Batteries, *Chem. Commun.* **54**(60), 8387–8390, (2018).
93. S. Komaba, W. Murata, T. Ishikawa, N. Yabuuchi, T. Ozeki, T. Nakayama, A. Ogata, K. Gotoh and K. Fujiwara, Electrochemical Na Insertion and Solid Electrolyte Interphase for Hard-Carbon Electrodes and Application to Na-Ion Batteries, *Adv. Funct. Mater.* **21**(20), 3859–3867, (2011).
94. S. Komaba, T. Ishikawa, N. Yabuuchi, W. Murata, A. Ito and Y. Ohsawa, Fluorinated Ethylene Carbonate as Electrolyte Additive for Rechargeable Na Batteries, *ACS Appl. Mater. Interfaces* **3**(11), 4165–4168, (2011).
95. E. Irisarri, A. Ponrouch and M. R. Palacin, Review—Hard Carbon Negative Electrode Materials for Sodium-Ion Batteries, *J. Electrochem. Soc.* **162**(14), A2476–A2482, (2015).

96. A. Ponrouch, A. R. Goñi and M. R. Palacín, High Capacity Hard Carbon Anodes for Sodium Ion Batteries in Additive Free Electrolyte, *Electrochem. commun.* **27**, 85–88, (2013).
97. L. F. Zhao, Z. Hu, W. H. Lai, Y. Tao, J. Peng, Z. C. Miao, Y. X. Wang, S. L. Chou, H. K. Liu and S. X. Dou, Hard Carbon Anodes: Fundamental Understanding and Commercial Perspectives for Na-Ion Batteries beyond Li-Ion and K-Ion Counterparts, *Adv. Energy Mater.* **11**(1), 1–28, (2021).
98. Y. Chu, Y. Shen, F. Guo, X. Zhao, Q. Dong, Q. Zhang, W. Li, H. Chen, Z. Luo and L. Chen, Advanced Characterizations of Solid Electrolyte Interphases in Lithium-Ion Batteries, *Electrochem. Energy Rev.* **3**(1), 187–219, (2020).
99. E. Peled and S. Menkin, Review—SEI: Past, Present and Future, *J. Electrochem. Soc.* **164**(7), A1703–A1719, (2017).
100. Y. Lei, L. Qin, R. Liu, K. C. Lau, Y. Wu, D. Zhai, B. Li and F. Kang, Exploring Stability of Nonaqueous Electrolytes for Potassium-Ion Batteries, *ACS Appl. Energy Mater.* **1**(5), 1828–1833, (2018).
101. F. A. Soto, A. Marzouk, F. El-Mellouhi and P. B. Balbuena, Understanding Ionic Diffusion through SEI Components for Lithium-Ion and Sodium-Ion Batteries: Insights from First-Principles Calculations, *Chem. Mater.* **30**(10), 3315–3322, (2018).
102. K. Xu, Nonaqueous Liquid Electrolytes for Lithium-Based Rechargeable Batteries, *Chem. Rev.* **104**(10), 4303–4418, (2004).
103. X. Wang, S. Tang, W. Guo, Y. Fu and A. Manthiram, Advances in Multimetallic Alloy-based Anodes for Alkali-ion and Alkali-metal Batteries, *Mater. Today* **50**, 259–275, (2021).
104. M. Dahbi, N. Yabuuchi, K. Kubota, K. Tokiwa and S. Komaba, Negative Electrodes for Na-ion Batteries, *Phys. Chem. Chem. Phys.* **16**(29), 15007–15028, (2014).
105. G. Suo, Y. Cheng, J. Zhang and S. M. Ahmed, Recent Progress and Perspectives on Alloying Anodes for Potassium-Ion Batteries, *ChemNanoMat* **7**(12), 1291–1308, (2021).
106. H. Li, T. Yamaguchi, S. Matsumoto, H. Hoshikawa, T. Kumagai, N. L. Okamoto and T. Ichitsubo, Circumventing Huge Volume Strain in Alloy Anodes of Lithium Batteries, *Nat. Commun.* **11**(1), 1–8, (2020).
107. A. Darwiche, C. Marino, M. T. Sougrati, B. Fraisse, L. Stievano and L. Monconduit, Better Cycling Performances of Bulk Sb in Na-ion Batteries Compared to Li-ion Systems: An Unexpected Electrochemical Mechanism, *J. Am. Chem. Soc.* **134**(51), 20805–20811, (2012).
108. M. He, K. Kravchyk, M. Walter and M. V. Kovalenko, Monodisperse Antimony Nanocrystals for High-rate Li-ion and Na-ion Battery Anodes: Nano Versus Bulk, *Nano Lett.* **14**(3), 1255–1262, (2014).
109. L. M. L. Fransson, J. T. Vaughey, R. Benedek, K. Edström, J. O. Thomas and M. M. Thackeray, Phase Transitions in Lithiated $Cu_2Sb$ Anodes for Lithium Batteries: An in situ X-ray Diffraction Study, *Electrochem. commun.* **3**(7), 317–323, (2001).

110. S. Sarkar, S. Roy, Y. Zhao and J. Zhang, Recent Advances in Semimetallic Pnictogen (As, Sb, Bi) based Anodes for Sodium-ion Batteries: Structural Design, Charge Storage Mechanisms, Key Challenges and Perspectives, *Nano Res.* **14**(11), 3690–3723, (2021).
111. Y. Zhao and A. Manthiram, High-Capacity, High-Rate Bi–Sb Alloy Anodes for Lithium-Ion and Sodium-Ion Batteries, *Chem. Mater.* **27**(8), 3096–3101, (2015).
112. L. Xiao, Y. Cao, J. Xiao, W. Wang, L. Kovarik, Z. Nie and J. Liu, High Capacity, Reversible Alloying Reactions in SnSb/C Nanocomposites for Na-Ion Battery Applications, *Chem. Commun.* **48**, 3321, (2012).
113. H. Xie, X. Tan, E. J. Luber, B. C. Olsen, W. P. Kalisvaart, K. L. Jungjohann, D. Mitlin and J. M. Buriak, β-SnSb for Sodium Ion Battery Anodes: Phase Transformations Responsible for Enhanced Cycling Stability Revealed by In Situ TEM, *ACS Energy Lett.* **3**(7), 1670–1676, (2018).
114. W. D. McCulloch, X. Ren, M. Yu, Z. Huang and Y. Wu, Potassium-Ion Oxygen Battery Based on a High Capacity Antimony Anode, *ACS Appl. Mater. Interfaces* **7**(47), 26158–26166, (2015).
115. J. Zheng, Y. Yang, X. Fan, G. Ji, X. Ji, H. Wang, S. Hou, M. R. Zachariah and C. Wang, Extremely Stable Antimony-carbon Composite Anodes for Potassium-ion Batteries, *Energy Environ. Sci.* **12**(2), 615–623, (2019).
116. T. Ramireddy, M. M. Rahman, T. Xing, Y. Chen and A. M. Glushenkov, Stable Anode Performance of an Sb-carbon Nanocomposite in Lithium-ion Batteries and the Effect of Ball Milling Mode in the Course of its Preparation, *J. Mater. Chem. A* **2**(12), 4282–4291, (2014).
117. K. Li, H. Liu and G. Wang, $Sb_2O_3$ Nanowires as Anode Material for Sodium-Ion Battery, *Arab. J. Sci. Eng.* **39**(9), 6589–6593, (2014).
118. H. Zhai, H. Jiang, Y. Qian, X. Cai, H. Liu, Y. Qiu, M. Jin, F. Xiu, X. Liu and L. Lai, $Sb_2S_3$ Nanocrystals Embedded in Multichannel N-doped Carbon Nanofiber for Ultralong Cycle Life Sodium-ion Batteries, *Mater. Chem. Phys.* **240**, 122139, (2020).
119. Y. Liu, Z. Tai, J. Zhang, W. K. Pang, Q. Zhang, H. Feng, K. Konstantinov, Z. Guo and H. K. Liu, Boosting Potassium-ion Batteries by Few-layered Composite Anodes Prepared via Solution-triggered One-step Shear Exfoliation, *Nat. Commun.* **9**(1), 1–10, (2018).
120. F. Yang, F. Yu, Z. Zhang, K. Zhang, Y. Lai and J. Li, Bismuth Nanoparticles Embedded in Carbon Spheres as Anode Materials for Sodium/Lithium-Ion Batteries, *Chem. – A Eur. J.* **22**(7), 2333–2338, (2016).
121. J. Sottmann, M. Herrmann, P. Vajeeston, Y. Hu, A. Ruud, C. Drathen, H. Emerich, H. Fjellvåg and D. S. Wragg, How Crystallite Size Controls the Reaction Path in Nonaqueous Metal Ion Batteries: The Example of Sodium Bismuth Alloying, *Chem. Mater.* **28**(8), 2750–2756, (2016).
122. C. Wang, D. Du, M. Song, Y. Wang and F. Li, A High-Power $Na_3V_2(PO_4)_3$-Bi Sodium-Ion Full Battery in a Wide Temperature Range, *Adv. Energy Mater.* **9**(16), 1900022, (2019).

123. J. Huang, X. Lin, H. Tan and B. Zhang, Bismuth Microparticles as Advanced Anodes for Potassium-Ion Battery, *Adv. Energy Mater.* **8**(19), 1703496, (2018).
124. P. Xiong, J. Wu, M. Zhou and Y. Xu, Bismuth–Antimony Alloy Nanoparticle@ Porous Carbon Nanosheet Composite Anode for High-Performance Potassium-Ion Batteries, *ACS Nano* **14**(1), 1018–1026, (2020).
125. J. Sun, M. Li, J. A. S. Oh, K. Zeng and L. Lu, Recent Advances of Bismuth Based Anode Materials for Sodium-ion Batteries, *Mater. Technol.* **33**(8), 563–573, (2018).
126. M.-K. Kim, S.-H. Yu, A. Jin, J. Kim, I.-H. Ko, K.-S. Lee, J. Mun and Y.-E. Sung, Bismuth Oxide as a High Capacity Anode Material for Sodium-ion Batteries, *Chem. Commun.* **52**(79), 11775–11778, (2016).
127. Z. Li, Y. Zhang, J. Zhang, Y. Cao, J. Chen, H. Liu and Y. Wang, Sodium-Ion Battery with a Wide Operation-Temperature Range from −70 to 100°C, *Angew. Chemie Int. Ed.* **61**(13), e202116930, (2022).
128. C. Wang, L. Wang, F. Li, F. Cheng and J. Chen, Bulk Bismuth as a High-Capacity and Ultralong Cycle-Life Anode for Sodium-Ion Batteries by Coupling with Glyme-Based Electrolytes, *Adv. Mater.* **29**, 1702212, (2017).
129. J. M. Stratford, M. Mayo, P. K. Allan, O. Pecher, O. J. Borkiewicz, K. M. Wiaderek, K. W. Chapman, C. J. Pickard, A. J. Morris and C. P. Grey, Investigating Sodium Storage Mechanisms in Tin Anodes: A Combined Pair Distribution Function Analysis, Density Functional Theory, and Solid-State NMR Approach, *J. Am. Chem. Soc.* **139**(21), 7273–7286, (2017).
130. F. Xin and M. S. Whittingham, Challenges and Development of Tin-Based Anode with High Volumetric Capacity for Li-Ion Batteries, *Electrochem. Energy Rev.* **3**(4), 643–655, (2020).
131. L. D. Ellis, B. N. Wilkes, T. D. Hatchard and M. N. Obrovac, In Situ XRD Study of Silicon, Lead and Bismuth Negative Electrodes in Nonaqueous Sodium Cells, *J. Electrochem. Soc.* **161**, A416, (2014).
132. I. Sultana, T. Ramireddy, M. M. Rahman, Y. Chen and A. M. Glushenkov, Tin-based Composite Anodes for Potassium-ion Batteries, *Chem. Commun.* **52**(59), 9279–9282, (2016).
133. F. Zoller, D. Böhm, T. Bein and D. Fattakhova-Rohlfing, Tin Oxide Based Nanomaterials and Their Application as Anodes in Lithium-Ion Batteries and Beyond, *ChemSusChem* **12**(18), 4140–4159, (2019).
134. L. Baggetto, P. Ganesh, R. P. Meisner, R. R. Unocic, J.-C. Jumas, C. A. Bridges and G. M. Veith, Characterization of Sodium Ion Electrochemical Reaction with Tin Anodes: Experiment and Theory, *J. Power Sources* **234**, 48–59, (2013).
135. Q. Wang, X. Zhao, C. Ni, H. Tian, J. Li, Z. Zhang, S. X. Mao, J. Wang and Y. Xu, Reaction and Capacity-Fading Mechanisms of Tin Nanoparticles in Potassium-Ion Batteries, *J. Phys. Chem. C* **121**(23), 12652–12657, (2017).
136. M. Shimizu, R. Yatsuzuka, T. Koya, T. Yamakami and S. Arai, Tin Oxides as a Negative Electrode Material for Potassium-Ion Batteries, *ACS Appl. Energy Mater.* **1**(12), 6865–6870, (2018).

137. T. Ramireddy, R. Kali, M. K. Jangid, V. Srihari, H. K. Poswal and A. Mukhopadhyay, Insights into Electrochemical Behavior, Phase Evolution and Stability of Sn upon K-alloying/de-alloying via In Situ Studies, *J. Electrochem. Soc.* **164**(12), A2360–A2367, (2017).
138. V. Gabaudan, R. Berthelot, M. T. Sougrati, P.-E. Lippens, L. Monconduit and L. Stievano, SnSb vs. Sn: Improving the Performance of Sn-based Anodes for K-ion Batteries by Synergetic Alloying with Sb, *J. Mater. Chem. A* **7**(25), 15262–15270, (2019).
139. W. Zhang, W. K. Pang, V. Sencadas and Z. Guo, Understanding High-Energy-Density $Sn_4P_3$ Anodes for Potassium-Ion Batteries, *Joule* **2**(8), 1534–1547, (2018).
140. V. Gabaudan, R. Berthelot, L. Stievano and L. Monconduit, Electrochemical Alloying of Lead in Potassium-Ion Batteries, *ACS Omega* **3**(9), 12195–12200, (2018).

https://doi.org/10.1142/9789811270956_0003

# Chapter 3

## High Performance Thermoelectric Materials

Subarna Das, Animesh Bhui and Kanishka Biswas*

*New Chemistry Unit, International Centre for Materials Science and School of Advanced Materials, Jawaharlal Nehru Centre for Advanced Scientific Research (JNCASR), Jakkur P.O., Bangalore 560 064, India*
**kanishka@jncasr.ac.in*

Thermoelectric materials, which have the potential to convert waste heat to electricity via solid-state thermoelectric devices, are going to be important in future energy management. Since 1950s, $Bi_2Te_3$ and its alloys are the most used thermoelectric materials in commercial applications, but its optimal operation has been limited only around room temperature which triggers researchers to look for new efficient thermoelectric materials for applications at the elevated temperatures. However, strong interdependency among the thermoelectric parameters makes it difficult to achieve ultra-high thermoelectric performance. Several innovative strategies have been employed to decouple the interrelated properties. In this chapter, we have discussed about the new and well-established high performance thermoelectric materials: metal chalcogenides, half-Heuslers, $Mg_3X_2$ (X = Bi/Sb) based compounds, skutterudites, clathrates, silicides and all-inorganic metal halide perovskites. A brief discussion about the chemical bonding, structural details, their intriguing chemical and physical properties, and thermoelectric performances has been presented. Suitable dopants to tune the electronic structure towards obtaining favourable thermoelectric properties and various phonon scattering mechanisms have also been reviewed. At the end, the challenges to improve thermoelectric performance further and possible strategies for future investigations have been discussed.

## 1. Introduction

The gradual decrease in the conventional energy sources has intensified interest among scientists to extensively search for alternative ways of energy generation and management. It is noteworthy to mention here that almost 65% of the total utilized energy is being dissipated away as waste heat to the environment. It would be both environmentally and economically beneficial if a part of this waste heat can be converted into useable energy which will provide one step forward towards solving the ongoing energy crisis. In this regard, thermoelectric (TE) materials, which have the potential to reversibly convert heat into electricity, can play a crucial role. It can be used for both power generation as well as for electronic refrigeration. Advances in TE materials could also end the monopoly of compression-based refrigeration systems where the harmful gases are used for refrigeration purpose.[1,2] The main bottleneck for large-scale commercial application of TE materials is its low efficiency. The efficiency of TE material is determined by dimensionless term *figure of merit* (zT), which can be expressed as:[1,2]

$$zT = \frac{\sigma S^2}{\kappa_{total}} T \tag{1}$$

Where $\sigma$, S, T and $\kappa_{total}$ are the electrical conductivity, Seebeck coefficient, absolute temperature and total thermal conductivity, respectively and $\sigma S^2$ is termed as power factor. While phonons, charge carriers, magnons, all can, in principle, contribute to heat transport, $\kappa_{total}$ mainly consists of electronic ($\kappa_{el}$), lattice ($\kappa_{lat}$) and bipolar ($\kappa_b$) thermal conductivity. With $\kappa_{lat}$ being the only independent parameter in the expression of zT, a highly efficient TE material requires a skilful optimization of interdependent material parameters *viz*, $\sigma$, S and $\kappa_{el}$ to obtain a high TE efficiency. While high $\sigma$ and S can be found in metals and semiconductors respectively, glass-like materials are required for low $\kappa_{total}$. High TE conversion efficiency ($\eta$) is necessary for application purpose, thereby, it is important to achieve high zT over a wide range of temperatures. The device figure of merit, denoted as ZT, is the weighted average of TE material zT across a finite temperature difference $\Delta T$ (= $T_{hot}$ - $T_{cold}$) and $\eta$ is expressed as:[1-3]

$$\eta = \frac{\Delta T}{T_{hot}} \frac{\sqrt{1+ZT}-1}{\sqrt{1+ZT}+\frac{T_{hot}}{T_{cold}}} \tag{2}$$

Recently, two main strategies are generally adopted for increasing zT: one is to increase power factor and the other involving efforts to reduce $\kappa_{lat}$. For increasing power factor, efforts are mainly given to decouple the TE

parameters via optimizing carrier concentration by suitable doping,[4-6] introducing carrier energy filtering effect,[7] engineering electronic band structure *viz.*, band convergence,[8-10] introduction of resonant impurity band,[10,11] band distortion,[12] band nestification[13] etc. On the other hand, for reduction of $\kappa_{lat}$, nanostructuring,[6,8,14] alloying,[15-17] introduction of dislocations,[18,19] and nanoprecipitates of second phase[8,15,20,21] have been proven to be successful techniques to scatter heat carrying phonons.
While the scientific and engineering challenges are many, the opportunity for scientists and industries in TE technology is also huge as the demand for electricity generation from waste heat is growing day by day. Since 1950s, solid solutions of bismuth telluride are the most established materials to be used in commercial TE devices.[18,19] However, with the peak TE performance for this material being only around 330–380 K, search for new materials is the need of the hour for application at elevated temperature. Intense investigation has been dedicated in lead chalcogenides as it is the champion TE material for mid-temperature applications.[22,23] However, tin chalcogenides[20,24] and germanium chalcogenides[11,16] give much promise to replace toxic lead-based materials and the search for highly efficient TE material is still ongoing. With a detailed comprehension of the correlation between electronic structure, chemical bonding and crystal chemistry, current research in TE materials revolves around metal chalcogenides, half-Heuslers, skutterudites, clathrates, Mg–Sb antimonides, silicides and all-inorganic metal halide perovskites.[3] In this chapter, attempt has been made to provide a brief discussion of the current material designing strategies and ongoing research status of high performance TE materials.

## 2. Different High Performance Thermoelectric Materials

In the following sections, different TE materials with promising TE performance will be introduced with their structural details. In addition, different strategies undertaken to enhance their zT and the present status of the materials in terms of TE application will be discussed.

## 2.1. *Metal chalcogenides*

### 2.1.1. *Tellurides*

Among the IV-V tellurides, SnTe,[25] PbTe[5,23,26] and GeTe[27] are the most popular TE materials for mid to high temperature applications. In ambient condition, both PbTe and SnTe crystallize in cubic rocksalt structure. Due to such simple high symmetric structure, electronic band structure shows multiple degenerate bands, which is favourable for obtaining high S.[5,20,25,28] However, at ambient temperature, GeTe possesses a rhombohedral crystal structure, and it changes its structure to cubic phase at T ~720 K.[11] Narrow band gap in the range of ~0.18–0.32 eV, along with heavy constituent elements, make these tellurides ideal for TE applications.[11,25,29]

Intrinsically low thermal conductivity stemming from large lattice anharmonicity and easily tuneable electronic band structure make PbTe a champion TE material. To increase the TE performance of these series of tellurides, two main strategies are used to introduce resonant level and band convergence, both of which modify electronic DOS significantly and help to increase S.[30] It is reported that a high zT ~1.5 can be obtained due to formation of resonant level near fermi surface through Tl doping in PbTe.[12] On the other hand, valence band convergence in $PbTe_{1-x}Se_x$:Na has resulted in zT ~1.8.[31] For lowering the $\kappa_{lat}$, manipulation of phonon transport through all-scale hierarchical architecture has been successfully demonstrated in PbTe, which lead to a very high zT ~2.2 at 915 K.[23] Furthermore, a high TE performance (zT ~2.57 at 850 K) has been obtained for Na-Eu-Sn doped PbTe, where $\kappa_{lat}$ is strongly suppressed due to lattice strain induced phonon scattering.[32] A close look into the electronic band structure of PbTe reveals that valence band consists of two neighbouring bands L and Σ, separated by a slight energy offset ($\Delta E_{L\text{-}\Sigma}$ = 0.15−0.20 eV).[33] Thus, it is feasible to converge these two bands to obtain high S and a high zT subsequently. However, near the fermi level, the conduction band lacks multiple bands and consists of only a single L band.[34] Thus, contrary to its p-type counterpart, obtaining a high S and hence high zT has always been a challenge for n-type PbTe material. In this regard, Dutta *et al.* have demonstrated that Gd doping in PbTe can flatten the conduction band, which improves the Seebeck coefficient

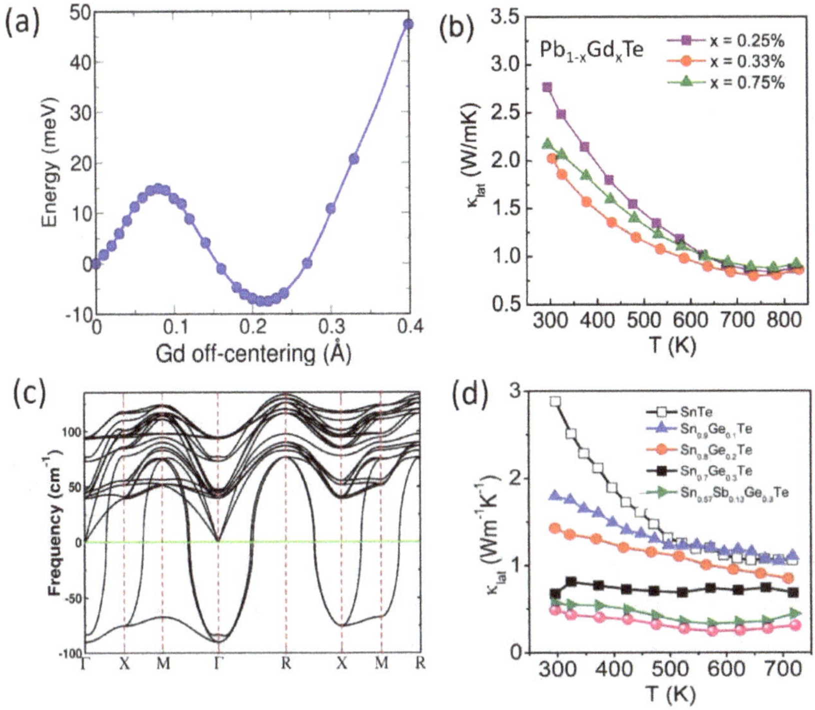


Figure 1. (a) Energy vs Gd off-centering showing that the structure is more stable when Gd is off-centered along the [111] direction by ~0.21 Å in the case of Gd doped PbTe. (b) Temperature-dependent lattice thermal conductivity ($\kappa_{lat}$) of $Pb_{1-x}Gd_xTe$ samples. (c) Phonon dispersion of $Sn_{0.75}Ge_{0.25}Te$ in cubic rocksalt-type structure showing unstable Γ-point transverse optical (TO) mode frequency at 93i $cm^{-1}$ due to ferroelectric instability. (d) Temperature variation of $\kappa_{lat}$ of Sb and Ge doped SnTe. Figures 1(a), 1(b) and 1(c), 1(d) are reproduced with permission from Ref. 5 © 2021 American Chemical Society and Ref. 9 © 2019 Royal Society of Chemistry respectively.

significantly through the increase of electron effective mass. In addition, an increased lattice anharmonicity leading to lowering of $\kappa_{lat}$ has also been observed in this compound due to off-centered position of Gd in the rock-salt PbTe lattice (Figures 1(a) and 1(b)).[5]

Despite being a reasonably efficient TE material, practical applicability of PbTe comes under question mark due to high toxicity of Pb. SnTe, structurally analogous to PbTe, has recently emerged with a promising alternative. However, due to inherent Sn-vacancies, p-type carrier concentration is quite high ($n \sim 10^{20}$–$10^{21}$ cm$^{-3}$) and thus the material suffers from a low S and high value of $\kappa_{el}$. Even though the electronic band structure is almost identical to that of PbTe, one of the striking differences is the relatively high energy offset ($\Delta E \sim 0.35$ eV) between the two topmost valance bands, which hinders SnTe to have a high S.[35] In addition, $\kappa_{lat}$ at room temperature (RT) is much higher (~3 Wm$^{-1}$K$^{-1}$) than the theoretical minimum thermal conductivity $\kappa_{min}$ (~0.44 Wm$^{-1}$K$^{-1}$), estimated using Cahill's model.[36] It is noteworthy to mention that to reduce the hole concentration, self-compensation in SnTe has been an effective way.[37] To overcome the low Seebeck coefficient issue, In doping has been proved to be effective by improving S significantly due to creation of resonant impurity levels near fermi level.[10,38] Furthermore, another strategy to increase S via convergence of valence bands has also been demonstrated by doping Sb,[9] Cd,[39] Ge, Mg,[20,40] Mn[37] in SnTe matrix. Synergistic effects of resonant level and valence band convergence have been observed to improve S via co-doping of In and Ag,[10] and In and Cd.[41] For lowering $\kappa_{lat}$ in SnTe, various commonly adopted strategies like alloying and nano-/meso-structures have been adopted.[20,37,42] Recently, a new strategy to lower $\kappa_{lat}$ in SnTe via engineering ferroelectric instability stemming from local structural distortion due to Ge alloying has been demonstrated by Banik *et al.*[9] It is revealed that ferroelectric instability induces the softening of low-lying polar transverse optical (TO) phonons, which again couple with strain to scatter heat-carrying acoustic phonons resulting in suppression of $\kappa_{lat}$ (Figures 1(c) and 1(d)).[9] Furthermore, chemical doping often exceeds its solubility limit and results in nanoprecipitates in SnTe. Nanoprecipitates of CdS,[39] MnTe[35] etc. have been reported to be efficient for phonon blocking and subsequently lower $\kappa_{lat}$. Recently, Pathak *et al.* have reported a high zT of ~1.55 at ~865 K in Mg and Ag co-doped SnTe, where Ag doping promotes valence band convergence and strong phonon scattering by MgTe nanoprecipitates resulting in ultra-low value of $\kappa_{lat}$ ~0.23 Wm$^{-1}$K$^{-1}$ at the temperature related to peak zT.[20]

Like PbTe and SnTe, in GeTe also, resonant state formation and valence band convergence has been one of the major strategies to increase its Seebeck coefficient. While In serves as the element to form resonant level in the electronic band structure of GeTe,[11] doping of Sb,[8] Mn,[43] Zn,[44] and Cd[45] have been proved to converge the valence bands. Recently, it has been observed in Sn -doped GeTe that Rashba spin splitting can also be effective in converging valence bands.[46] By co-doping In and Bi in GeTe, Perumal *et al.* have reported a high zT ~2.1 at 723 K, where In forms the resonant level and Bi doping increasing the point defect induced phonon scattering resulting in an ultralow $\kappa_{lat}$ of ~0.27 $Wm^{-1}K^{-1}$ at 625 K.[11] Synergistic effects of lowering $\kappa_{total}$ and increase S via valence band convergence have been obtained by co-doping Sb and Bi in GeTe.[8] It is observed that doping of Se and S in Te site modulates the electronic transport properties of GeTe and reduces $\kappa_{lat}$ due to point defect induced phonon scattering. A further reduction of $\kappa_{lat}$ of $(GeTe)_{0.9}(GeSe)_{0.05}(GeS)_{0.05}$ has been obtained by Sb doping followed by spark plasma sintering (SPS).[15] However, it should be mentioned here that among GeTe-based TE material, $(GeTe)_x(AgSbSe_2)_{100-x}$ alloys, well known as TAGS-x, are the most recognized due to their high performance.[47] Like SnTe, GeTe also suffers from intrinsic Ge vacancy, which makes it difficult to obtain a n-type GeTe. In this regard, Samanta *et al.* have demonstrated that gradual alloying $AgBiSe_2$ in GeTe causes a transition from p-type rhombohedral phase to n-type cubic phase (Figure 2(a)). In the cubic phase, the authors have managed to achieve zT ~0.6 at 500 K in n-type $(GeTe)_{50}(AgBiSe_2)_{50}$.[48] Electronic band structure calculation reveals that band degeneracy is higher in cubic phase of GeTe than in its rhombohedral phase. To be precise, in cubic phase, GeTe possesses 12-fold degenerate heavy band at lower energy along with 4-fold degenerate light band at higher energy. In rhombohedral phase, the polar distortion along the [111] crystallographic direction splits both the light and heavy hole bands and band degeneracy is partially lifted. So, efforts have been made to stabilize cubic phase of GeTe at lower temperature so that higher TE performance can be achieved. In this context, alloying MnTe,[43] Bi and Mn co-doping[49] and In and Sb co-doping[50] has been proven to be effective. However, Li *et al.* have demonstrated that effective band convergence of higher degree can be obtained by precisely controlling the rhombohedral distortion and a high zT~2.4 at 600 K is achieved in Pb and Bi co-doped GeTe.[51] Figure 2(b) shows comparison plot of temperature dependence of zT for different high performance GeTe based TE materials.[8,11,15,47-48,50-51]

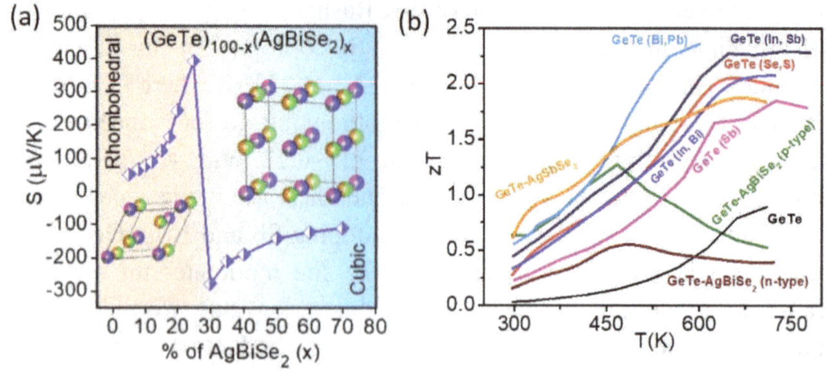


Figure 2. (a) Seebeck coefficient as a function of $AgBiSe_2$ percentage (x) for $(GeTe)_{100-x}(AgBiSe_2)_x$ exhibiting that carrier type changes from p-type to n-type when the crystal structure transforms from rhombohedral to cubic phase. Color codes are represented as follows: Ge, blue; Ag, grey; Bi, purple; Te, green; S, yellow. (b) Temperature dependence of thermoelectric figure of merit, *zT*, of selected high-performance GeTe based materials: Ge-In-Sb-Te,[50] Ge-In-Bi-Te,[11] GeTe-$AgSbSe_2$,[47] Ge-Pb-Bi-Te,[51] GeTe-$AgBiSe_2$,[48] Ge-Pb-Te,[45] Ge-Sb-Te,[8] Ge-Te-Se-S.[15] Figure 2(a) has been reproduced with permission from Ref. 48 © 2019 American Chemical Society.

Recently, Roychowdhury *et al.* have achieved a very high zT of ~2.6 at 573 K in $AgSbTe_2$ having a rocksalt-type cubic crystal structure (at RT) by Cd doping. It is noteworthy to mention that inherent atomic disorder in $AgSbTe_2$ restricts it to achieve high zT due to a low σ. Cd doping tunes the disorder induced localization and improves the electronic transport properties via enhancing cationic ordering. At the same time, reduction of $\kappa_{lat}$ is also observed due to formation of nanoscale superstructures (~2–4 nm).[52] Such synergistic effect of increasing σ without lowering S significantly and decreasing $\kappa_{total}$ produces high TE performance in Cd doped $AgSbTe_2$.

Bismuth telluride ($Bi_2Te_3$) and its alloy with $Sb_2Te_3$, $Bi_2Se_3$ are well-known TE materials due to their potential for near room temperature applications. From 1950s till now, it has been the mostly used TE material for commercial application.[18] Their unique properties *e.g.* small band gap ($E_g$ ~0.15 eV), small effective mass, high mobility, and poor $\kappa_{total}$ made them fascinating as one of the promising TE materials. Under ambient conditions, $Bi_2Te_3$ crystallizes in tetradymite structure. The Bravais lattice of bulk $Bi_2Te_3$ has rhombohedral crystal structure with space group *R*-3*m* having 5 atoms in the unit cell. Along the crystallographic c-direction, three quintuple layers of Te(1)–Bi–Te(2)–Bi–Te(1) are stacked with two neighbouring Te(1) layers which are separated via weak van der Waals' (vdW) interactions. Due to such layered structure, measurement of thermal and electrical transport properties along parallel and perpendicular to layered plane show anisotropic behaviour.[53] While a highly degenerate band structure helps to obtain a high Seebeck coefficient, a significant reduction of $\kappa_{lat}$ can also be achieved by alloying $Bi_2Te_3$ with $Sb_2Te_3$ due to point defects scattering stemming from Bi and Sb disorder in $Bi_{0.5}Sb_{1.5}Te_3$.[18] Over the years, different techniques like microstructure engineering and nano structuring have been employed to reduce $\kappa_{lat}$ further while maintaining a relatively high mobility. In $Bi_xSb_{2-x}Te_3$ alloys, a p-type zT ~1.4 was achieved at 373 K by hot pressing ball-milled nano powders.[54] In spite of introducing point defects and grain boundaries through alloying and nano structuring to scatter the low and high frequency phonons, suitable strategy to scatter mid-frequency phonons in this alloy was missing in the literature until 2015. In this regard, Kim *et al.* introduced the idea of synthesising the material via liquid phase sintering technique. He demonstrated that such synthesis process produces dense dislocation arrays in the low energy grain boundaries. Such dislocations proved to be highly effective in scattering mid frequency phonons and drag $\kappa_{lat}$ down to ~0.33 $Wm^{-1}K^{-1}$ at 320 K. A very high zT of 1.86 at 320 K was obtained.[18] Recently, Zhu *et al.* have applied similar technique of liquid phase sintering in Sb doped $Bi_2Te_{2.7}Se_{0.3}$ to investigate the effectiveness of the synthesis process in n-type material and obtained a high zT of ~1.4 at 425 K.[19] Microstructure engineering can also be an effective tool to optimize thermal transport. Hong *et al.* have reported a n-type zT of ~1.23 at 480 K in $Bi_2Te_{2.7}Se_{0.3}$ nanoplates synthesised via a microwave-assisted synthesis route.[55]

### 2.1.2. *Selenides*

Among the selenides, SnSe is the most popular TE material and recent research has put up some promising TE performances for this material. SnSe is a narrow band gap ($E_g$ ~0.90 eV) semiconductor with presence of multiple band maximums near fermi level, which gives the scope to tune their positions to manipulate the electronic density of states and hence obtain high Seebeck coefficient.[56-57] At ambient condition, SnSe has a layered orthorhombic structure with *Pnma* space group. However, SnSe undergoes a displacive or shear phase transition to the high symmetric *Cmcm* phase from a low symmetric *Pnma* phase. Such phase transition increases lattice strain and creates extended defects resulting in poor mechanical strength of the system.[58] As far as the thermal transport properties are concerned, due to high anharmonicity in the lattice, an ultralow $\kappa_{lat}$ of ~0.20 $Wm^{-1}K^{-1}$ at 973 K is observed along crystallographic a-direction, which make the single crystalline SnSe a promising candidate to study for high TE performance. Furthermore, a complex electronic band structure having 6 degenerate valence bands ensures a high Seebeck coefficient and hence a high power factor.[14,58] As a result, while pristine SnSe single crystals exhibit a high p-type zT ~2.6 at 923 K along b-direction,[58] a high n-type zT of 2.8 at 773 K is observed along the a-axis in Br doped SnSe crystals.[59] Despite having such high zT, high fragility and poor mechanical stability along with a lengthy and expensive synthesis procedure restricts single crystalline SnSe to be used for commercial applications.[60] To overcome this problem, attempts have been made to prepare nanocrystals/nanostructures of SnSe via various solution-phase synthesis techniques *viz.,* solvothermal/hydrothermal method. Chandra *et al.* have successfully synthesized two dimensional (2D) nanoplates of Ge doped SnSe by hydrothermal route and densified the synthesized power via SPS for transport measurements. Ge doping enhances p-type carrier concentration significantly resulting in a high electrical conductivity and power factor. In addition, high lattice anharmonicity, nanoscale grain boundary and Ge precipitates in SnSe matrix cause strong scattering of heat carrying phonons resulting in an ultralow $\kappa_{lat}$ of ~0.18 $Wm^{-1}K^{-1}$ and zT~2.1 at 873 K in the perpendicular pressing direction of SPS.[24] In an attempt to investigate the effect of rare earth element doping in SnSe, it is observed that Gd doping causes a phase separation of $Gd_2Se_3$ nanoprecipitates (2–5 nm) and hence improved hole concentration due to $Sn^{2+}$ vacancy formation. Enhanced magnetic fluctuations due to presence of paramagnetic nanoprecipitates of $Gd_2Se_3$ lead to scattering of low and mid-frequency phonons. As a result, a low $\kappa_{lat}$ ~0.41 $Wm^{-1}K^{-1}$ at 868 K was obtained for 3% Gd doping in SnSe.[61]

Furthermore, attempt has also been made to synthesize ultrathin (1.2–3 nm thick) n-type few layered nanosheets of Bi doped SnSe. An ultralow $\kappa_{lat}$ (~0.3 $Wm^{-1}K^{-1}$) in 300–720 K temperature range was observed in these Bi doped nanosheets due to scattering of phonons by nanoscale grain boundaries, interface of SnSe layers and point defects.[6] However, polycrystals and nano powders of SnSe suffer from rapid surface oxidation. As a result, thermally conductive layers of SnO and $SnO_2$ restrict achieving ultralow $\kappa_{lat}$ and also result in degraded electrical conductivity due to poor carrier mobility, which deteriorates the overall TE performance of SnSe.[62-63] In this regard, Zhou *et al.* synthesized

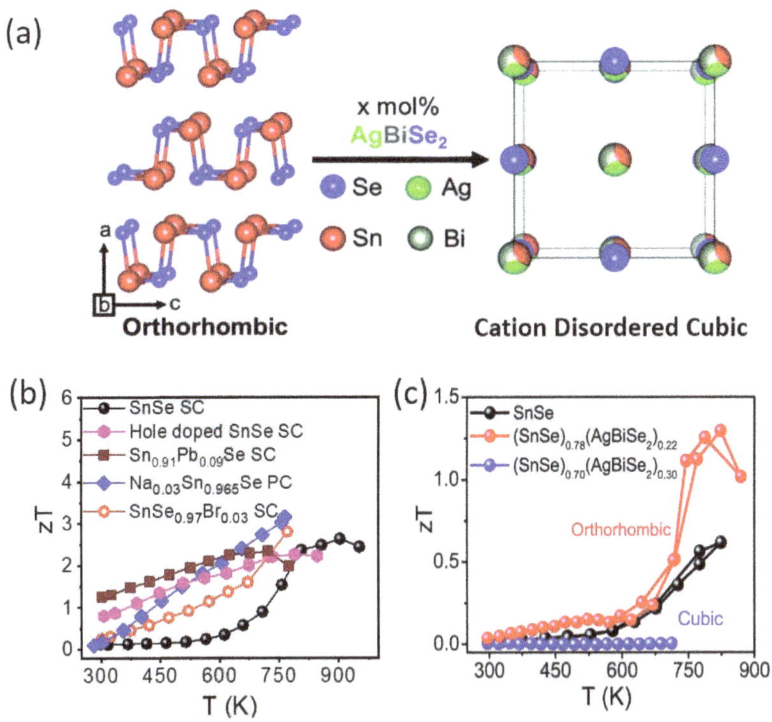


Figure 3. (a) Structural transition of SnSe from the orthorhombic to the rock-salt cubic structure upon $AgBiSe_2$ alloying. (b) Thermoelectric figure of merit (zT) of various SnSe based high performance thermoelectric materials (SC and PC denote single crystalline and polycrystalline sample respectively).[56-59,63] (c) Temperature-dependent zT of polycrystalline $(SnSe)_{1-x}(AgBiSe_2)_x$ measured parallel to the SPS pressing direction (black and blue symbols are orthorhombic phase and cubic phase). Figures 3(a) and 3(c) are reproduced with permission from Ref. 64 © 2020, American Chemical Society.

polycrystalline SnSe using tin oxide free Sn by chemically reducing it in $H_2$/Ar flow at elevated temperature and achieved an ultralow $\kappa_{lat}$ ~0.07 $Wm^{-1}K^{-1}$ at 783 K and a very high zT ~3.1 in polycrystalline $Na_{0.03}Sn_{0.965}Se$ (Figure 3(b)).[63] Recently, Chandra *et al.* have effectively stabilized the high-pressure cubic rock salt structure ($Fm\bar{3}m$) of SnSe by providing chemical pressure to the system via $AgBiSe_2$ alloying. For $(SnSe)_{1-x}(AgBiSe_2)_x$ system, while narrow band gap orthorhombic phase is stable for $0.00 \leq x < 0.28$, band gap closes and system goes to a cubic rock salt phase under further introduction of $AgBiSe_2$ ($0.30 \leq x < 0.70$) (Figure 3(a)). Cubic $(SnSe)_{0.70}(AgBiSe_2)_{0.30}$ shows n-type nature at 300 K. However, a highest zT ~1.3 at 823 K was achieved in this series for orthorhombic p-type $(SnSe)_{0.78}(AgBiSe_2)_{0.22}$ sample due to optimized electronic structure (Figure 3(c)).[64]

#### 2.1.3. *Sulphides*

Over the past few decades, TE community has evidenced improvement in performance of telluride and selenide-based TE materials. But TE technology demands low-cost and earth-abundant constituent elements besides having a high TE performance. In the last decade, metal-based sulphides gained popularity with the finding of several novel inexpensive, and environmentally sustainable TE materials.[65] However, a lighter mass of S compared to Se and Te leads to comparatively higher $\kappa_{lat}$ in crystalline metal sulphides. Simultaneously, lower carrier mobility results in a poor $\sigma$ and a low power factor and hence a lower zT.[65-66] Therefore, the TE study of sulphides had been somewhat overlooked with respect to selenide and tellurides, yet the diversity of the structure is more significant in sulphides, and S is much more earth-abundant and cheaper than Se and Te. In this part we discuss about few recently discovered metal sulphides showing ultralow $\kappa_{lat}$ and a decent high TE performance.

Although few metal sulphides have recently shown promising TE properties *viz.* $CuFeS_2$,[67] $AgBi_3S_5$,[68] $Cu_{12}Sb_4S_{13}$[69] and $Cu_2S$,[70] performance of n-type sulphides is still lower in comparison to the p-type metal sulphide thermoelectrics. Recently, Rathore *et al.* have explored the intrinsic ultralow $\kappa_{lat}$ of ~0.68–0.48 $Wm^{-1}K^{-1}$ in 298–820 K temperature range (Figure 4(a)) and a maximum zT of ~0.7 at ~823 K (Figure 4(b)) in the

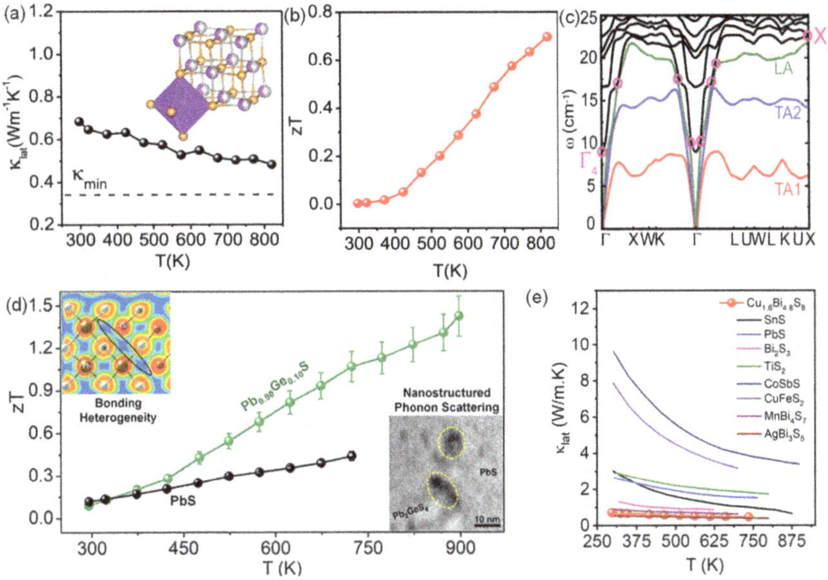


Figure 4. (a) Lattice thermal conductivity of $AgBiS_2$ measured in the temperature range 298–820 K. Inset of (a) shows refined $AgBiS_2$ with small distortions of the central cations away from the octahedral center in space group *P*1 (gray: Ag, violet: Bi, and orange: S). (b) Temperature dependent thermoelectric figure of merit (zT) of bulk $AgBiS_2$. (c) Calculated phonon dispersion for $AgBiS_2$ in the low-frequency region. (d) zT of $Pb_{1-x}Ge_xS$ (x = 0 and 0.10) samples. Inset of (d) shows electron localization function for $Pb_{0.89}Ge_{0.11}S$ along [110] plane (left side) and presence of $Pb_2GeS_4$ nanoprecipitates in the PbS matrix (right side). (e) Comparison of $\kappa_{lat}$ of $Cu_{1.6}Bi_{4.8}S_8$ with different state-of-the-art metal sulfides with low thermal conductivity. Figures 4(a), 4(b), and 4(c), are reproduced with permission from Ref. 71 © 2019 American Chemical Society. Figures 4(d) and 4(e) are reproduced with permission from Ref. 79 © 2022, Elsevier and Ref. 66 © 2021 American Chemical Society.

pristine n-type metal-sulfide $AgBiS_2$.[71] $AgBiS_2$, a member of the I-V-$VI_2$ compounds family (where I = Ag/Au/Cu; V = Sb/Bi/As; and VI = S/Se/Te), having a hexagonal crystal structure (*P*-3*m*1) at RT. At 475 K the hexagonal crystal structure undergoes transition into a cation-disordered rock salt cubic (*Fm*-3*m*) crystal structure. Because of the presence of stereochemical active lone pair in the $Bi^{3+}$ centre along the crystallographic [011] direction, $AgBiS_2$ possesses a locally distorted structure although the global structure remains cubic (inset of Figure 4(a)). Distortion leads to a low longitudinal and transverse sound velocity of ~3047 ± 85 m/s and

~1476 ± 13 m/s due to soft bonding interactions related to interactions and propagation of phonons in the crystals.[71] The local distortion induces significant anharmonicity in the crystal lattice, specified by a high anharmonicity index, Grüneisen parameter ($\gamma$) of 2.1. Additionally, theoretical study shows that anharmonicity generates from Bi lone pair and soft vibrations of Ag atoms create low-lying soft optical phonon modes which can scatter the heat carrying acoustic phonons (Figure 4(c)) significantly and reduce the $\kappa_{lat}$ to an ultralow value in globally cubic $AgBiS_2$. Eventually, a noteworthy *figure-of-merit* (zT) of ~0.7 (Figure 4(b)) was achieved in n-type metal sulfide pristine-$AgBiS_2$, which can be further improved via different doping and structural modulations.[71]

PbTe[5] and PbSe[72] have been deemed as potential thermoelectric materials with their high zT value, but the high cost and scarcity of Te and Se motivate the scientific community to study their low-cost and earth-abundant sulfide analogues, PbS. However, PbS has been rarely explored previously in terms of TE properties due to high $\kappa_{lat}$ (~2.75 $Wm^{-1}$ $K^{-1}$ at RT)[73] of PbS which leads to a low zT although it has a rock salt (*Fm-3m*) crystal structure similar to PbTe and PbSe. Additionally, higher deformation potential of PbS compared to PbTe and PbSe lowers the performance in PbS, which leads to a very small change in $\sigma$ at comparable doping concentration because of the better electron-phonon interaction in PbS.[73] However, instigating new strategies for tailoring the electronic band structure and nano/microstructures of PbS, enhanced zT has been achieved. Recently, Zhao's group have reduced $\kappa_{lat}$ by ~50% at 723 K for p-type PbS via incorporating up to 4.0 at. % of SrS or CaS in the PbS matrix.[74] CaS or SrS addition leads to the formation of both nanostructured and solid solution material which has combination of both point defects and nanoscale precipitates.[74] The $\kappa_{lat}$ is reduced from ~2.34 $Wm^{-1}K^{-1}$ (pristine PbS) to ~0.83 $Wm^{-1}K^{-1}$ for $Pb_{0.975}Na_{0.025}S$ + 4.0% CaS. Recently, $\kappa_{lat}$ of PbS has been further reduced by nanostructuring in p-type materials, for example, in PbS-(Zn,Cd)S,[75] PbS-(Sr,Ca)S,[74] and in n-type PbS-Sb,[76] PbS-Sn-PbTe,[77] PbS-Cu.[78] These nanoprecipitates and point defects play a crucial role in decreasing the $\kappa_{lat}$ by scattering phonons. However, a maximum zT of ~1.3 has been realized at 923 K for p-type $Pb_{0.975}Na_{0.025}S$-3.0% CdS,[75] whereas a similar zT of ~1.3 has been reported in n-type $Pb_{0.93}Sb_{0.05}S$[76] and $Pb_{0.94}Sn_{0.06}S$-8% PbTe.[77] Hence, the expedition to enhance the TE performance in earth-abundant PbS still remains an open opportunity for further exploration, which can be accomplished via concurrently lowering the $\kappa_{lat}$ and improving the electronic transport. Both the thermal and electronic transport of PbS can be fine-tuned by nano-structuring and chemical bond tailoring.

Motivated by the previous work, Rathore *et al.* alloyed 10 mol% Ge in PbS, which significantly lowered the $\kappa_{lat}$ (~0.56 $Wm^{-1}K^{-1}$ at 900 K) which reached closer to the theoretical $\kappa$ minima ($\kappa_{min}$) of ~0.45 $Wm^{-1}K^{-1}$ for pristine-PbS.[79] The presence of 5–10 nm sized inclusions of $Pb_2GeS_4$ in $Pb_{1-x}Ge_xS$ (x > 0.04) has been directly observed from high-resolution transmission electron microscopy (TEM) analysis (Figure 4(d)). These nano-inclusions of $Pb_2GeS_4$ significantly scatter the heat-carrying acoustic phonons.[79] The $\kappa_{lat}$ reduced by ~40% for 10% Ge doped $Pb_{0.90}Ge_{0.10}S$ in comparison to that of pristine PbS at 298 K. Moreover, Ge doping reduces the energy and lifetime of transverse acoustic (TA) and longitudinal acoustic (LA) phonon modes to an ultrashort value of ~1–4.5 ps, which was observed experimentally from inelastic neutron scattering (INS) measurement. Softening of phonon modes and ultrashort lifetime indicated the presence of enhanced scattering which resulted in ultra-low $\kappa_{lat}$. Enhanced covalency with increased heterogeneity in chemical bonding via Ge doping in PbS has also been observed from electron localization functional (ELF), COHP analysis and deformation potential calculations. The reduced electron-phonon coupling upon Ge-doping enhanced the electrical conductivity. This combined effects of (a) improved electrical transport and (b) supressed $\kappa_{lat}$ induced by $Pb_2GeS_4$ nano-inclusion and low-lying anharmonic phonons lift the zT to ~1.45 in the n-type Ge-substituted PbS at 900 K, which could become a promising low-cost substitute for PbSe- and PbTe-based TE materials.[79]

Metal sulphides generally show higher $\kappa_{lat}$ in comparison to their selenides and tellurides counterparts because of the lighter atomic weight of sulphur. But inherently, ultra-low $\kappa_{lat}$ in the crystalline metal sulphides grows much interest due to high natural abundance and low cost of elemental sulphur. Recently Bhui *et al.*, have investigated the structural origin of the low $\kappa_{lat}$ (0.71–0.44 $Wm^{-1}K^{-1}$ in 296–736 K range of temperature) of an environment-friendly and earth-rich *n*-type $Cu_{1.6}Bi_{4.8}S_8$ having a zT of ~0.26 at 736 K.[66] Ultralow $\kappa_{lat}$ measured in $Cu_{1.6}Bi_{4.8}S_8$ is indeed lower than that of the many state-of-the-art metal sulphides (Figure 4(e)) *viz.* SnS,[80] PbS,[79] $Bi_2S_3$,[81] $TiS_2$,[82] CoSbS,[83] $CuFeS_2$,[67] $MnBi_4S_7$,[84] and $AgBi_3S_5$.[68] $Cu_{1.6}Bi_{4.8}S_8$ exhibits fascinating structural features comprising of significant bonding heterogeneity and interstitial Cu-rattlers.[66] These rattling atoms vibrate independently akin to Einstein rattlers and induce low-lying optical phonon modes which was evidenced from the temperature-dependent Raman spectroscopy and heat capacity study. The presence of chemical bonding hierarchy in $Cu_{1.6}Bi_{4.8}S_8$ also observed from the ELF study which indicates weak bonding interactions of Cu-atoms to the covalently bonded Bi-S lattice framework. Density functional theoretical (DFT) calculations of phonon dispersion showed that substantial anharmonicity in the crystal lattice with the presence of low-lying optical phonons. Flat potential well of interstitial Cu-atoms compared to the deeper potential well of Bi and S indicates facile rattling of these Cu atoms. Complex structure and bonding hierarchy together with inherent rattler atoms considerably supress the phonon propagation and reduce the phonon lifetime to an ultrashort value of ~2–4.5 ps and result in inherently ultralow $\kappa_{lat}$ in $Cu_{1.6}Bi_{4.8}S_8$.[66]

## 2.2. *Half-Heusler*

Half-Heusler (HH) compounds have a typical composition consisting of XYZ (X = Sc, Ti, V group transition metal, Y = Fe, Co, Ni group transition metal and Z = Sn, Sb, Ge, Pb or Bi). Out of these compounds, three typical systems, *viz*., MCoSb, MNiSn (M = Ti, Zr, Hf), and XFeSb (X = V, Nb, Ta) attract attention of researchers in the thermoelectric community. The crystal structure of HH alloys can be treated as combination of rock-salt (XZ) and zinc blende (YZ) structures with cubic space group $F\bar{4}3m$ (216). HH alloys having 18 valence shell electronic count are, in general, semiconducting in nature. DFT calculation shows a small value of band gap ranging from 0.5–1 eV.[85] Highly symmetric cubic structure of these compounds results in multiple carrier pockets along with high band degeneracy ($N_v$) near fermi level. For example, for p-type NbFeSb and ZrCoSb, $N_v$ is 8.[86-87] Such high band degeneracy leads to large DOS effective mass and hence large Seebeck coefficient. In addition, in heavy band HH system, high carrier mobility is ensured by weak scattering of charge carriers by phonons due to low value of deformation potential.[88] One of the main disadvantages of HH thermoelectric compound is having high lattice $\kappa_{lat}$, which restricts it from achieving high zT. Such high $\kappa_{lat}$ can largely be suppressed to 2–3 $Wm^{-1}K^{-1}$ at temperature related to peak zT through nanostructuring, alloying, phase separation and formation of nanocomposites. However, $\kappa_{lat}$ still remains much higher than the theoretically minimum $\kappa_{lat}$ ~1 $Wm^{-1}K^{-1}$, estimated through Cahill's model.[36]

In the HH family, best n-type TE material is MNiSn based compound with zT > 1.[89-91] It is noteworthy to mention here that, recently, MCoSb has been found to exhibit high zT value as it has both p and n-type forms. Through proper compositional design and doping optimization, a maximum zT of greater than 1.5 has also been found for XFeSb-based heavy-band HH system.[88,92-93] Even though nominal 19-electron HH system are expected to show metal-like behavior, a moderate zT ~0.4 was found in NbCoSb.[94] Detailed investigation revealed that XCoSb is a defective HH compound having cationic vacancies as large as 20%. Such cation vacancies play a pivotal role in suppressing $\kappa_{lat}$. Through tuning the vacancy content, a high zT ~0.9 has been achieved for $Nb_{1-x}CoSb$ system.[95] This paves a new route for exploring 19-electron HH systems for high performance TE material by investigating the correlation between the vacancies and the transport properties. Recently, power density of 3.1 $Wcm^{-2}$ and a maximum conversion efficiency ~10.5% was achieved using p-type XFeSb and n type MNiSn under a temperature difference of 680 K.[96-97]

## 2.3. *$Mg_3X_2$ based compounds*

n-type $Mg_3X_2$, where X = Sb, Bi or mixture of both, is considered to be one of the promising TE materials in the room temperature to mid-temperature (300–700 K) range. This material can be treated as a subgroup of $AM_2X_2$ Zintl phases where both A and M sites are occupied by Mg atoms. The crystal structure is also identical to the $AM_2X_2$ Zintl phases having space group $P\bar{3}m1$ (No. 164).[98] In stoichiometric $Mg_3X_2$, Mg vacancies are formed, which kills electron. To supress such Mg vacancy formation and ensure n-type conduction via Mg-excess thermodynamic condition, an excess of Mg is required so that the nominal composition is $Mg_{3+x}X_2$. However, high vapor pressure along with high reactivity of Mg at an elevated temperature leads to loss of Mg and consequently loss of n-type conduction during high temperature (T > 700 K) synthesis.[99-101] As far as the electronic band structure is concerned, highly degenerate conduction band exists having valley degeneracy $N_v = 6$. In addition to a high Seebeck coefficient due to highly degenerate valley, an ultralow thermal conductivity ($\kappa_{lat}$ ~0.5 $Wm^{-1}K^{-1}$) leads to high zT values (> 1.5) at 700 K and ~1 around 300 K in $Mg_3Sb_2$.[102] After realization of such high TE performance, efforts have been made to optimize zT by chemical doping[100,103] and manipulating the electronic band structure.[104] Grain boundary engineering has a pivotal role to play in determining TE performance of these compounds. Charged defects at grain boundaries trap and scatter the mobile charge carriers leading to low electrical conductivity in polycrystalline samples. Such grain boundary effect can be minimized by increasing the size of the grains through various techniques.[100,103-105]

As $Mg_3Sb_2$ and $Mg_3Bi_2$ have identical crystal structure, they can make a solid solution of $Mg_3Sb_{2-x}Bi_x$ for the entire range of x from 0 to 1. With alloying, $\kappa_{lat}$ is massively decreased (~50%) and electronic band structure can be tuned by varying x. With increasing Bi content, band gap (0.5–0.6 eV for pure $Mg_3Sb_2$) decreases and effective mass reduces resulting in an increased weighted mobility.[106] It has been observed that 25% doping of Bi, i.e., $Mg_3Sb_{1.5}Bi_{0.5}$ is an optimal composition for mid temperature (~700 K) TE application. Increased Bi doping (> 50 %) results in a shift of zT peak towards lower temperature. However, doping in Mg site disrupts the high conduction band degeneracy and decreases carrier mobility, which leads to a significant reduction of thermoelectric performance.[107-108] Effective control of charge carriers can be achieved by doping Te in the anion sites of these compounds.[98,109-110] Some recent reports showed that while coupled with p- type MgAgSb[111] and p-type $Bi_2Te_3$,[112] these materials can achieve a high conversion efficiency (> 7%) under 300 K temperature difference with hot side temperature ~573 K.

## 2.4. *Skutterudites*

In terms of TE applications, in the intermediate temperature, skutterudites are one of the best TE materials discovered. Binary skutterudites having a common chemical formulation $MX_3$ (M = Rh, Co or Ir; X = Sb, As or P) usually crystallizes in a bcc structure (*Im*-3*m*). The structure of representative cobalt antimonide ($CoSb_3$) is shown in Figure 5(a) displaying the lattice framework composed of corner-shared $CoSb_6$ octahedra.[113] This type of framework leaves two icosahedral voids per unit $M_8X_{24}$, which can be made accessible by different types of guest atoms *viz.* rare earth, alkaline earth, alkalis, and others, to form filled skutterudites. Filled skutterudites are generally represented as $G_yM_4X_{12}$, where G stand for the filler and y stand for filling fractions. Interestingly, filler G atom remains weakly connected to the skutterudite lattice framework and undergoes independent vibration as an Einstein rattler, which can significantly reduce the thermal conductivity.[114] Filled skutterudites satisfy the concept of phonon-glass electron-crystal (PGEC) by maintaining low $\kappa_{lat}$ and decent electrical transport properties.[115]

Among the reported skutterudites family, filled-$CoSb_3$ based skutterudites exhibits superb TE performance due to its low band gap of ~0.2 eV with high carrier mobility.[116] When alkali metals, alkaline earth or rare earth metals are acclimatised in the void of Sb-icosahedral without any extra charge compensation, the resultant filled-skutterudites typically exhibit n-type electrical conductivity because of the extra introduced electrons in the lattice framework. Gainza *et al.* have recently reported a detailed TE property study of high-pressure synthesized $M_xCo_4Sb_{12}$ (M = Y, K, Sr) skutterudites.[117] The evaluated structural parameters imply primary dependence on the filling fraction, irrespective of the size of cations. Among all the fillers, Sr-filled $CoSb_3$ sample shows least distorted [$Sb_4$] rings, related to a higher filling fraction. Due to the higher filling of Sr, $Sr_{0.48}Co_4Sb_{12}$ sample shows a low $\kappa_{lat}$, resulting in a high zT of ~0.80 at 720 K (Figure 5(b)).[117]

DFT calculations show that the vibrational frequencies of the different filler from separate chemical groups are significantly different. It has been observed that phonon frequencies near the frequencies of filler atoms get scattered significantly to reduce the thermal conductivity. Therefore,

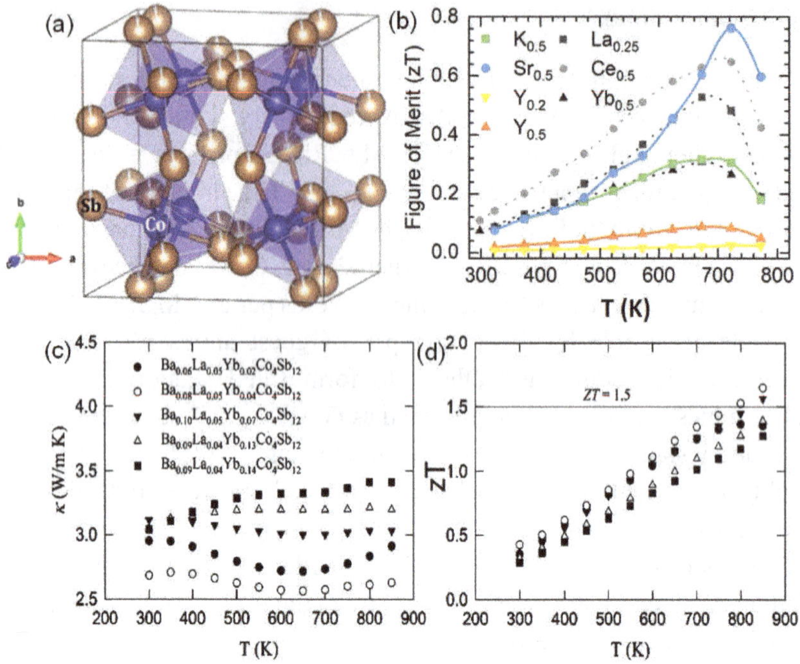


Figure 5. (a) Skutterudite crystal structure of $CoSb_3$ indicating the presence of $CoSb_6$ octahedra. (b) zT of $M_xCo_4Sb_{12}$ (M = Y, K, Sr, La, Ce, Yb) prepared under identical conditions. (c) Temperature dependence of $\kappa_{tot}$, and (b) zT for $Ba_uLa_vYb_wCo_4Sb_{12}$ multiple-filled skutterudites. Figures 5(a) and 5(b) are reproduced with permission from Ref. 113 © 2022 American Chemical Society and Ref. 117 © 2020 John Wiley and Sons, respectively. Figures 5(c) and 5(d) are reproduced with permission from Ref. 116 © 2011, American Chemical Society.

incorporation of two different filler atoms from separate chemical groups into the icosahedral voids of $CoSb_3$ could lead to generation of two characteristics filler vibrational frequencies to scatter wider range of phonons and further reduce the $\kappa_{lat}$.[116,118] In this regard, Shi *et al.* have studied co-filling of three different elements (Ba, La and Yb) simultaneously into the icosahedral nanocage of $CoSb_3$.[116] The lattice framework remains unperturbed and well-maintained with excellent electrical conductivity. The total thermal conductivities (Figure 5(c)) of the multiple-filled skutterudites are shown in the range of 300 to 850 K. The $\kappa_{lat}$ decreases with increasing the total filling fraction. In the multiple filled

$Ba_{0.08}La_{0.05}Yb_{0.04}Co_4Sb_{12}$, the maximum zT obtained reaches to a very high value of ~1.7 at 850 K (Figure 5(d)).[116]

## 2.5. *Clathrates*

Inorganic clathrates are well-known as a potential TE material with phonon-glass electron-crystal (PGEC) behavior having low $\kappa_{lat}$ coupled with high electrical conductivity. These materials are typically composed of elements from group 13, 14 and 15 with a low band gap and behave like semiconductors.[119] Compounds having Type-I clathrate structure possess a typical formula of $A_8E_{46}$, (A = Ba, K, Na; E = Si, Ge, Sn, Al, Ga, In) and have an open-framework cubic crystal structure with *Pm*-3*n* space group (Figure 6(a)).[120-121] Whereas clathrates II and III adopt a general formula of $A_{24}E_{136}$ and $A_{30}E_{172}$, respectively, with similar A and E atoms, but there are very rare examples of such structures.[119] While type-I clathrates show promising TE properties, reports on the TE property study of type-II and type-III clathrates are rare.

Sui *et al.* have reported TE studies of Ba-doped type-I clathrate, $K_8Al_8Si_{38}$, which exhibits high electrical resistivity and high Seebeck coefficients in its pristine form.[121] In comparison to the pristine $K_8Al_8Si_{38}$, $K_{8-x}Ba_xAl_{8+x}Si_{38-x}$ shows enhanced electrical conductivity owing to increase in n-type charge carrier concentration with Ba-substitution. The low $\kappa_{lat}$ observed in the $K_{8-x}Ba_xAl_{8+x}Si_{38-x}$ is due to a higher scattering of acoustic phonons with incorporation of dopants. Overall, an improved zT of ~0.4 at 873 K has been obtained for the composition of $K_{6.5}Ba_{1.5}Al_9Si_{37}$ (Figure 6(b)).[121] In another study, Zhang *et al.* have reported a series of type-I inorganic clathrates having composition of $Ba_8(Al_xGa_{1-x})_{16}Ge_{30}$, synthesized using $Ba_8Al_{16}Ge_{30}$ and $Ba_8Ga_{16}Ge_{30}$ via ball milling followed by SPS.[122] It is observed that for samples with x = 0.20, 0.23, and 0.25, electrical conductivity increases significantly, whereas Seebeck coefficient reduces very slightly (Figure 6(c)), which results in highest power factor value of ~1.89 mW $m^{-1}$ $K^{-2}$ in $Ba_8(Al_{0.25}Ga_{0.75})Ge_{30}$. Microscopic studies revealed that the material is comprised of Al particles/nano-inclusions in the Ga-rich clathrate matrix. These aggregated particles or inclusions on the grain boundaries further reduce the thermal conductivity.

Therefore, a maximum zT of ~0.93 at 1073 K has been achieved in the sample $Ba_8(Al_{0.25}Ga_{0.75})Ge_{30}$ (Figure 6(d)).[122]

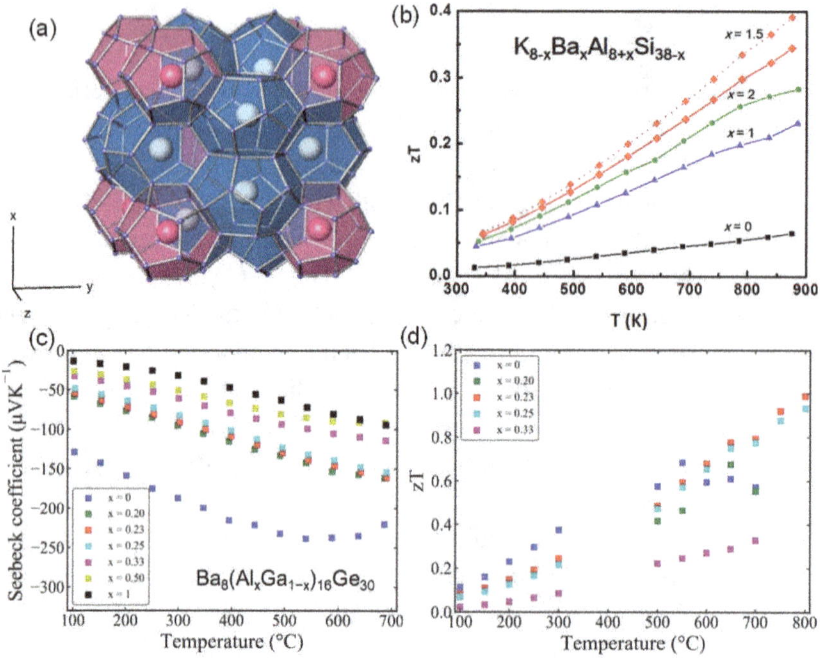


Figure 6. (a) Representation of the Type I clathrate structure having two different types of cages. The 20-vortex cage is shown by red colour and the 24-vertex cage is shown in blue colour. (b) zT versus temperature plot for $K_{8-x}Ba_xAl_{8+x}Si_{38-x}$ (x = 0, 1, 1.5, and 2). (c) Seebeck coefficient of $Ba_8(Al_xGa_{1-x})_{16}Ge_{30}$ pellets prepared by SPS. (d) Temperature-dependent zT of $Ba_8(Al_xGa_{1-x})_{16}Ge_{30}$ sintered by SPS. Figures 6(a) and 6(b) are reproduced with permission from Ref. 121 © 2016 American Chemical Society. Figures 6(c) and 6(d) are reproduced with permission from Ref. 122 © 2021, John Wiley and Sons.

## 2.6. *Silicides*

Since the 1960s, Si-based TE materials have been investigated because of their potential TE properties together with their low-cost and environmentally benign constituent materials. Furthermore, combination of Si with any other elements/metals gives an excellent platform for TE property studies. Among several Si-based TE materials, mostly $Mg_2Si$, $MnSi_x$, $ReSi_{1.75}$, $CrSi_2$ and β-$FeSi_2$, have attracted considerable attention of the thermoelectric society due to their high TE performances.[123] Among these identified high-performance silicides, mostly $Mg_2Si$ and associated solid solutions show very promising zT.

Efforts have been made to optimize $Mg_2Si$-based TE materials regarding synthesis, stability, and performance over time. $Mg_2Si$ has a cubic (*Fm*-3*m*) antifluorite type structure (Figure 7(a)), and it is a n-type semiconductor.[124] Liu *et al.* have prepared a series of compounds, $Mg_{2(1+z)}Si_{0.5-y}Sn_{0.5}S$ ($0 \leq z \leq 0.15$ and $0 \leq y \leq 0.015$) and optimised the TE properties with controlled Mg content and variable Sb doping.[125] With a

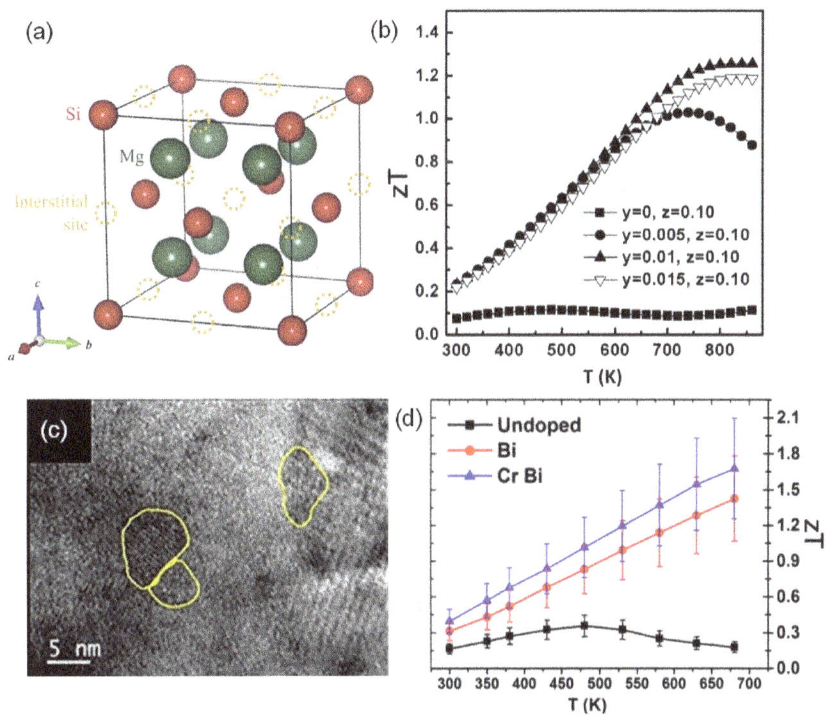


Figure 7. (a) Antifluorite crystal structure of $Mg_2Si$. (b) Temperature variation of zT of $Mg_{2(1+z)}Si_{0.5-y}Sn_{0.5}Sb_y$ ($z = 0.1$) compounds for different Sb doping ($0 \leq y \leq 0.015$). (c) HRTEM images of $Mg_{1.98}Cr_{0.02}(Si_{0.3}Sn_{0.7})_{0.98}Bi_{0.02}$ showing presence of scattered precipitates in the matrix. (d) Temperature-dependent zT of $Mg_2Si_{0.3}Sn_{0.7}$, $Mg_2(Si_{0.3}Sn_{0.7})_{0.98}Bi_{0.02}$ and $Mg_{1.98}Cr_{0.02}(Si_{0.3}Sn_{0.7})_{0.98}Bi_{0.02}$. Figures 7(a) and 7(b) are reproduced with permission from Ref. 124 © 2020 AIP Publishing LLC and Ref. 125 © 2011 American Chemical Society. Figures 7(c) and 7(d) are reproduced with permission from Ref. 126 © 2019, American Chemical Society.

sub-stoichiometric Mg content, the point defect only consists of Mg-vacancies which counteract the impact of Sb doping. Whereas, with over-stoichiometric Mg content in $Mg_{2(1+z)}Si_{0.49}Sn_{0.5}Sb_{0.01}$, the primary point defects are Si/Sn vacancies and interstitial Mg.[125] Excess Mg content simultaneously enhances the electron carrier concentration. Thus, excess Mg doped sample, $Mg_{2.10}Si_{0.49}Sn_{0.5}Sb_{0.01}$, has much higher electron carrier concentration compared to the $Mg_2Si_{0.49}Sn_{0.5}Sb_{0.01}$. Moreover, a small excess over-stoichiometric Mg content boosts the power factor and zT. A maximum zT of ~1.25 (Figure 7(b)) has been achieved at 800 K for $Mg_{2.20}Si_{0.49}Sn_{0.5}Sb_{0.01}$ which is 60% higher than the nominal stoichiometric compositions, $Mg_2Si_{0.49}Sn_{0.5}Sb_{0.01}$.[125]

The best results are achieved in this compound via a combination of reduced $\kappa_{lat}$ (improved phonon scattering) and enhanced electrical transport (band convergence). Goyal *et al.* have obtained highest zT of ~1.7 at 673 K via co-doping of Bi and Cr in $Mg_2Si_{0.3}Sn_{0.7}$.[126] While Cr-doping leads to formation of nanoprecipitates (Figure 7(c)), Bi-doping optimizes the electrical conductivity. The embedded nanoprecipitates, rich in elemental Cr and Sn, create an additional strain in the precipitate-matrix interface and enhance the phonon scattering to further reduce the $\kappa_{lat}$ and to boost the zT (Figure 7(d)) to a record high value of ~1.7 at 673 K.[126]

## 2.7. *All-inorganic metal halide perovskites*

Recently, the study of thermal transport property of metal halide perovskites is growing to develop new materials for hot-phonon bottle-neck in photovoltaics, thermoelectrics, and refractories.[127] The ultralow thermal conductivity of the halide perovskites mainly arises interest to attain a phonon blockage for thermal insulation purposes and thermo-electricity. The study of thermal transport in crystalline perovskite halides is still in its infancy and requires exploration of the broad family of perovskite and related structures of halides for fundamental understanding of the correlation of phonon transport, chemical bonding and solid-state structure. Although most of the experimental and theoretical studies have been performed based on transport properties of hybrid perovskites ($MAPbX_3$ (X = I, Br and Cl), $FAPbBr_3$ and $(MA)_3Bi_2I_9$, where MA and FA stands for methylammonium and formamidinium respectively),[128-131] recent studies on all inorganic halide perovskite are gaining more significance over hybrid halide perovskites due to their superior thermal stability.[132-133] Halide perovskites have a good Seebeck coefficient at RT which together with ultralow thermal conductivity make these systems interesting for TE transport properties investigation.[134]

Fundamental comprehension of thermal conductivity and its correlation with the nature of chemical bonding, crystal structure, and lattice dynamics is essential to design on-demand materials, such as crystalline materials with low thermal conductivity to achieve high thermoelectric performance and hot-phonon bottleneck in optoelectronics. Because of much lower toxicity of Sn over Pb, Sn-based all-inorganic perovskites are recently considered as potential alternative over Pb-based halides. $Cs_2SnI_6$, exhibiting vacancy ordered double perovskite structure (*Fm*-3*m*) (Figure 8(a)), have a comparatively low band gap ~1.25 eV, which make them attractive for studying thermoelectric properties.[133] Thereby, exploring the thermoelectric properties and thermal transport of $Cs_2SnI_6$ with intriguing defect-variant perovskite crystal structure and bonding hierarchy in combination with the co-existence of ionic and covalent bonding.

Recently Bhui *et al.*, have demonstrated ultralow $\kappa_{lat}$ of ~0.29–0.22 $Wm^{-1}K^{-1}$ (Figure 8(b)) in the air-stable and Pb-free vacancy-ordered double perovskite $Cs_2SnI_6$.[133] $\kappa_{lat}$ value of crystalline $Cs_2SnI_6$ is lesser than that of several state-of-the-art low thermal conductive halide perovskites and metal chalcogenides. Unlike perovskites $CsSnX_3$ (X = Br, I) based sample where degenerate hole doping makes them *p*-type semiconductor, $Cs_2SnI_6$ exhibits a rare *n*-type thermoelectric conduction with high Seebeck coefficient (> −600 μV/K in the 296–423 K temperature range). DFT calculation predicted small values of shear and bulk modulus originating from elastically soft lattice structure of $Cs_2SnI_6$ with weakly connected $[SnI_6]^{2-}$ octahedra.[133] The abundance of low-lying optical phonons originate from localized anharmonic Cs atom rattling in the lattice framework constructed by $[SnI_6]^{2-}$ together with the dynamic rotation of the $SnI_6$ octahedra (Figure 8(a)). These low-lying optical phonon modes couple with heat-transporting acoustic phonon modes and cause significant dampening. The soft nature of the acoustic phonons and their strong scattering from low-lying highly anharmonic optical phonons results in a very short phonon lifetime ~1 ps and consequently leads to the ultralow $\kappa_{lat}$. Also, an improved power factor of ~9.6 μW/m·K$^2$ has been achieved in 5 mol % Ge-alloyed $Cs_2SnI_6$, which is 3 times more in comparison to the pristine $Cs_2SnI_6$.[133]

Two-dimensional (2D) metal halide perovskites have recently created sensation for their intriguing optoelectronic applications. Thermal conductivity is one of the most critical parameters for the success of these materials. In a recent work, Acharyya *et al.* have grown single crystal of 2D Ruddlesden-Popper (RP) halide perovskite, $Cs_2PbI_2Cl_2$ (*I4/mmm* space group, Figure 8(c)) and demonstrated ultralow $\kappa_{lat}$ of ~0.37–0.28 $Wm^{-1}K^{-1}$ (Figure 8(d)) in the 295–523 K along the crystallographic c-axis and perpendicular (⊥) to Bridgman growth direction.[132] Anisotropic $\kappa_{lat}$ value along both the directions of Bridgman growth (Figure 8(d)) are extremely low for a crystalline material and being single crystal, the observed low $\kappa_{lat}$ seems intrinsic to the system. Presence of mixed covalent (strong) and ionic (weak) bonding play an intriguing role in lowering the $\kappa_{lat}$. DFT phonon calculations found low elastic moduli for this system together with dispersion less flat low energy optical phonon branches.[132] Lone-pair induced anharmonic vibration of halide atoms distorts the Pb-octahedra, which, in turn, results in these low energy optical phonons and low thermal conductivity. Temperature-dependent heat capacity and Raman data also experimentally confirm the presence of low-lying flat phonon bands. Such novel thermal or phonon transport in this low-dimensional perovskite halide is due to dynamic octahedral rotation/tilting driven by lone pair and anharmonic lattice dynamics.[132]

Thermoelectric property studies of $CsPbX_3$ (X = Br, I) and $CH_3NH_3PbX_3$ (X = Cl, Br, I) have revealed a notable Seebeck coefficient (300–5500 $\mu VK^{-1}$) with an ultralow thermal conductivity.[135-136] However, most of these studies are mainly focused on the nanowires or films and meticulous study of thermoelectric transport properties on the bulk polycrystalline sample is demanding. The major difference between the Sn and Pb-based halide perovskite is the capability to dope. In comparison to the Pb-based perovskite halides, $ASnX_3$ undergoes degenerate self-doping of $Sn^{4+}$ via the oxidation of $Sn^{2+}$ and can exhibit a controllable electrical conductivity. In this regard, Xie *et al.* have recently studied the thermoelectric properties and thermal transport of the $CsSnBr_{3-x}I_x$ perovskites.[134] Among the

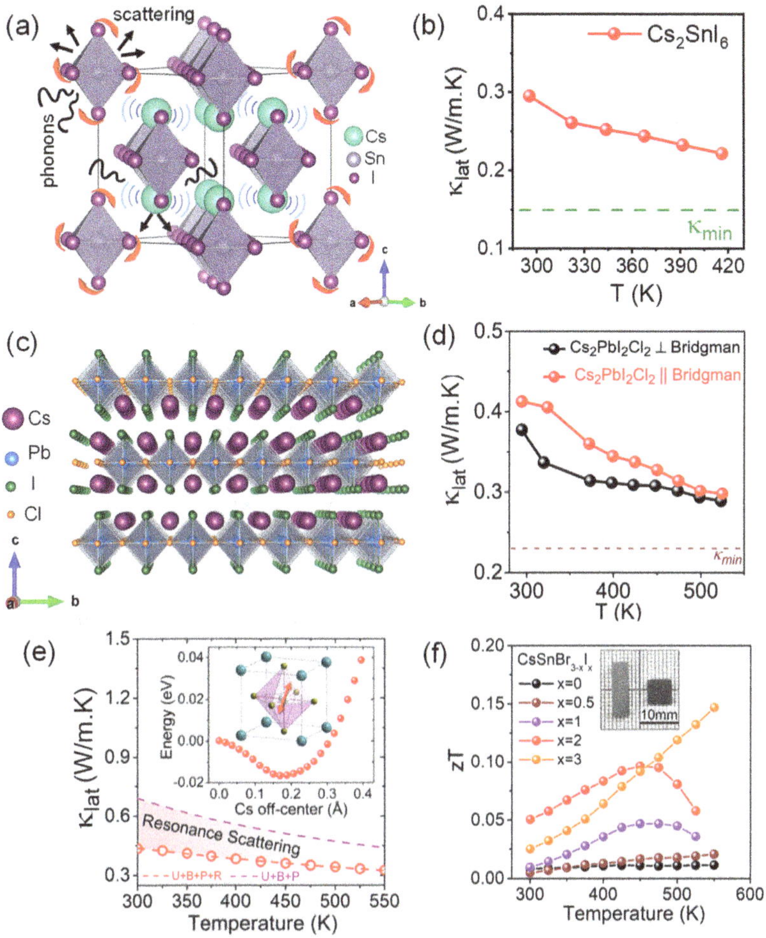


Figure 8. (a) Crystal structure of $Cs_2SnI_6$ showing the presence of phonon scattering centers. (b) Temperature-dependent $\kappa_{lat}$ of $Cs_2SnI_6$. (c) Crystal structure and (d) temperature-dependent $\kappa_{lat}$ in parallel (∥) and perpendicular (⊥) with respect to the Bridgman growth directions of layered Ruddlesden−Popper (RP) $Cs_2PbI_2Cl_2$. (e) $\kappa_{lat}$ of $CsSnBrI_2$ with contribution of various scattering processes; here, the R, P, B and U represent phonon resonance scattering, point defect scattering, grain boundaries scattering and Umklapp scattering respectively. Inset figure shows off-centered Cs is shifted from center towards (111) direction by 0.2 Å. (f) zT of $CsSnBr_{3-x}I_x$ (x = 0, 0.5, 1, 2 and 3). The inset images display the sample specimens used for the measurement of transport properties. Figures 8(a) and 8(b) are reproduced with permission from Ref. 133 © 2022 American Chemical Society. Figures 8(c) and 8(d) are reproduced with permission from Ref. 132 © 2020, American Chemical Society. Figures 8(e) and 8(f) are reproduced with permission from Ref. 134 © 2020, American Chemical Society.

polycrystalline mixed halide ($CsSnBr_{3-x}I_x$) series of perovskite materials, $CsSnBr_2I$ has an ultralow $\kappa_{lat}$ of ~0.32 $Wm^{-1}K^{-1}$ at 550 K (Figure 8(e)). Pristine $CsSnI_3$ exhibits metal like transport properties with a considerable high $\sigma$ of about 300 S $cm^{-1}$ at ~298 K whereas pristine $CsSnBr_3$ is having a large S of ~325 $\mu VK^{-1}$ which further increases up to ~560 $\mu VK^{-1}$ with increasing temperature to 550 K.[134] The alloying of both $CsSnI_3$ and $CsSnBr_3$ shows an efficient way of modulating charge carrier transport while maintaining the ultralow $\kappa_{lat}$. Large Umklapp scattering in combination with phonon resonance scattering, point defect scattering and grain boundary scattering induce an exceptionally low $\kappa_{lat}$ in the mixed halide $CsSnBrI_2$ (Figure 8(e)). The weak chemical bonding, primarily among Cs-X (ionic bonding) bonds, leads to a larger vibration of the Cs atoms. Theoretical calculations and the experimental results show that Cs-atoms are in an off-centred site (inset of Figure 8(e)) in the cuboidal cage and behave as a heavy rattler. The Cs-rattler atoms together with bonding heterogeneity generate low-lying optical phonons that can scatter the heat-carrying acoustic phonons, lower the phonon velocity and Debye temperature, and lead to ultralow $\kappa_{lat}$. With good electrical conductivity, decent Seebeck coefficient and ultra-low $\kappa_{lat}$, $CsSnBr_{3-x}I_x$ shows a decent zT of ~0.15 (Figure 8(f)) which is the highest among the perovskite halides.[134]

## 3. Conclusions and Outlook

A rapid progression in the thermoelectric performance of existing materials have been witnessed in the last two decades with a simultaneous discovery of new thermoelectric materials. New material design strategies have been applied to decouple the correlated transport parameters. zT values in the third-generation thermoelectric material crosses the threshold of 2 and it is reaching up to 3. New strategies like removing oxide layer in SnSe, atomic order-disorder optimization in $AgSbTe_2$ and valence band convergence in GeTe push the thermoelectric performances upwards. Several fascinating materials such as $AgBiS_2$, $AgBi_3S_5$, $Cu_{1.6}Bi_{4.8}S_8$ showing intriguing low thermal conductivity and promising thermo-electric performance have been discovered. All-inorganic metal halide perovskites viz. $CsSnI_3$, $Cs_2PbI_2Cl_2$ and $Cs_2SnI_6$ offer high Seebeck coefficients with intrinsically ultralow lattice thermal conductivity, paving a new direction to design novel thermoelectric materials. We believe that there is enormous scope to achieve high-performance thermoelectrics either via modulating the electronic structure and manipulating nano-structures, or via exploring new materials with intriguing structure and chemical bonding. The finding of new compounds can be boosted further by applying high-throughput computational method followed by detailed experimental investigations. Total scattering probe, such as synchrotron X-ray pair distribution function, is being used to understand local structure and thermoelectric property relationships. Inelastic neutron scattering can provide experimental phonon spectra which is a valuable tool to understand thermal conductivity. Key chemical intuition, understanding of bonding and structure, synthesis and materials processing and transport measurements are the key to create new high performance thermoelectric materials.

## Acknowledgments

S.D. and A.B. contributed equally in writing this chapter. K.B. would like to acknowledge the financial support from Swarnajayanti Fellowship, Science and Engineering Research Board (SERB) (SB/SJF/2019-20/06). A.B. is thankful to the CSIR for the fellowship.

## References

1. G. J. Snyder and E. S. Toberer, Complex Thermoelectric Materials, *Nat. Mater.* **7**(2), 105-114, (2008).
2. X.-L. Shi, J. Zou and Z.-G. Chen, Advanced Thermoelectric Design: From Materials and Structures to Devices, *Chem. Rev.* **120**(15), 7399-7515, (2020).
3. R. Freer, D. Ekren, T. Ghosh, K. Biswas, P. Qiu, S. Wan, L. Chen, S. Han, C. Fu, T. Zhu, A. K. M. Ashiquzzaman Shawon, A. Zevalkink, K. Imasato, G. J. Snyder, M. Ozen, K. Saglik, U. Aydemir, R. Cardoso-Gil, E. Svanidze, R. Funahashi, A. V. Powell, S. Mukherjee, S. Tippireddy, P. Vaqueiro, F. Gascoin, T. Kyratsi, P. Sauerschnig and T. Mori, Key Properties of Inorganic Thermoelectric Materials — Tables (version 1), *J. Phys. Energy* **4**(2), 022002, (2022).

4. S. Perumal, S. Roychowdhury, D. S. Negi, R. Datta and K. Biswas, High Thermoelectric Performance and Enhanced Mechanical Stability of p-Type $Ge_{1-x}Sb_xTe$, *Chem. Mater.* **27**(20), 7171-7178, (2015).
5. M. Dutta, R. K. Biswas, S. K. Pati and K. Biswas, Discordant Gd and Electronic Band Flattening Synergistically Induce High Thermoelectric Performance in n-type PbTe, *ACS Energy Lett.* **6**(4), 1625-1632, (2021).
6. S. Chandra, A. Banik and K. Biswas, n-Type Ultrathin Few-Layer Nanosheets of Bi-Doped SnSe: Synthesis and Thermoelectric Properties, *ACS Energy Lett.* **3**(5), 1153-1158, (2018).
7. C. Gayner and Y. Amouyal, Energy Filtering of Charge Carriers: Current Trends, Challenges, and Prospects for Thermoelectric Materials, *Adv. Funct. Mater.* **30**, 1901789, (2020).
8. S. Perumal, P. Bellare, U. S. Shenoy, U. V. Waghmare and K. Biswas, Low Thermal Conductivity and High Thermoelectric Performance in Sb and Bi Codoped GeTe: Complementary Effect of Band Convergence and Nanostructuring, *Chem. Mater.* **29**(24), 10426-10435, (2017).
9. A. Banik, T. Ghosh, R. Arora, M. Dutta, J. Pandey, S. Acharya, A. Soni, U. V. Waghmare and K. Biswas, Engineering Ferroelectric Instability to Achieve Ultralow Thermal Conductivity and High Thermoelectric Performance in $Sn_{1-x}Ge_xTe$, *Energy Environ. Sci.* **12**(2), 589-595, (2019).
10. A. Banik, U. S. Shenoy, S. Saha, U. V. Waghmare and K. Biswas, High Power Factor and Enhanced Thermoelectric Performance of SnTe-$AgInTe_2$: Synergistic Effect of Resonance Level and Valence Band Convergence, *J. Am. Chem. Soc.* **138**(39), 13068-13075, (2016).
11. S. Perumal, M. Samanta, T. Ghosh, U. S. Shenoy, A. K. Bohra, S. Bhattacharya, A. Singh, U. V. Waghmare and K. Biswas, Realization of High Thermoelectric Figure of Merit in GeTe by Complementary Co-doping of Bi and In, *Joule* **3**(10), 2565-2580, (2019).
12. J. P. Heremans, V. Jovovic, E. S. Toberer, A. Saramat, K. Kurosaki, A. Charoenphakdee, S. Yamanaka and G. J. Snyder, Enhancement of Thermoelectric Efficiency in PbTe by Distortion of the Electronic Density of States, *Science* **321**(5888), 554-557, (2008).
13. H. Jang, S. Abbey, W. H. Nam, B. Frimpong, C. V. Nguyen, S.-J. Joo, H. S. Shin, J. Y. Song, E. N. Cho, M. Kim, Y. S. Jung and M.-W. Oh, Order-disorder Transition-induced Band Nestification in $AgBiSe_2$–$CuBiSe_2$ Solid Solutions for Superior Thermoelectric Performance, *J. Mater. Chem. A* **9**(8), 4648-4657, (2021).
14. S. Chandra, P. Dutta and K. Biswas, High-Performance Thermoelectrics Based on Solution-Grown SnSe Nanostructures, *ACS Nano* **16**(1), 7-14, (2022).
15. M. Samanta and K. Biswas, Low Thermal Conductivity and High Thermoelectric Performance in $(GeTe)_{1-2x}(GeSe)_x(GeS)_x$: Competition between Solid Solution and Phase Separation, *J. Am. Chem. Soc.* **139**(27), 9382-9391, (2017).

16. P. Acharyya, S. Roychowdhury, M. Samanta and K. Biswas, Ultralow Thermal Conductivity, Enhanced Mechanical Stability, and High Thermoelectric Performance in $(GeTe)_{1-2x}(SnSe)_x(SnS)_x$, *J. Am. Chem. Soc.* **142**(48), 20502-20508, (2020).
17. M. Dutta, K. Pal, M. Etter, U. V. Waghmare and K. Biswas, Emphanisis in Cubic $(SnSe)_{0.5}(AgSbSe_2)_{0.5}$: Dynamical Off-Centering of Anion Leads to Low Thermal Conductivity and High Thermoelectric Performance, *J. Am. Chem. Soc.* **143**(40), 16839-16848, (2021).
18. S. Kim, K. H. Lee, H. A. Mun, H. S. Kim, S. W. Hwang, J. W. Roh, D. J. Yang, W. H. Shin, X. S. Li and Y. H. Lee, Dense Dislocation Arrays Embedded in Grain Boundaries for High-performance Bulk Thermoelectrics, *Science* **348**(6230), 109-114, (2015).
19. B. Zhu, X. Liu, Q. Wang, Y. Qiu, Z. Shu, Z. Guo, Y. Tong, J. Cui, M. Gu and J. He, Realizing Record High Performance in n-type $Bi_2Te_3$-based Thermoelectric Materials, *Energy Environ. Sci.* **13**(7), 2106-2114, (2020).
20. R. Pathak, D. Sarkar and K. Biswas, Enhanced Band Convergence and Ultra-Low Thermal Conductivity Lead to High Thermoelectric Performance in SnTe, *Angew. Chem. Int. Ed.* **60**(32), 17686-17692, (2021).
21. D. Sarkar, S. Das and K. Biswas, Valence Band Convergence and Nanostructured Phonon Scattering Trigger High Thermoelectric Performance in SnTe, *Appl. Phys. Lett.* **119**(25), 253901, (2021).
22. K. Biswas, J. He, Q. Zhang, G. Wang, C. Uher, V. P. Dravid and M. G. Kanatzidis, Strained Endotaxial Nanostructures with High Thermoelectric Figure of Merit, *Nat. Chem.* **3**, 160-166, (2011).
23. K. Biswas, J. He, I. D. Blum, C.-I. Wu, T. P. Hogan, D. N. Seidman, V. P. Dravid and M. G. Kanatzidis, High-performance Bulk Thermoelectrics with All-scale Hierarchical Architectures, *Nature* **489**(7416), 414-418, (2012).
24. S. Chandra and K. Biswas, Realization of High Thermoelectric Figure of Merit in Solution Synthesized 2D SnSe Nanoplates via Ge alloying, *J. Am. Chem. Soc.* **141**(15), 6141-6145, (2019).
25. A. Banik, S. Roychowdhury and K. Biswas, The Journey of Tin Chalcogenides towards High-performance Thermoelectrics and Topological Materials, *Chem. Commun.* **54**(50), 6573-6590, (2018).
26. L. D. Zhao, H. J. Wu, S. Q. Hao, C. I. Wu, X. Y. Zhou, K. Biswas, J. Q. He, T. P. Hogan, C. Uher, C. Wolverton, V. P. Dravid and M. G. Kanatzidis, All-scale Hierarchical Thermoelectrics: MgTe in PbTe Facilitates Valence Band Convergence and Suppresses Bipolar Thermal Transport for High Performance, *Energy Environ. Sci.* **6**(11), 3346-3355, (2013).
27. S. Perumal, S. Roychowdhury and K. Biswas, High Performances Thermoelectric Materials and Devices Based on GeTe, *J. Mater. Chem. C* **4**(32), 7520-7536, (2016).
28. S. Roychowdhury, U. S. Shenoy, U. V. Waghmare and K. Biswas, An Enhanced Seebeck Coefficient and High Thermoelectric Performance in p-type In and Mg

Co-doped $Sn_{1-x}Pb_xTe$ via the Co-adjuvant Effect of the Resonance Level and Heavy Hole Valence Band, *J. Mater. Chem. C* **5**(23), 5737-5748, (2017).

29. S. Roychowdhury, M. Dutta and K. Biswas, Enhanced Thermoelectric Performance in Topological Crystalline Insulator n-type $Pb_{0.6}Sn_{0.4}Te$ by Simultaneous Tuning of the Band Gap and Chemical Potential, *J. Mater. Chem. A* **6**(47), 24216-24223, (2018).
30. S. Ahmad, A. Singh, S. Bhattacharya, M. Navaneethan, R. Basu, R. Bhatt, P. Sarkar, K. N. Meshram, A. K. Debnath, K. P. Muthe and D. K. Aswal, Band Convergence and Phonon Scattering Mediated Improved Thermoelectric Performance of SnTe–PbTe Nanocomposites, *ACS Appl. Energy Mater.* **3**(9), 8882-8891, (2020).
31. Y. Pei, X. Shi, A. LaLonde, H. Wang, L. Chen and G. J. Snyder, Convergence of Electronic Bands for High Performance Bulk Thermoelectrics, *Nature* **473**, 66-69, (2011).
32. Y. Wu, Z. Chen, P. Nan, F. Xiong, S. Lin, X. Zhang, Y. Chen, L. Chen, B. Ge and Y. Pei, Lattice Strain Advances Thermoelectrics, *Joule* **3**(5), 1276-1288, (2019).
33. Y. Xiao and L. D. Zhao, Charge and Phonon Transport in PbTe-based Thermoelectric Materials, *npj Quantum Mater.* **3**, 55, (2018).
34. Y. Zhong, J. Tang, H. Liu, Z. Chen, L. Lin, D. Ren, B. Liu and R. Ang, Optimized Strategies for Advancing n-Type PbTe Thermoelectrics: A Review, *ACS Appl. Mater. Interfaces* **12**(44), 49323–49334, (2020).
35. G. Tan, F. Shi, S. Hao, H. Chi, T. P. Bailey, L. D. Zhao, C. Uher, C. Wolverton, V. P. Dravid and M. G. Kanatzidis, Valence Band Modification and High Thermoelectric Performance in SnTe Heavily Alloyed with MnTe, *J. Am. Chem. Soc.* **137**(35), 11507-11516, (2015).
36. D. G. Cahill, S. K. Watson and R. O. Pohl, Lower Limit to the Thermal Conductivity of Disordered Crystals, *Phys. Rev. B: Condens. Matter Mater. Phys.* **46**(10), 6131, (1992).
37. D. Sarkar, T. Ghosh, A. Banik, S. Roychowdhury, D. Sanyal and K. Biswas, Highly Converged Valence Bands and Ultralow Lattice Thermal Conductivity for High-Performance SnTe Thermoelectrics, *Angew. Chem., Int. Ed.* **59**(27), 11115-11215, (2020).
38. Q. Zhang, B. Liao, Y. Lan, K. Lukas, W. Liu, K. Esfarjani, C. Opeil, D. Broido, G. Chen and Z. Ren, High Thermoelectric Performance by Resonant Dopant Indium in Nanostructured SnTe, *Proc. Natl. Acad. Sci. U. S. A.* **110**(33), 13261-13266, (2013).
39. G. Tan, L. D. Zhao, F. Shi, J. W. Doak, S. H. Lo, H. Sun, C. Wolverton, V. P. Dravid, C. Uher and M. G. Kanatzidis, High Thermoelectric Performance of p-Type SnTe via a Synergistic Band Engineering and Nanostructuring Approach, *J. Am. Chem. Soc.* **136**(19), 7006-7017, (2014).
40. A. Banik, U. S. Shenoy, S. Anand, U. V. Waghmare and K. Biswas, Mg Alloying in SnTe Facilitates Valence Band Convergence and Optimizes Thermoelectric Properties, *Chem. Mater.* **27**(2), 581-587, (2015).
41. G. Tan, F. Shi, S. Hao, H. Chi, L. D. Zhao, C. Uher, C. Wolverton, V. P. Dravid and M. G. Kanatzidis, Codoping in SnTe: Enhancement of Thermoelectric Performance

through Synergy of Resonance Levels and Band Convergence, *J. Am. Chem. Soc.* **137**(15), 5100-5112, (2015).

42. S. Roychowdhury, R. K. Biswas, M. Dutta, S. K. Pati and K. Biswas, Phonon Localization and Entropy-Driven Point Defects Lead to Ultralow Thermal Conductivity and Enhanced Thermoelectric Performance in $(SnTe)_{1-2x}(SnSe)_x(SnS)_x$, *ACS Energy Lett.* **4**(7), 1658-1662, (2019).
43. Z. Zheng, X. Su, R. Deng, C. Stoumpos, H. Xie, W. Liu, Y. Yan, S. Hao, C. Uher, C. Wolverton, M. Kanatzidis and X. Tang, Rhombohedral to Cubic Conversion of GeTe via MnTe Alloying Leads to Ultralow Thermal Conductivity, Electronic Band Convergence, and High Thermoelectric Performance, *J. Am. Chem. Soc.* **140**(7), 2673-2686, (2018).
44. M. Hong, Y. Wang, T. Feng, Q. Sun, S. Xu, S. Matsumura, S. T. Pantelides, J. Zou and Z.-G. Chen, Strong Phonon–Phonon Interactions Securing Extraordinary Thermoelectric $Ge_{1-x}Sb_xTe$ with Zn-Alloying-Induced Band Alignment, *J. Am. Chem. Soc.* **141**(4), 1742-1748, (2019).
45. M. Hong, Y. Wang, W. Liu, S. Matsumura, H. Wang, J. Zou and Z.-G. Chen, Arrays of Planar Vacancies in Superior Thermoelectric $Ge_{1-x-y}Cd_xBi_yTe$ with Band Convergence, *Adv. Energy Mater.* **8**(30), 1801837, (2018).
46. M. Hong, W. Lyv, M. Li, S. Xu, Q. Sun, J. Zou and Z. G. Chen, Rashba Effect Maximizes Thermoelectric Performance of GeTe Derivatives, *Joule* **4**(9), 2030-2043, (2020).
47. M. Samanta, S. Roychowdhury, J. Ghatak, S. Perumal and K. Biswas, Ultrahigh Average Thermoelectric Figure of Merit, Low Lattice Thermal Conductivity and Enhanced Microhardness in Nanostructured $(GeTe)_x(AgSbSe_2)_{100-x}$, *Chem. - Eur. J.* **23**(31), 7438-7443, (2017).
48. M. Samanta, T. Ghosh, R. Arora, U. V. Waghmare and K. Biswas, Realization of Both n- and p-Type GeTe Thermoelectrics: Electronic Structure Modulation by $AgBiSe_2$ Alloying, *J. Am. Chem. Soc.* **141**(49), 19505-19512, (2019).
49. Z. Liu, J. Sun, J. Mao, H. Zhu, W. Ren, J. Zhou, Z. Wang, D. J. Singh, J. Sui, C. W. Chu and Z. Ren, Phase-Transition Temperature Suppression to Achieve Cubic GeTe and High Thermoelectric Performance by Bi and Mn Codoping, *Proc. Natl. Acad. Sci. U. S. A.* **115**(21), 5332-5337, (2018).
50. M. Hong, Z. G. Chen, L. Yang, Y. C. Zou, M. S. Dargusch, H. Wang and J. Zou, Realizing zT of 2.3 in $Ge_{1-x-y}Sb_xIn_yTe$ via Reducing the Phase-Transition Temperature and Introducing Resonant Energy Doping, *Adv. Mater.* **30**(11), 1705942, (2018).
51. J. Li, X. Zhang, Z. Chen, S. Lin, W. Li, J. Shen, I. T. Witting, A. Faghaninia, Y. Chen and A. Jain, Low-Symmetry Rhombohedral GeTe Thermoelectrics, *Joule* **2**(5), 976-987, (2018).
52. S. Roychowdhury, T. Ghosh, R. Arora, M. Samanta, L. Xie, K. Singh Niraj, A. Soni, J. He, V. Waghmare Umesh and K. Biswas, Enhanced Atomic Ordering Leads to High Thermoelectric Performance in $AgSbTe_2$, *Science* **371**(6530), 722-727, (2021).

53. H. J. Goldsmid and R. W. Douglas, The Use of Semiconductors in Thermoelectric Refrigeration, *Br. J. Appl. Phys.* **5**(11), 386-390, (1954).
54. B. Poudel, Q. Hao, Y. Ma, Y. Lan, A. Minnich, B. Yu, X. Yan, D. Wang, A. Muto and D. Vashaee, High-Thermoelectric Performance of Nanostructured Bismuth Antimony Telluride Bulk Alloys, *Science* **320**(5876), 634-638, (2008).
55. M. Hong, T. C. Chasapis, Z.-G. Chen, L. Yang, M. G. Kanatzidis, G. J. Snyder and J. Zou, n-Type $Bi_2Te_{3-x}Se_x$ Nanoplates with Enhanced Thermoelectric Efficiency Driven by Wide-Frequency Phonon Scatterings and Synergistic Carrier Scatterings, *ACS Nano* **10**(4), 4719-4727, (2016).
56. L. D. Zhao, G. Tan, S. Hao, J. He, Y. Pei, H. Chi, H. Wang, S. Gong, H. Xu, V. P. Dravid, C. Uher, G. J. Snyder, C. Wolverton and M. G. Kanatzidis, Ultrahigh Power Factor and Thermoelectric Performance in Hole-Doped Single-Crystal SnSe, *Science* **351**(6269), 141-144, (2016).
57. B. Qin, D. Wang, X. Liu, Y. Qin, J. F. Dong, J. Luo, J. W. Li, W. Liu, G. Tan, X. Tang, J. F. Li, J. He and L. D. Zhao, Power Generation and Thermoelectric Cooling Enabled by Momentum and Energy Multiband Alignments, *Science* **373**(6554), 556-561, (2021).
58. L. D. Zhao, S. H. Lo, Y. Zhang, H. Sun, G. Tan, C. Uher, C. Wolverton, V. P. Dravid and M. G. Kanatzidis, Ultralow Thermal Conductivity and High Thermoelectric Figure of Merit in SnSe Crystals, *Nature* **508**, 373–377, (2014).
59. C. Chang, M. Wu, D. He, Y. Pei, C. F. Wu, X. Wu, H. Yu, F. Zhu, K. Wang and Y. Chen, 3D Charge and 2D Phonon Transports Leading to High Out-of-Plane ZT in n-Type SnSe Crystals, *Science* **360**(6390), 778-783, (2018).
60. P. C. Wei, S. Bhattacharya, J. He, S. Neeleshwar, R. Podila, Y. Chen and A. Rao, The Intrinsic Thermal Conductivity of SnSe, *Nature* **539**, E1–E2, (2016).
61. S. Chandra, P. Dutta and K. Biswas, Enhancement of the Thermoelectric Performance of 2D SnSe Nanoplates through Incorporation of Magnetic Nanoprecipitates, *ACS Appl. Energy Mater.* **3**(9), 9051-9057, (2020).
62. Y. K. Lee, Z. Luo, S. P. Cho, M. G. Kanatzidis and I. Chung, Surface Oxide Removal for Polycrystalline SnSe Reveals Near-Single-Crystal Thermoelectric Performance, *Joule* **3**(3), 719-731, (2019).
63. C. Zhou, Y. K. Lee, Y. Yu, S. Byun, Z. Luo, H. Lee, B. Ge, Y. L. Lee, X. Chen, J. Y. Lee, O. Cojocaru-Mirédin, H. Chang, J. Im, S. Cho, M. Wuttig, V. P. Dravid, M. G. Kanatzidis and I. Chung, Polycrystalline SnSe with a Thermoelectric Figure of Merit Greater than the Single Crystal, *Nat. Mater.* **20**, 1378, (2021).
64. S. Chandra, R. Arora, U. V. Waghmare and K. Biswas, Modulation of the Electronic Structure and Thermoelectric Properties of Orthorhombic and Cubic SnSe by $AgBiSe_2$ Alloying, *Chem. Sci.* **12**, 13074-13082, (2021).
65. Z.-H. Ge, L.-D. Zhao, D. Wu, X. Liu, B.-P. Zhang, J.-F. Li and J. He, Low-cost, abundant binary sulfides as promising thermoelectric materials, *Mater. Today* **19**(4), 227-239, (2016).

66. A. Bhui, M. Dutta, M. Mukherjee, K. S. Rana, A. K. Singh, A. Soni and K. Biswas, Ultralow Thermal Conductivity in Earth-Abundant $Cu_{1.6}Bi_{4.8}S_8$: Anharmonic Rattling of Interstitial Cu, *Chem. Mater.* **33**(8), 2993-3001, (2021).
67. S. Tippireddy, F. Azough, Vikram, F. T. Tompkins, A. Bhui, R. Freer, R. Grau-Crespo, K. Biswas, P. Vaqueiro and A. V. Powell, Tin-Substituted Chalcopyrite: An n-Type Sulfide with Enhanced Thermoelectric Performance, *Chem. Mater.* **34**(13), 5860-5873, (2022).
68. G. Tan, S. Hao, J. Zhao, C. Wolverton and M. G. Kanatzidis, High Thermoelectric Performance in Electron-doped $AgBi_3S_5$ with Ultralow Thermal Conductivity, *J. Am. Chem. Soc.* **139**(18), 6467-6473, (2017).
69. X. Lu, D. T. Morelli, Y. Wang, W. Lai, Y. Xia and V. Ozolins, Phase Stability, Crystal Structure, and Thermoelectric Properties of $Cu_{12}Sb_4S_{13-x}Se_x$ Solid Solutions, *Chem. Mater.* **28**(6), 1781-1786, (2016).
70. Y. He, T. Day, T. Zhang, H. Liu, X. Shi, L. Chen and G. J. Snyder, High Thermoelectric Performance in Non-toxic Earth-abundant Copper Sulfide, *Adv. Mater.* **26**, 3974, (2014).
71. E. Rathore, R. Juneja, S. P. Culver, N. Minafra, A. K. Singh, W. G. Zeier and K. Biswas, Origin of Ultralow Thermal Conductivity in n-Type Cubic Bulk $AgBiS_2$: Soft Ag Vibrations and Local Structural Distortion Induced by the Bi $6s^2$ Lone Pair, *Chem. Mater.* **31**(6), 2106-2113, (2019).
72. H. Wang, Z. M. Gibbs, Y. Takagiwa and G. J. Snyder, Tuning bands of PbSe for Better Thermoelectric Efficiency, *Energy Environ. Sci.* **7**(2), 804-811, (2014).
73. Z.-Z. Luo, S. Hao, S. Cai, T. P. Bailey, G. Tan, Y. Luo, I. Spanopoulos, C. Uher, C. Wolverton, V. P. Dravid, Q. Yan and M. G. Kanatzidis, Enhancement of Thermoelectric Performance for n-Type PbS through Synergy of Gap State and Fermi Level Pinning, *J. Am. Chem. Soc.* **141**(15), 6403-6412, (2019).
74. L.-D. Zhao, S.-H. Lo, J. He, H. Li, K. Biswas, J. Androulakis, C.-I. Wu, T. P. Hogan, D.-Y. Chung, V. P. Dravid and M. G. Kanatzidis, High Performance Thermoelectrics from Earth-Abundant Materials: Enhanced Figure of Merit in PbS by Second Phase Nanostructures, *J. Am. Chem. Soc.* **133**(50), 20476-20487, (2011).
75. L.-D. Zhao, J. He, S. Hao, C.-I. Wu, T. P. Hogan, C. Wolverton, V. P. Dravid and M. G. Kanatzidis, Raising the Thermoelectric Performance of p-Type PbS with Endotaxial Nanostructuring and Valence-Band Offset Engineering Using CdS and ZnS, *J. Am. Chem. Soc.* **134**(39), 16327-16336, (2012).
76. B. Jiang, X. Liu, Q. Wang, J. Cui, B. Jia, Y. Zhu, J. Feng, Y. Qiu, M. Gu, Z. Ge and J. He, Realizing High-efficiency Power Generation in Low-cost PbS-based Thermoelectric Materials, *Energy Environ. Sci.* **13**(2), 579-591, (2020).
77. Y. Xiao, D. Wang, Y. Zhang, C. Chen, S. Zhang, K. Wang, G. Wang, S. J. Pennycook, G. J. Snyder, H. Wu and L.-D. Zhao, Band Sharpening and Band Alignment Enable High Quality Factor to Enhance Thermoelectric Performance in n-Type PbS, *J. Am. Chem. Soc.* **142**(8), 4051-4060, (2020).

78. M. Zhao, C. Chang, Y. Xiao, R. Gu, J. He and L.-D. Zhao, Investigations on Distinct Thermoelectric Transport Behaviors of Cu in n-type PbS, *J. Alloys Compd.* **781**, 820-830, (2019).
79. E. Rathore, R. Juneja, D. Sarkar, S. Roychowdhury, M. Kofu, K. Nakajima, A. K. Singh and K. Biswas, Enhanced Covalency and Nanostructured-phonon Scattering Lead to High Thermoelectric Performance in n-type PbS, *Mater. Today Energy* **24**, 100953, (2022).
80. W. He, D. Wang, H. Wu, Y. Xiao, Y. Zhang, D. He, Y. Feng, Y.-J. Hao, J.-F. Dong, R. Chetty, L. Hao, D. Chen, J. Qin, Q. Yang, X. Li, J.-M. Song, Y. Zhu, W. Xu, C. Niu, X. Li, G. Wang, C. Liu, M. Ohta, S. J. Pennycook, J. He, J.-F. Li and L.-D. Zhao, High Thermoelectric Performance in Low-cost $SnS_{0.91}Se_{0.09}$ Crystals, *Science* **365**(6460), 1418-1424, (2019).
81. K. Biswas, L.-D. Zhao and M. G. Kanatzidis, Tellurium-Free Thermoelectric: The Anisotropic n-Type Semiconductor $Bi_2S_3$, *Adv. Energy Mater.* **2**(6), 634-638, (2012).
82. E. Guilmeau, Y. Bréard and A. Maignan, Transport and Thermoelectric Properties in Copper Intercalated $TiS_2$ Chalcogenide, *Appl. Phys. Lett.* **99**(5), 052107, (2011).
83. Y. You, X. Su, S. Hao, W. Liu, Y. Yan, T. Zhang, M. Zhang, C. Wolverton, M. G. Kanatzidis and X. Tang, Ni and Se Co-doping Increases the Power Factor and Thermoelectric Performance of CoSbS, *J. Mater. Chem. A* **6**(31), 15123-15131, (2018).
84. J.-B. Labégorre, A. Virfeu, A. Bourhim, H. Willeman, T. Barbier, F. Appert, J. Juraszek, B. Malaman, A. Huguenot, R. Gautier, V. Nassif, P. Lemoine, C. Prestipino, E. Elkaim, L. Pautrot-d'Alençon, T. Le Mercier, A. Maignan, R. Al Rahal Al Orabi and E. Guilmeau, $XBi_4S_7$ (X = Mn, Fe): New Cost-Efficient Layered n-Type Thermoelectric Sulfides with Ultralow Thermal Conductivity, *Adv. Funct. Mater.* **29**(48), 1904112, (2019).
85. T. Graf, C. Felser and S. S. P. Parkin, Simple Rules for the Understanding of Heusler Compounds, *Prog. Solid. State Chem.* **39**(1), 1-50, (2011).
86. C. Fu, T. Zhu, Y. Pei, H. Xie, H. Wang, G. J. Snyder, Y. Liu, Y. Liu and X. Zhao, High Band Degeneracy Contributes to High Thermoelectric Performance in p-Type Half-Heusler Compounds, *Adv. Energy Mater.* **4**(18), 1400600, (2014).
87. M.-S. Lee, F. P. Poudeu and S. D. Mahanti, Electronic Structure and Thermoelectric Properties of Sb-based Semiconducting Half-Heusler Compounds, *Phys. Rev. B* **83**(8), 085204, (2011).
88. C. Fu, S. Bai, Y. Liu, Y. Tang, L. Chen, X. Zhao and T. Zhu, Realizing High Figure of Merit in Heavy-Band p-Type Half-Heusler Thermoelectric Materials, *Nat. Commun.* **6**, 8144, (2015).
89. Q. Shen, L. Chen, T. Goto, T. Hirai, J. Yang, G. P. Meisner and C. Uher, Effects of Partial Substitution of Ni by Pd on the Thermoelectric Properties of ZrNiSn-Based Half-Heusler Compounds, *Appl. Phys. Lett.* **79**, 4165, (2001).

90. S. R. Culp, S. J. Poon, N. Hickman, T. M. Tritt and J. Blumm, Effect of Substitutions on the Thermoelectric Figure of Merit of Half-Heusler Phases at 800°C, *Appl. Phys. Lett.* **88**(4), 042106, (2006).
91. C. Yu, T.-J. Zhu, R.-Z. Shi, Y. Zhang, X.-B. Zhao and J. He, High-performance Half-Heusler Thermoelectric Materials $Hf_{1-x}Zr_xNiSn_{1-y}S$ Prepared by Levitation Melting and Spark Plasma Sintering, *Acta Mater.* **57**(9), 2757-2764, (2009).
92. R. He, D. Kraemer, J. Mao, L. Zeng, Q. Jie, Y. Lan, C. Li, J. Shuai, S. Kim Hee, Y. Liu, D. Broido, C.-W. Chu, G. Chen and Z. Ren, Achieving High Power Factor and Output Power Density in p-type Half-Heuslers $Nb_{1-x}Ti_xFeSb$, *Proc. Natl. Acad. Sci. U. S. A.* **113**(48), 13576-13581, (2016).
93. H. Zhu, J. Mao, Y. Li, J. Sun, Y. Wang, Q. Zhu, G. Li, Q. Song, J. Zhou, Y. Fu, R. He, T. Tong, Z. Liu, W. Ren, L. You, Z. Wang, J. Luo, A. Sotnikov, J. Bao, K. Nielsch, G. Chen, D. J. Singh and Z. Ren, Discovery of TaFeSb-based Half-Heuslers with High Thermoelectric Performance, *Nat. Commun.* **10**(1), 270, (2019).
94. L. Huang, R. He, S. Chen, H. Zhang, K. Dahal, H. Zhou, H. Wang, Q. Zhang and Z. Ren, A New n-type Half-Heusler Thermoelectric Material NbCoSb, *Mater. Res. Bull.* **70**, 773-778, (2015).
95. K. Xia, Y. Liu, S. Anand, G. J. Snyder, J. Xin, J. Yu, X. Zhao and T. Zhu, Enhanced Thermoelectric Performance in 18-Electron $Nb_{0.8}CoSb$ Half-Heusler Compound with Intrinsic Nb Vacancies, *Adv. Funct. Mater.* **28**(9), 1705845, (2018).
96. J. Yu, Y. Xing, C. Hu, Z. Huang, Q. Qiu, C. Wang, K. Xia, Z. Wang, S. Bai, X. Zhao, L. Chen and T. Zhu, Half-Heusler Thermoelectric Module with High Conversion Efficiency and High Power Density, *Adv. Energy Mater.* **10**(25), 2000888, (2020).
97. Y. Xing, R. Liu, J. Liao, C. Wang, Q. Zhang, Q. Song, X. Xia, T. Zhu, S. Bai and L. Chen, A Device-to-Material Strategy Guiding the "Double-High" Thermoelectric Module, *Joule* **4**(11), 2475-2483, (2020).
98. H. Tamaki, H. K. Sato and T. Kanno, Isotropic Conduction Network and Defect Chemistry in $Mg_{3+\delta}Sb_2$-Based Layered Zintl Compounds with High Thermoelectric Performance, *Adv. Mater.* **28**, 10182, (2016).
99. S. Ohno, K. Imasato, S. Anand, H. Tamaki, S. D. Kang, P. Gorai, H. K. Sato, E. S. Toberer, T. Kanno and G. J. Snyder, Phase Boundary Mapping to Obtain n-type $Mg_3Sb_2$-Based Thermoelectrics, *Joule* **2**(1), 141-154, (2018).
100. X. Shi, T. Zhao, X. Zhang, C. Sun, Z. Chen, S. Lin, W. Li, H. Gu and Y. Pei, Extraordinary n-Type $Mg_3SbBi$ Thermoelectrics Enabled by Yttrium Doping, *Adv. Mater.* **31**(36), 1903387, (2019).
101. J. Shuai, J. Mao, S. Song, Q. Zhu, J. Sun, Y. Wang, R. He, J. Zhou, G. Chen, D. J. Singh and Z. Ren, Tuning the Carrier Scattering Mechanism to Effectively Improve the Thermoelectric Properties, *Energy Environ. Sci.* **10**(3), 799-807, (2017).
102. W. Peng, G. Petretto, G.-M. Rignanese, G. Hautier and A. Zevalkink, An Unlikely Route to Low Lattice Thermal Conductivity: Small Atoms in a Simple Layered Structure, *Joule* **2**(9), 1879-1893, (2018).

103. X. Shi, C. Sun, X. Zhang, Z. Chen, S. Lin, W. Li and Y. Pei, Efficient Sc-Doped $Mg_{3.05-x}Sc_xSbBi$ Thermoelectrics Near Room Temperature, *Chem. Mater.* **31**(21), 8987-8994, (2019).
104. K. Imasato, S. D. Kang and G. J. Snyder, Exceptional Thermoelectric Performance in $Mg_3Sb_{0.6}Bi_{1.4}$ for Low-grade Waste Heat Recovery, *Energy Environ. Sci.* (3), 965-971, (2019).
105. J. Mao, H. Zhu, Z. Ding, Z. Liu, G. A. Gamage, G. Chen and Z. Ren, High Thermoelectric Cooling Performance of n-type $Mg_3Bi_2$-based Materials, *Science* (6452), 495-498, (2019).
106. G. J. Snyder, A. H. Snyder, M. Wood, R. Gurunathan, B. H. Snyder and C. Niu, Weighted Mobility, *Adv. Mater.* **32**(25), 2001537, (2020).
107. M. Wood, K. Imasato, S. Anand, J. Yang and G. J. Snyder, The Importance of the Mg–Mg Interaction in $Mg_3Sb_2$–$Mg_3Bi_2$ Shown Through Cation Site Alloying, *J. Mater. Chem. A* **8**(4), 2033-2038, (2020).
108. K. Imasato, M. Wood, J. J. Kuo and G. J. Snyder, Improved Stability and High Thermoelectric Performance Through Cation Site Doping in n-type La-doped $Mg_3Sb_{1.5}Bi_{0.5}$, *J. Mater. Chem. A* **6**(41), 19941-19946, (2018).
109. J. Zhang, L. Song, S. H. Pedersen, H. Yin, L. T. Hung and B. B. Iversen, Discovery of High-Performance Low-Cost n-Type $Mg_3Sb_2$-Based Thermoelectric Materials with Multi-Valley Conduction Bands, *Nat. Commun.* **8**, 13901, (2017).
110. K. Imasato, S. D. Kang, S. Ohno and G. J. Snyder, Band Engineering in $Mg_3Sb_2$ by Alloying with $Mg_3Bi_2$ for Enhanced Thermoelectric Performance, *Mater. Horiz.* **5**(1), 59-64, (2018).
111. Z. Liu, N. Sato, W. Gao, K. Yubuta, N. Kawamoto, M. Mitome, K. Kurashima, Y. Owada, K. Nagase, C.-H. Lee, J. Yi, K. Tsuchiya and T. Mori, Demonstration of Ultrahigh Thermoelectric Efficiency of ~7.3% in $Mg_3Sb_2$/MgAgSb Module for Low-temperature Energy Harvesting, *Joule* **5**(5), 1196-1208, (2021).
112. Z. Liang, C. Xu, H. Shang, Q. Zhu, F. Ding, J. Mao and Z. Ren, High Thermoelectric Energy Conversion Efficiency of a Unicouple of n-type $Mg_3Bi_2$ and p-type $Bi_2Te_3$, *Mater. Today Phys.* **19**, 100413, (2021).
113. J. E. F. S. Rodrigues, J. Gainza, F. Serrano-Sánchez, C. Marini, Y. Huttel, N. M. Nemes, J. L. Martínez and J. A. Alonso, Atomic Structure and Lattice Dynamics of $CoSb_3$ Skutterudite-Based Thermoelectrics, *Chem. Mater.* **34**(3), 1213-1224, (2022).
114. V. Keppens, D. Mandrus, B. C. Sales, B. C. Chakoumakos, P. Dai, R. Coldea, M. B. Maple, D. A. Gajewski, E. J. Freeman and S. Bennington, Localized Vibrational Modes in Metallic Solids, *Nature* **395**(6705), 876-878, (1998).
115. B. C. Sales, D. Mandrus, B. C. Chakoumakos, V. Keppens and J. R. Thompson, Filled Skutterudite Antimonides: Electron Crystals and Phonon Glasses, *Phys. Rev. B* **56**(23), 15081-15089, (1997).
116. X. Shi, J. Yang, J. R. Salvador, M. Chi, J. Y. Cho, H. Wang, S. Bai, J. Yang, W. Zhang and L. Chen, Multiple-filled Skutterudites: High Thermoelectric Figure of Merit

Through Separately Optimizing Electrical and Thermal Transports, *J. Am. Chem. Soc.* **133**(20), 7837-7846, (2011).

117. J. Gainza, F. Serrano-Sánchez, J. E. Rodrigues, J. Prado-Gonjal, N. M. Nemes, N. Biskup, O. J. Dura, J. L. Martínez, F. Fauth and J. A. Alonso, Unveiling the Correlation between the Crystalline Structure of M-Filled $CoSb_3$ (M = Y, K, Sr) Skutterudites and Their Thermoelectric Transport Properties, *Adv. Funct. Mater.* **30**(36), 2001651, (2020).
118. X. Shi, H. Kong, C. P. Li, C. Uher, J. Yang, J. R. Salvador, H. Wang, L. Chen and W. Zhang, Low Thermal Conductivity and High Thermoelectric Figure of Merit in n-type $Ba_xYb_yCo_4Sb_{12}$ Double-filled Skutterudites, *Appl. Phys. Lett.* **92**(18), 182101, (2008).
119. H. Kleinke, New Bulk Materials for Thermoelectric Power Generation: Clathrates and Complex Antimonides, *Chem. Mater.* **22**(3), 604-611, (2010).
120. B. B. Iversen, A. E. C. Palmqvist, D. E. Cox, G. S. Nolas, G. D. Stucky, N. P. Blake and H. Metiu, Why are Clathrates Good Candidates for Thermoelectric Materials?, *J. Solid State Chem.* **149**(2), 455-458, (2000).
121. F. Sui and S. M. Kauzlarich, Tuning Thermoelectric Properties of Type I Clathrate $K_{8-x}Ba_xAl_{8+x}Si_{38-x}$ through Barium Substitution, *Chem. Mater.* **28**(9), 3099-3107, (2016).
122. Y. Zhang, J. Brorsson, R. Qiu and A. E. C. Palmqvist, Enhanced Thermoelectric Performance of $Ba_8Ga_{16}Ge_{30}$ Clathrate by Modulation Doping and Improved Carrier Mobility, *Adv. Electron. Mater.* **7**(2), 2000782, (2021).
123. A. Nozariasbmarz, A. Agarwal, Z. A. Coutant, M. J. Hall, J. Liu, R. Liu, A. Malhotra, P. Norouzzadeh, M. C. Öztürk, V. P. Ramesh, Y. Sargolzaeiaval, F. Suarez and D. Vashaee, Thermoelectric Silicides: A Review, *Jpn. J. Appl. Phys.* **56**(5S1), 05DA04, (2017).
124. K. Hayashi, W. Saito, K. Sugimoto, K. Ohoyama, K. Hayashi, N. Happo, M. Harada, K. Oikawa, Y. Inamura and Y. Miyazaki, Preparation, Thermoelectric Properties, and Crystal Structure of Boron-doped $Mg_2Si$ Single Crystals, *AIP Adv.* **10**(3), 035115, (2020).
125. W. Liu, X. Tang, H. Li, J. Sharp, X. Zhou and C. Uher, Optimized Thermoelectric Properties of Sb-Doped $Mg_{2(1+z)}Si_{0.5-y}Sn_{0.5}Sb_y$ through Adjustment of the Mg Content, *Chem. Mater.* **23**(23), 5256-5263, (2011).
126. G. K. Goyal, S. Mukherjee, R. C. Mallik, S. Vitta, I. Samajdar and T. Dasgupta, High Thermoelectric Performance in $Mg_2(Si_{0.3}Sn_{0.7})$ by Enhanced Phonon Scattering, *ACS Appl. Energy Mater.* **2**(3), 2129-2137, (2019).
127. M. A. Haque, S. Kee, D. R. Villalva, W.-L. Ong and D. Baran, Halide Perovskites: Thermal Transport and Prospects for Thermoelectricity, *Adv. Sci.* **7**(10), 1903389, (2020).
128. T. Ye, X. Wang, X. Li, A. Q. Yan, S. Ramakrishna and J. Xu, Ultra-high Seebeck Coefficient and Low Thermal Conductivity of a Centimeter-sized Perovskite Single

Crystal Acquired by a Modified Fast Growth Method, *J. Mater. Chem. C* **5**(5), 1255-1260, (2017).

129. C. Ge, M. Hu, P. Wu, Q. Tan, Z. Chen, Y. Wang, J. Shi and J. Feng, Ultralow Thermal Conductivity and Ultrahigh Thermal Expansion of Single-Crystal Organic–Inorganic Hybrid Perovskite $CH_3NH_3PbX_3$ (X = Cl, Br, I), *J. Phys. Chem. C* **122**(28), 15973-15978, (2018).
130. G. A. Elbaz, W.-L. Ong, E. A. Doud, P. Kim, D. W. Paley, X. Roy and J. A. Malen, Phonon Speed, Not Scattering, Differentiates Thermal Transport in Lead Halide Perovskites, Nano Lett. 17(9), 5734-5739, (2017).
131. X. Long, Z. Pan, Z. Zhang, J. J. Urban and H. Wang, Solvent-free Synthesis of Organometallic Halides $CH_3NH_3PbI_3$ and $(CH_3NH_3)_3Bi_2I_9$ and Their Thermoelectric Transport Properties, *Appl. Phys. Lett.* **115**(7), 072104, (2019).
132. P. Acharyya, T. Ghosh, K. Pal, K. Kundu, K. Singh Rana, J. Pandey, A. Soni, U. V. Waghmare and K. Biswas, Intrinsically Ultralow Thermal Conductivity in Ruddlesden-Popper 2D Perovskite $Cs_2PbI_2Cl_2$: Localized Anharmonic Vibrations and Dynamic Octahedral Distortions, *J. Am. Chem. Soc.* **142**(36), 15595-15603, (2020).
133. A. Bhui, T. Ghosh, K. Pal, K. Singh Rana, K. Kundu, A. Soni and K. Biswas, Intrinsically Low Thermal Conductivity in the n-Type Vacancy-Ordered Double Perovskite $Cs_2SnI_6$: Octahedral Rotation and Anharmonic Rattling, *Chem. Mater.* **34**(7), 3301-3310, (2022).
134. H. Xie, S. Hao, J. Bao, T. J. Slade, G. J. Snyder, C. Wolverton and M. G. Kanatzidis, All-Inorganic Halide Perovskites as Potential Thermoelectric Materials: Dynamic Cation off-Centering Induces Ultralow Thermal Conductivity, *J. Am. Chem. Soc.* **142**(20), 9553-9563, (2020).
135. W. Lee, H. Li, A. B. Wong, D. Zhang, M. Lai, Y. Yu, Q. Kong, E. Lin, J. J. Urban, J. C. Grossman and P. Yang, Ultralow Thermal Conductivity in All-inorganic Halide Perovskites, *Proc. Natl. Acad. Sci. U. S. A.* **114**(33), 8693-8697, (2017).
136. Y. Xiong, L. Xu, P. Wu, L. Sun, G. Xie and B. Hu, Bismuth Doping–Induced Stable Seebeck Effect Based on $MAPbI_3$ Polycrystalline Thin Films, *Adv. Funct. Mater.* **29**(16), 1900615, (2019).

https://doi.org/10.1142/9789811270956_0004

# Chapter 4

# Two-dimensional Materials for Supercapacitor Applications

Bhaskar Kakoty, Premkumar Senguttuvan* and C.N.R. Rao†

*Jawaharlal Nehru Centre for Advanced Scientific Research*
*Jakkur, Bangalore 560064, India*
**prem@jncasr.ac.in, †cnrrao@jncasr.ac.in*

The high-power density and fast charge/discharge ability of supercapacitor established it as a potent candidate among sustainable and efficient energy storage devices. The bottleneck for their avenue is the lower energy density compared to that of batteries and fuel cells. 2D materials have the scope to design different materials with enhanced surface area that may help to tackle these issues. In this chapter, we have highlighted recent progress in some of these 2D materials such as graphene with transition metal oxides, chalcogenides and MXenes.

## 1. Introduction

The development of sustainable and clean energy technologies is crucial to address the raising global warming and $CO_2$ footprint. At present, renewable sources such as wind and solar have been utilized.[1–3] However, they suffer from intermittent energy production which requires robust energy storage technologies.[4] Electrochemical energy storage devices (batteries, fuel cells and supercapacitors) play a remarkable role in the electrification of portable devices, automobiles, and large-scale storage. They offer high round-trip efficiency, low cost, and maintenance when compared to other technologies.[5]

Whilst batteries are known for their high energy density (e.g., lithium-ion; ~300 Wh $kg^{-1}$), supercapacitors are preferred for their high power and extraordinary cycle life (> 100,000 cycles).[6] The difference in their performance metrics can be ascribed to their charge storage mechanisms. While the charge storage in batteries is dominated by sluggish diffusion of ions in the bulk electrodes, faster electrostatic or Faradaic process, which occur at the electrode-electrolyte interface of active materials, enable supercapacitor to deliver faster discharge and charge rates within seconds or minutes but lower specific energy (5–10 Wh $kg^{-1}$).[6] The first supercapacitor, otherwise known as "Leyden jar", was developed by German scientist Ewald Georg Von Kleis and Dutch scientist Pieter Van Musschenbroek in 1745.[7] Their design involved two metal foils which were separated by water inside a glass jar and by rotating the jar, static electricity was produced. Since then, several theories on charge storage mechanism across the electrical double layer have been proposed by Helmoltz, Gouy, Chapman, Stern, and Grahame.[8–11] In 1978, a Japanese company, Nippon Electric Crop., was the first to commercialize supercapacitor for its application in back-up power for clock ships and complementary metal-oxide-semiconductor (CMOS) memories. At present, several business houses, such as Maxwell (USA), ELTON (Russia), Nesscap (Korea) and Nippon (Japan), produce different types of supercapacitors for various applications.[12]

In this book chapter, we will discuss the application of two-dimensional materials (2D) in supercapacitors. In general, 2D materials such as graphene, MXenes, transition metal oxides, hydroxides and dichalcogenides (TMO, TMOH and TMD) have received significant interest in the discipline of energy storage and conversion.[13,14] They offer greater tunability of their chemical and physical properties which impart high capacitive storage performances, flexibility and stability.

## 2. Fundamentals of Supercapacitor

Supercapacitors, like any other electrochemical devices, consist of two electrodes (positive and negative) separated by an ionic conducting and electronically insulating membrane or a separator.[15] Although it shares similar operating principle with conventional capacitors, it can store thousands of Farads per device compared to milli-Farad storage of conventional capacitors. It is due to higher surface area, nanoscale dielectric distances and faster Faradaic pseudocapacitive reactions present in supercapacitor electrodes. Supercapacitors have the potential to be a bridge between the conventional capacitors and batteries.[6] Based on the mechanism of charge storage, these supercapacitors are divided into three categories, namely, electrical double layer capacitor, pseudo capacitor and hybrid capacitor.

Electrical double layer supercapacitor (EDLC), is a kind of an electrochemical capacitor in which the charge storage mechanism is solely dependent on the formation of electrochemical double layer at the interphase.[16,17] These electrochemical double layers are formed by electrostatic interactions between the charged electrode and the ions. Thus, there is no charge transfer reaction or Faradaic reaction occurring between the electrode and the electrolyte at the interphase. The capacitance of these capacitors is higher than the normal electrochemical capacitors due to larger surface area of the electrode. The capacitance of these capacitors is calculated from the following equation:[18]

$$C = \frac{\varepsilon_r \varepsilon_0}{d}$$

where $\varepsilon_0$ is relative permittivity in vacuum, $\varepsilon_r$ is relative permittivity of double layer, d is the double layer thickness and S is the surface area of the electrode. The time of formation of a double layer is about ~$10^{-8}$ s, which is much higher than the time constant for a redox reaction $10^{-2}$–$10^{-4}$ s to occur.[6] As the double layer capacitor involves charge rearrangement rather than redox reaction, the response time of capacitors is much higher than that of a battery. Materials that are used as electrodes for this kind of capacitor are activated carbon,[19] carbon aerogel,[20] carbon xerogel,[21] porous carbon derived from metal oxide or polymer,[21] or carbon nanotubes. High surface area of an electrode is one of the prerequisites but there are other parameters to be considered to attain high performance supercapacitor. For instance, the pore size distribution of these electrodes as well as the size of the ions in electrolyte plays a crucial role in forming a double layer. Carbon based materials can lead to a capacitance of 300–500 F $g^{-1}$ but experimentally only a capacity of ~250 F $g^{-1}$ has been achieved till date.[22] Despite experimental and theoretical advances, there is more to comprehend about the charge storage mechanism of EDLC in nano confined spaces.

In 1971, pseudo capacitance was first observed in $RuO_2$.[23] Unlike EDLC, pseudo capacitance occurs due to fast and reversible faradic reaction occurring at the electrode surface. Despite the occurrence of charge transfer reaction, the cyclic voltammogram of a pseudo capacitor is rectangular in shape,[24] which is the signature of an EDLC. Materials known for their pseudo capacitance behavior are ruthenium oxide,[25] manganese oxide,[26] vanadium nitride,[27] $Co_3O_4$[28] etc. $RuO_2$ is an ideal candidate to understand the phenomena of pseudocapacitance. The presence of structural water in ruthenium oxide ($RuO_2.xH_2O$) plays a crucial role in increasing the capacitance than the amorphous $RuO_2$[29] (720 *vs.* 380 F $g^{-1}$). The storage of proton by hydrated $RuO_2$ is shown below:

$$RuO_a(OH)_b + \delta H^+ + \delta e^- \rightarrow RuO_{a-\delta}(OH)_{b+\delta}$$

where $RuO_a(OH)_b$ and $RuO_{a-\delta}(OH)_{b+\delta}$ are the interfacial oxyruthenium species at higher and lower oxidation states, respectively. Pure bulk — nanostructured $RuO_2.xH_2O$ is observed to display a specific capacitance of around 1300 F $g^{-1}$.[30] The pseudo capacitive behaviour of $RuO_2.xH_2O$ is due to the existence of $Ru^{4+}$ as redox active species, its porous structure and metallic conductivity which assist in the rapid electron transport, the hydrous nature of $RuO_2$ which enables the rapid proton transport, and a large outer surface area that decreases diffusion distances.

The charge storage mechanism of pseudo capacitor can be categorized into three types which occur due to different physical processes of different materials: (a) underpotential deposition, where metal ions are adsorbed on the metal surface, positive to their redox potential (e.g., lead on surface of gold or proton on surface of platinum)[31,32] (b) redox pseudo-capacitance, where ions are adsorbed on the surface of metal or near to the metal surface with simultaneous faradaic reaction (e.g., $RuO_2$[33] or $MnO_2$[34]) (c) intercalation pseudo-capacitance where intercalation of ions occur with concomitant faradaic charge transfer, provided no crystallographic phase change (e.g., $Nb_2O_5$[35]). As pseudo capacitance has the contribution from both Faradaic current and capacitive current, the current, *i*, obtained from CV analysis of a capacitor will have both the components.

$$i\,(V) = k_1\nu + k_2\nu^{1/2}$$

where $k_1\nu$ and $k_2\nu^{1/2}$ are contributions from both capacitive process and charge transfer process.[36] The capacitance value of a pseudocapacitor is much higher than that of electrochemical double layer capacitance (~10–100 times), whilst it suffers from lower power density due to slower faradaic reaction. Further instability of the cathode during cycling is also an unavoidable issue.

Hybrid capacitors are the combination of electrochemical double layer capacitor (e.g., graphene, activated carbon) and pseudocapacitor (e.g., $RuO_2$, $MnO_2$, conducting polymer).[37,38] The effort of coupling the battery-like electrode (Faradaic) and capacitor-like electrode (non-faradaic) leads to an increase in capacity of the system by two to three folds the individual EDLC and pseudocapacitor. Depending on the configuration of assembly, a hybrid supercapacitor can be divided into symmetric and asymmetric supercapacitor capacitor.

## 3. Two-dimensional Electrode Materials used in Supercapacitors

### 3.1. *Graphene-based materials*

Graphene is one of the 'researcher's choice' materials since its discovery back in 2004. This honeycomb like material has astonishing properties like superior electron mobility (2.5×105 $cm^2V^{-1}s^{-1}$),[39] excellent thermal conductivity[40] (> 3000 W $mK^{-1}$), high theoretical surface area 2630 $m^2$ $g^{-1}$,[41] exceptional intrinsic strength (130 GPa), Youngs modulus(~1 TPa) and elastic stiffness (−2.0 TPa).[42] These novel qualities of graphene make it a potential candidate for a high energy density supercapacitor with a theoretical capacitance of 550 F $g^{-1}$.[43] Graphene can be produced by various methods such as self-assembly, chemical activation, and light treatment, however, cost and scalability limits its uses.[43] Rao and his coworkers[44] in 2008 demonstrated that graphene developed from thermally exfoliated graphitic oxide at 1050°C and nanodiamond heated at 1650°C were able to deliver a specific capacitance of 117 F $g^{-1}$ and 75 F $g^{-1}$ in aqueous and ionic liquid electrolyte, respectively. This was the highest reported value among the other allotropes of carbon. In the same year, Zhu *et al.*[45] reported a specific capacitance of 135 and 99 F $g^{-1}$ by chemically reducing graphene oxide using hydrazine hydrate, in aqueous and organic electrolyte, respectively. The surface area of the resultant graphene was 705 $m^2$ $g^{-1}$.

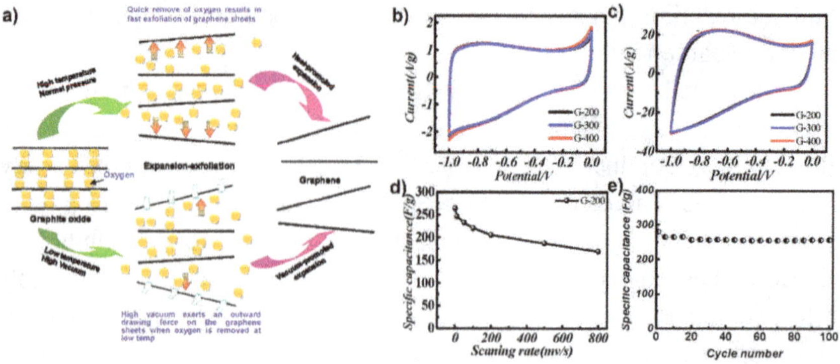


Figure 1. (a) Schematic depiction of chemical exfoliation of graphenes: (Top) At high temperature (above 1000°C) and normal pressure; (bottom) at low temperature (~200°C) and high vacuum; cyclic voltammetry at scan rate (b) 10 mV $s^{-1}$, (c) 200 mV $s^{-1}$; (d) Specific capacitance of G-200 at different scan rate; (e) cycling performance of G-200 at current density 100 mA $g^{-1}$. Reprinted with permission from Ref 48. Copyright 2009 American Chemical Society.

The BET surface area of graphene can be increased to 3100 $m^2$ $g^{-1}$ by exfoliating graphite oxide with microwave, followed by KOH activation. As a result, the specific capacitance of KOH activated microwave exfoliated graphite oxide electrode is enhanced to ~170 F $g^{-1}$.[46]

Restacking of graphene sheets, during its preparation, is a major problem reducing its effective surface area. In 2010, curved graphene sheets were prepared which not only mitigate the issue of restacking of graphene sheets but also enhanced the accessibility and wettability of the pores by large and viscous ionic liquids.[47] Supercapacitor fabricated from these graphene sheets electrode delivered an enhanced electrochemical double layer capacitance of 154.1 F $g^{-1}$ at room temperature. Apart from chemical treatment, graphite oxide is also reduced by physical treatment like low-temperature vacuum treatment, laser treatment, thermal treatment etc. Lv *et al.*[48] prepared graphene by exfoliating graphite oxide at 200°C (G-200) accompanied by high vacuum as shown in Figure 1a. The CV curve measured in aqueous solution (5.5 M KOH) at both low (10 mV $s^{-1}$) and high (200 mV $s^{-1}$) scan rates exhibited a typical rectangular curve with a minimal distortion at high sweep rate (Figures 1b and 1c). This indicates low resistance and facile double layer charge storage mechanism in the porous texture of the material. Figure 1d displays the specific capacitance of G-200 at various sweep rates. At a high scan rate of 800 mV $s^{-1}$, 65% of the initial specific capacitance (262 F $g^{-1}$ at 1 mV $s^{-1}$) was retained which indicated the decent power capability of the supercapacitor. The supercapacitor delivered a good cyclability with a stable specific capacitance of 256 F $g^{-1}$ for 100 cycles (Figure 1e).

Gao *et al.*[49] proposed a very facile and non-toxic way of building electrical double layer capacitor by reducing graphite oxide via laser writing technique. They made use of entrapped water molecules inside the layers of graphene oxide, which in turn act as electrolyte and electrode separator. The ionic conductivity ($1.1\times10^{-5}$ S $cm^{-1}$ for the sandwich geometry and $2.8\times10^{-3}$ S $cm^{-1}$ for the in-plane configuration) was observed due to movement of protons arising from hydrolysis of different functional groups present in graphene oxide. The capacitance of in-plane geometry (0.51 mF $cm^{-2}$) was higher than the sandwich structure. However, due to eventual water loss, the capacitance of both in-plane circular device and sandwich devise dropped to 30% of its original capacitance. Kady *et al.*[50] developed laser-scribed high-performance supercapacitor by reduction of graphite oxide using LightScribe DVD optical drive. The robust graphene films displayed high electrical conductivity (1738 S $m^{-1}$) and a surface area (1520 $m^2$ $g^{-1}$) with excellent areal capacitance of 4.04 mF $cm^{-2}$ in aqueous 1M $H_2SO_4$.

Huge van der Waals attraction, among the layers of graphene, force the graphene layers to restack. This self-restacking of graphene layers can also be prevented by introducing spacers between the layers of graphene. These spacers can be carbon materials, solvents, polymers or metals.[51] For example, introduction of small amount of 2D carbon black nanoparticle by vacuum filtration open up the structure for effective ion transport and charge storage. The pillared graphene electrode showed stable specific capacitance of 138 F $g^{-1}$ in aqueous electrolyte (6 M KOH) and 83.2 F $g^{-1}$ in organic electrolyte (1 M $LiPF_6$) for long number of cycles.[52]

Graphene nanofibres and nano palettes have attracted huge attraction due their potential application in electronic wearables. Qu and his coworker[53] developed vertically oriented graphene nanoribbon fibre which has high surface area and electrical conductivity. These vertically oriented graphene showed high areal capacitance of 3.2 mF $cm^{-1}$ when compared to that of pure graphene fibre 0.2 mF $cm^{-1}$. The synthesis of two-dimensional graphene nanoribbon was also performed by thermal transformation of metal organic frameworks nanorods (MOF). A specific capacitance of 168 F $g^{-1}$ was observed for these nanoribbons at a current density of 1 A $g^{-1}$ in 1 M $H_2SO_4$ electrolyte.[54] The basal planes in graphene nanoribbon remain inert and do not engage in the electrochemical process. To mitigate this problem, hetero-atom doping came into picture. Doping graphene with hetero atoms, such as nitrogen, sulphur, iodine or boron, altered the electrode electrolyte interphase, which results in increase in the electrochemical performance of the capacitor.[55] Several different synthesis strategies has been adopted to synthesize nitrogen-doped (N-doped) graphene. Hydrothermal reaction between graphene oxide and urea resulted in formation of nitrogen doped graphene yielding 10.13% nitrogen.[56] The ammonia released in process of decomposition of graphene reacted with oxygen of different functional group present in the graphene oxide resulting in N-doped graphene sheet. The electrodes exhibited excellent electrochemical capacitance of 326 F $g^{-1}$ at 0.2 A $g^{-1}$ current density and good cycling stability for about 2000 cycles in 6 M KOH solution. Moreover, the graphite oxide was treated with cyanamide, yielding $C_3N_4$ polymer on top of graphene surface. These highly porous (3.42 $cm^3$ $g^{-1}$) and crumpled nitrogen-doped graphene exhibited specific capacitance of 248 F $g^{-1}$ at 5 mV $s^{-1}$.[57] The capacitance of electrical double layer in pyrrolic structure enriched nitrogen-doped (N-doped) graphene layers were further observed to be 194 F $g^{-1}$.[58]

Graphene forms are 3D interconnected network of graphene with porous structures having high electronic conductivity and high surface area. However, these materials suffer from low wettability and a smaller number of active sites, which limits their direct use in supercapacitor. N-doped graphene are known for its improved wettability, accessibility of active sites and good conductivity. Combining these two materials, Zhang *et al.*[59] reported N-superdoped 3D graphene network structure where graphene form acts as a skeleton nested with N-doped reduced graphene oxide aerogels. This material combines the advantages of the material yielding high conductivity (3.33 S $cm^{-1}$), large specific surface area (583 $m^2$ $g^{-1}$), enhanced wettability, and more accessible active sites. Due to these multiple features, the 3D superdoped structure displayed remarkable specific capacitance in alkaline (380 F $g^{-1}$), acidic (332 F $g^{-1}$) and neutral electrolytes (245 F $g^{-1}$).

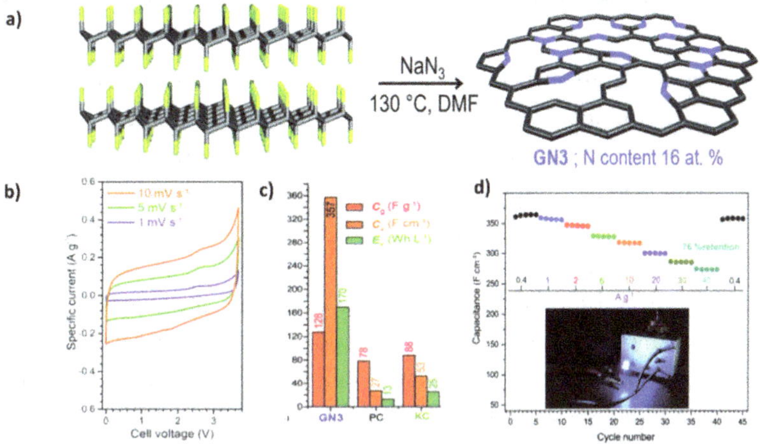


Figure 2. (a) Schematic representation of the synthesis of GN3. (b) CV of GN3 at different sweep rate. (c) Comparison of performance of GN3 with that of porous carbon (PC) and Kuraray carbon (KC). (d) Rate performance of GN3 at various current density. Reproduced from Ref. 61 with permission from the Royal Society of Chemistry.

Cheap and environmentally friendly one-step production of N-doped graphene 3D sponge from waste polyethylene terephthalate (PET) bottles opened up the wide opportunity for reuse of waste. These PET bottles were mixed with urea and treated at different temperatures to produce N-doped graphene.[60] These cost-effective N-doped graphene displayed an outstanding performance with specific capacitance of 405 F $g^{-1}$ at 1 A $g^{-1}$ in 6 M KOH solution as electrolytes. Veronica and her co-workers[61] recently showed N doped-graphene with diamond-like bond. They prepared their material by reacting fluorographene with sodium azide ion which acts as defluorinating agent and introduced nitrogen into the structure (Figure 2a). The novel combination of $sp^2$ and $sp^3$ carbons in the same structure with high nitrogen doping and vacancies displayed efficient charge transport and ion -diffusion properties. The cyclic voltammetry curve (CV) of GN3 exhibited quasi rectangular shape with small faradaic peak (Figure 2b). The shape of the curve maintained at high sweep rate indicated fast charge transfer at the interface. Compared to benchmark compound such as porous carbon (PC) and Kuraray carbon (KC), GN3 performed far better in both volumetric and gravimetric terms (Figure 2c). The rate capability of the cell is examined at various current densities and 76% of the initial capacity is preserved at a current density of 40 A $g^{-1}$ (Figure 2d).

Like N-doped graphene, boron-doped (B-doped) graphene is also known for its extraordinary electrochemical performance. Thermal annealing of graphene oxide and $B_2O_3$ at 1000°C led to high boron concentration of around 6.04% in the graphene sheets.[62] These B-doped graphene electrode increased the electronic properties and the surface area of the graphene and exhibited specific capacitance of 448 F $g^{-1}$ in aqueous electrolyte. Pandurangan and co-workers[63] developed copper-anchored boron-doped graphene nanosheets via chemical vapour deposition technique using $CuSO_4$ an $B_2O_3$. These B-doped graphene sheets act as negative electrode in quasi solid electrolyte of $H_2SO_4$/PVA to give a capacitance of 483 F $g^{-1}$ at 1 A $g^{-1}$ with 96 % retention after 5000 cycles. Symmetrical supercapacitor using Boron doped graphene/boron-doped diamond displayed potential application in energy storage system.[64] These boron-doped (B-doped) graphene and Boron-doped diamond delivered high energy density of 30.7 Wh $kg^{-1}$ and retained its capacitance to 99.6% in 3.2 A $g^{-1}$ current density for 10,000 cycles.

### 3.2. *Graphene composites with transition metal oxide and chalcogenides*

Transition metal oxides are attractive due to their superior redox activity. The capacitance or the energy density of transition metal oxide pseudocapacitors are about 10–100 times higher than that of electrochemical double layer capacitors.[65] However, the poor electronic conductivity and low surface area limits their uses. Therefore, combining the advantage of both pseudocapacitor with electrochemical double layer capacitor can give rise to supercapacitors with enhanced power density and energy density. Zhang *et al.*[66] fabricated an asymmetric supercapacitor with $RuO_2$/graphene electrode and PANi/graphene electrode which exhibited significantly improved capacitance than symmetrically reduced graphene oxide (RGO)-$RuO_2$ or RGO-PANi supercapacitor. This enhancement in performance was due to broadening of potential window (1.4 V) which resulted in double the energy density (26.3 Wh $kg^{-1}$) than that of its symmetrical counterpart. An advanced rational design of $Mn_3O_4/MnO_2$ hierarchical nanoarchitecture anchored in 3D N-doped graphene showed superior electrochemical capacitance (739 F $g^{-1}$ at 0.5 A $g^{-1}$) and rate capabilities and stability (93.4% capacitance retention after 10000 cycles) than that of similar graphene/$MnO_2$[67,68] composites. The improvement in performance of this electrode was due to enhanced available electrochemical active sites in 4 nm thick nanosheet $MnO_2$ and the presence of mixed- valence cations ($Mn^{2+}$, $Mn^{3+}$, and $Mn^{4+}$) led to the formation of ionic defects and electronic defects.[69]

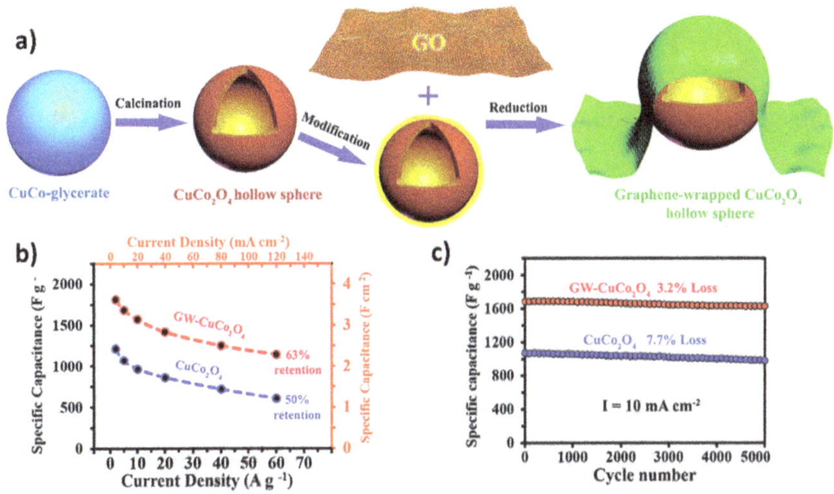


Figure 3. (a) Schematic of the process of formation of graphene-wrapped (GW) $CuCo_2O_4$ hollow spheres. (b, c) Electrochemical performance of the GW $CuCo_2O_4$ and $CuCo_2O_4$. Reproduced from Ref. 71 with permission from the Royal Society of Chemistry.

Among different transition metal oxides, ternary metal oxides are known for their high electrical conductivity and electrochemical activity due to faster electron transfer between the cations. Sethi *et al.*[70] have reported the preparation of porous graphene and $NiFe_2O_4$ nanocomposite. The stronger bond between oxygen and metal ions and multiple oxidation states of both Ni and Fe make this a suitable candidate for electrochemical devices. With a specific surface area of 107 $m^2$ $g^{-1}$, this electrode displayed a specific capacitance of 1465 F $g^{-1}$ with 94% of capacitive retention even after

10,000 cycles in 6 M KOH solution. Kaverlavani *et al.*[71] developed graphene wrapped $CuCo_2O_4$ nanostructure composite having a well-defined hollow sphere nanostructure (Figure 3a). This unique nanoporous hollow sphere electrode consists of large surface area (106.2 $m^2$ $g^{-1}$) and displayed a high capacitance of 1813 F $g^{-1}$ at 2 A $g^{-1}$ compared to that of pure $CuCo_2O_4$ (1214 F $g^{-1}$) (Figures 3b and 3c). The high surface area, porous shell, hollow interior, and surface permeability, which play supreme roles in facile electron and ion transport, fast kinetics, and more accessible active sites, led to outstanding performance of the super-capacitor.

Transition metal dichalcogenides ($MX_2$ where M and X refers to transition metals and chalcogen anions e.g. S or Se, respectively)[72] are a group of 2D compounds with layer of metal atoms sandwiched between two layers of chalcogenide *via* covalent interaction and every layer of $MX_2$ is held *via* van der Waals interaction.[73] Such layers can be exfoliating to $MX_2$ nano-sheets exposing their large surface area and electrochemical potential.[74] The exfoliation of $MoS_2$ can also be done in aqueous electrolyte ($Na_2SO_4$) where graphene is added to further enhance the conductivity of the material and hence, increase the electrochemical capacitance of the $MoS_2$/graphene electrode.[75] Highly flexible and stretchable all-solid state supercapacitor has been prepared from $MoS_2$/graphene composite. The individual layer structure of $MoS_2$ and graphene tends to get damaged upon little stress. However, by growing layered $MoS_2$ nanosheets into the holes and surface of graphene foam, 100% stretchability of the compact graphene/$MoS_2$ film can be achieved. These electrodes can bear a tensile strength of 60% and show a volumetric capacity of 19.44 F $cm^{-3}$ (70.00 mF $cm^{-2}$).[76] Apart from $MoS_2$, other transition metal chalcogenides were also used. $WS_2$ is a naturally occurring transition metal sulfide. The availability of a wide range of oxidation state from +2 to +6 in tungsten made it a potential candidate for pseudo capacitors. Rout *et al.*[77] developed $WS_2$/graphene hybrid electrode which exhibits 5 and 2.5 times higher specific capacitance (350 F $g^{-1}$) than that of its individual components ($WS_2$ and reduced graphene oxide sheets, respectively).

### 3.3. *MXenes*

Naguib *et al.*,[78] first reported the synthesis of 2D MXenes material from laminar MAX parent phases, where M donates the transition metal (e.g., Ti, V, Cr, Mo), X corresponds to nitrogen and/or carbon, and A is an element from group 13 or 14 (such as Sn, Al, and Si). Upon selective etching of A from interlayer, these MAX phases are converted to MXene-$M_{n+1}X_nT_x$ (n = 1–3) wherein $T_x$ denotes the surface terminal group (O, OH and/or F) and x is the number of terminal groups (Figure 4).

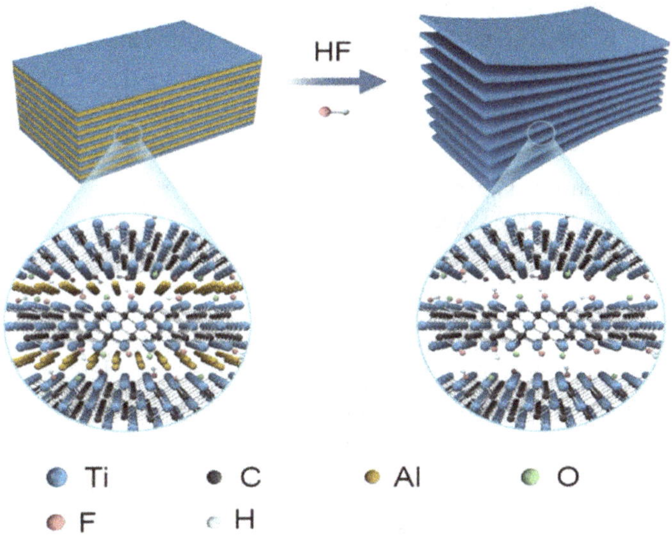


Figure 4. Schematic illustration of MXenes precursor and its synthesis. Reprinted with permission from Ref. 90. Copyright 2021 American Chemical Society.

MXenes offer larger interlayer spacing and surface area which provide faster storage kinetics and extraordinary chemical and thermal stabilities. Several compositions of MXenes such as $Ti_2CT_x$, $Ti_3C_2T_x$, $V_2CT_x$, $Nb_2CT_x$, $Nb_4C_3T_x$ and$Ta_4C_3T_x$ were explored as the electrodes in supercapacitors.[79] The 2D morphology with interlayer spacing provides room for fast ion-transport the outer transition metal oxide layer on the surface generates redox active sites and the inner layer of transition metal carbide facilitate fast electron transfer.[80] This unique design of MXenes makes it a potential candidate for both electrochemical double layer capacitors and the pseoudocapacitors.

The 2D MXenes can electrochemically intercalate various monovalent and multivalent cations such as $Mg^{2+}$, $Na^{+}$, $K^{+}$, $Al^{3+}$ and $NH^{4+}$. The pH of the solution and nature of intercalating ions affect the capacitance of the material. Gogotsi and his coworkers[81] reported capacitance of 350 F $cm^{-3}$ for $Ti_3C_2T_x$ paper in basic potassium hydroxide solution. The microporous $Ti_3C_2T_x$ film delivered a specific capacitance of 210 F $g^{-1}$ at high sweep rate of 10 V $s^{-1}$ outperforming most of the carbon-based supercapacitors. In acidic electrolyte such as $H_2SO_4$, the capacitance of $Ti_3C_2T_x$ was found to be 520 F $cm^{-3}$ at a sweep rate of 2 mV $s^{-1}$. This enhancement in capacitance was due to change in the surface chemistry of MXenes such as intercalation of potassium salt, delamination of $Ti_3C_2T_x$ with DMSO intercalation and replacement of terminal fluorine with oxygen containing functional group.[82] The mechanism of charge storage in acidic solution was further probed by XANES studies which revealed that with change in potential the oxidation state of Ti changes linearly:[83]

$$Ti_3C_2O_x(OH)_yF_{2-x-y} + \delta e^- + \delta H^+ \rightarrow Ti_3C_2O_{x-d}\,(OH)_{y+\delta}\,F_{2-x-y}$$

Supercapacitors are known for their quick storage and delivery of charge due to fast ion-adsorption mechanism. Using battery material in super-capacitor results in large charge storge capacity due to intercalation of ion, but at the cost of slower diffusion limited process. Intercalation of ions in MXenes occurs at a very high rate resulting in high capacitance, and leading to enigma among the researchers. To shed a light into this paradox, Levi *et al.*[84] characterized the mechanical deformations of these electrodes at different stages via electrochemical quartz crystal admittance coupled with *in-situ* electronic conductance and electrochemical impedance spectroscopy. They demonstrated that existence of water molecules between the MXenes layers facilitate the fast 2D adsorption of ions at the interface. Moreover, the ion intercalation occurs as fast as ion adsorption accompanied by exfoliation or deformation of the MXene particles. This exfoliation/breaking up of individual MXene flakes leads to higher capacitance at high cycling rates.

Like other 2D materials, face to face restacking is a serious problem lowering the electrochemical capacitance of MXenes. To mitigate this issue, Zhang *et al.*[85] synthesized 3D $Ti_3C_2$ aerogel electrodes. The unique channel and pores of the aerogel avoid restacking of the MXenes layers. With a high surface area of 176.3 $m^2$ $g^{-1}$ these supercapacitors exhibit high areal capacitance of 1012.5 mF $cm^{-1}$ at scan rate of 2 mV $s^{-2}$ in KOH solution. Further, stable 3D MXenes hydrogel was prepared by freeze drying method. Here, gelation was done by dispersing the MXene in aqueous solution of $FeCl_2$.[86] The $Fe^{2+}$ ion in the solution impair the repulsive forces between the MXene nanosheets and can interact with the $OH^-$ groups present on its surface and act as a linker to form a stable 3D MXene network. This hydrogel suppressed the restacking of the MXene nanosheet and gave excellent performance of 226 F $g^{-2}$ at 1 V $s^{-1}$.

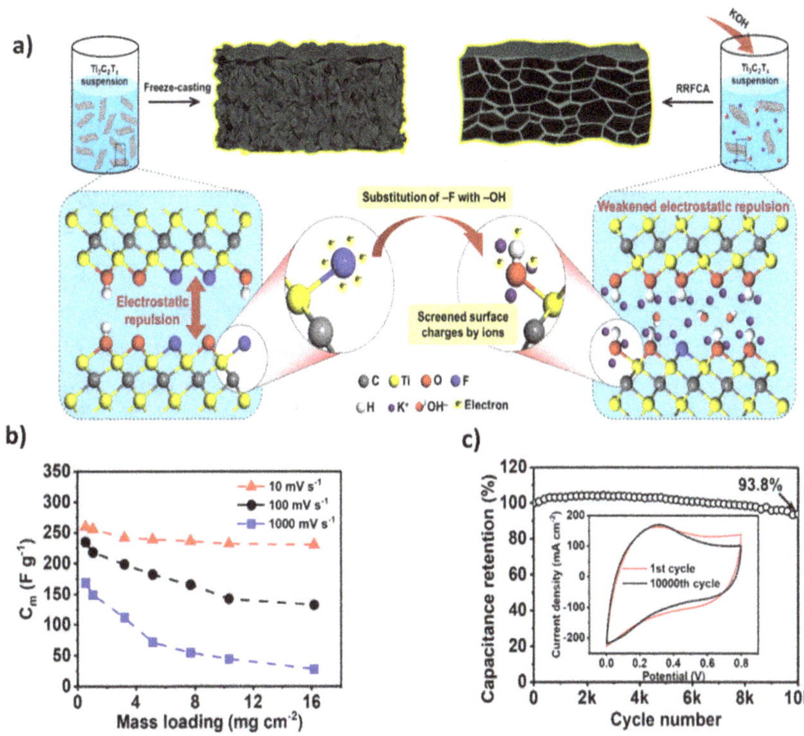


Figure 5. (a) Schematic representation of conventional freeze casting (left) and reduced-repulsion freeze-casting assembly (RRFCA) process (right), (b) Gravimetric capacitance of the electrode with various mass loading at different scan rate, (c) Cycling performance of porous $Ti_3C_2T_x$ electrode at sweep rate 100 mV $s^{-1}$. The inlet displays the 1st and 10000th cycle of CV. Reprinted with permission from Ref. 87. Copyright 2020 American Chemical Society.

For mass production of 3D porous electrode for supercapacitor problems related to high mass loading must be solved. In this direction, Ostrikov and his coworkers[87] suggested the production of 3D porous $Ti_3C_2T_x$ by reduced-repulsion freeze-casting assembly (RRFCA) without addition of any gelation agents (Figure 5a). To diminish the repulsive forces between the layers of MXenes layers, potassium hydroxide solution was added to the $Ti_3C_2T_x$ suspension, which in turn, substituted the electronegative -F functional group with less electronegative -OH groups. The addition of these potassium and hydroxyl ions also helped in screening the surface charge by forming an electrical double layer at the interface. As shown in Figure 5b, the porous structure delivered 88.8% retention of gravimetric capacitance while increasing the mass loading from 0.53 to 16.19 mg $cm^{-2}$ at 10 mV $s^{-1}$ sweep rate. This signifies the steady electron transfer in the 3D porous structure. However, at higher sweep rates (100 and 1000 mV $s^{-1}$) the capacitance faded rapidly with increase in mass loading due to increase in ion diffusion distance in thicker electrode. The cell also exhibited a stable capacitance retention of 93.8% for 10000 cycles with a mass loading of 16.18 mg $cm^{-2}$ at scan rate 100 mV $s^{-1}$ (Figure 5c). Moreover, the problem of restacking was also tackled by vertically orienting the 2D flakes. The vertical alignment of 2D $Ti_3C_2T_x$ was achieved under an external mechanical shearing force of discotic lamellar liquid crystal phase of MXene.[88] The vertical alignment of these inorganic nanosheets was executed in presence of non-ionic surfactant, hexaethylene glycol mono-dodecyl ether, $C_{12}E_6$, to improve the molecular interaction (hydrogen bonding between the OH group of $C_{12}E_6$ and F/O group on the MXene surface) among the nanosheets and thereby enhanced the systems packing symmetry. The MXene sheets were reoriented into vertical direction due to the torque generated from the mechanical shearing of discotic lamellar phase. The film displayed an extraordinary performance where irrespective of film thickness (up to 200 μm) above 200 F $g^{-1}$ capacitance was retained at high scan rate of 2000 mV $s^{-1}$. The vertical alignment of electrodes structure was so stable that it retained almost 100% capacitance after cycling for 20,000 cycles at current density 20 A $g^{-1}$.

The performance of supercapacitors is not solely dependent on the electrode material. It can be enhanced by modifying the surface of the current collector or by changing the electrolyte. Kumar *et al.*[89]

demonstrated a way of passivating the current collector by using graphene, which improved the capacitance of the supercapacitor. The graphene was grown on top of Ni surface by chemical vapour deposition techniques which was used as a current collector to coat nano-layered $Ti_3C_2T_x$. The graphene-based supercapacitor outperformed the supercapacitors without graphene by 1.5 times. The graphene-based supercapacitor facilitated the charge transfer mechanism and was able to yield a specific capacitance of ~542 F $g^{-1}$ at a scan rate of 5 mV $s^{-1}$.

## 4. Conclusion

To sum up, here we briefly discussed about recent progress in some selected 2D materials that are being used in supercapacitors. Despite the enhanced specific surface area and the exceptional mechanical and electrochemical performance of these materials, there are various hurdles that need to be addressed to enjoy the benefits of supercapacitor to the fullest. The use of Faradaic and capacitive properties of materials is a novel idea but a deep insight of the synergistic interactions is still lacking. Though 2D materials, like graphene, opened the door to tune and achieve the high surface area, restacking remained a serious problem after few cycles. Moreover, in aqueous electrolyte, MXenes are prone to oxidation which increase the resistance and decrease the capacitance over time. Therefore, electrolyte optimisation is equally significant to achieve the stability of the electrode for longer duration. Besides using *in-situ* and *ex-situ* experimental tools, theoretical modelling and computational stimulation can help in better understanding of charge storage mechanism.

## References

1. A. Hagfeldt, G. Boschloo, L. Sun, L. Kloo and H. Pettersson, Dye-sensitized Solar Cells, *Chem. Rev.* **110**(11), 6595–6663 (2010).
2. A. C. Dillon, Carbon Nanotubes for Photoconversion and Electrical Energy Storage, *Chem. Rev.* **110**(11), 6856–6872 (2010).
3. T. R. Cook, D. K. Dogutan, S. Y. Reece, Y. Surendranath, T. S. Teets and D. G. Nocera, Solar Energy Supply and Storage for the Legacy and Nonlegacy Worlds, *Chem. Rev.* **110**(11), 6474–6502 (2010).

4. Z. Yang, J. Zhang, M. C. W. Kintner-Meyer, X. Lu, D. Choi, J. P. Lemmon and J. Liu, Electrochemical Energy Storage for Green Grid, *Chem. Rev.* **111**(5), 3577–3613 (2011).
5. B. Dunn, H. Kamath and J. M. Tarascon, Electrical Energy Storage for the Grid: A Battery of Choices, *Science* **334** (6058), 928–935 (2011).
6. M. Winter and R. J. Brodd, What are Batteries, Fuel Cells, and Supercapacitors?, *Chem. Rev.* **105**, 1021 (2005).
7. T. Pandolfo, V. Ruiz, S. Sivakkumar and J. Nerkar, *In Supercapacitors: Materials, Systems, and Applications*. Wiley-VCH Verlag GmbH & Co. KGaA (2013).
8. D. L. Chapman, A Contribution to the Theory of Electrocapillarity, *Phylosophycal Magazine* **25**(148), 475–481 (2010).
9. D. C. Grahame, The Electrical Double Layer and the Theory of Electrocapilarity, *Chem. Rev.* **41**(3), 441–501 (1947).
10. O. Z. Stern, Theory of The Electrical Double Layer, *Elektrochem.* **30**, 508–508 (1924).
11. G. Gouy, Constitution of the Electric Charge at The Surface of an Electrolyte. *Comptes Rendus* **149**, 654–654 (1910).
12. J. R. Miller, *Market and Applications of Electrochemical Capacitors. Supercapacitors.* Wiley-VCH Verlag GmbH & Co. KGaA (2013).
13. P. Forouzandeh and S. C. Pillai, S. C. Two-dimensional (2D) Electrode Materials for Supercapacitors. *Mater. Today Proc.* **41**(3), 498–505 (2021).
14. B. Raj, A. K. Padhy, S. Basu and M. Mohapatra, Review-Futuristic Direction for R&D Challenges to Develop 2D Advanced Materials Based Supercapacitors, *J. Electrochem. Soc.* **167**, 136501 (2020).
15. Y. Shao, M. F. El-Kady, J. Sun, Y. Li, Q. Zhang, M. Zhu, H. Wang, B. Dunn and R. B. Kaner, Design and Mechanisms of Asymmetric Supercapacitors. *Chem. Rev.* **118**(18), 9233–9280 (2018).
16. Y. Zhai, Y. Dou, D. Zhao, P. F. Fulvio, R. T. Mayes and S. Dai, Carbon Materials for Chemical Capacitive Energy Storage. *Adv. Mater.* **23**(42), 4828–4850 (2011).
17. R. Kötz and M. Carlen, Principles and Applications of Electrochemical Capacitors. *Electrochimica Acta* **45**(15–16), 2483-2498 (2000).
18. L. L. Zhang and X. S. Zhao, Carbon-based Materials as Supercapacitor Electrodes. *Chem. Soc. Rev.* **38**, 2520–2531 (2009).
19. E. Frackowiak and F. Béguin, Electrochemical Storage of Energy in Carbon Nanotubes and Nanostructured Carbons. *Carbon,* **40**, 1775–1787 (2002).
20. S. T. Mayer, R. W. Pekala and J. L. Kaschmitter, The Aerocapacitor: An Electrochemical Double-Layer Energy-Storage Device. *J. Electrochem. Soc.* **140**, 446–451 (1993).
21. S. Yoon, J. Lee, T. Hyeon and S. M. Oh, Electric Double-Layer Capacitor Performance of a New Mesoporous Carbon. *J. Electrochem. Soc.* **147**, 2507 (2000).
22. M. F. El-Kady, V. Strong, S. Dubin and R. B. Kaner, Laser Scribing of High-performance and Flexible Graphene-based Electrochemical Capacitors. *Science (1979)* **335**(6074), 1326–1330 (2012).

23. S. Trasatti and G. Buzzanca, Ruthenium Dioxide: A New Interesting Electrode Material. Solid State Structure and Electrochemical Behaviour. *J. Electroanal. Chem. Interfacial Electrochem.* **29**(2), A1–A5 (1971).
24. V. Augustyn, P. Simon and B. Dunn, Pseudocapacitive Oxide Materials for High-rate Electrochemical Energy Storage. *Energy Environ. Sci.* **7**, 1597–1614 (2014).
25. N. Yoshida, Y. Yamada, S. Nishimura, Y. Oba, M. Ohnuma and A. Yamada, Unveiling the Origin of Unusual Pseudocapacitance of RuO2·nH2O from Its Hierarchical Nanostructure by Small-Angle X-ray Scattering. *J. Phys. Chem. C.* **117**(23), 12003-12009 (2013).
26. H. Zhang, G. Cao, Z. Wang, Y. Yang, Z. Shi and Z. Gu, Growth of Manganese Oxide Nanoflowers on Vertically-Aligned Carbon Nanotube Arrays for High-Rate Electrochemical Capacitive Energy Storage. *Nano Lett.* **8**(9), 2664–2668 (2008).
27. D. Choi, G. E. Blomgren and P. N. Kumta, Fast and Reversible Surface Redox Reaction in Nanocrystalline Vanadium Nitride Supercapacitors. *Adv. Mater.* 18, 1178–1182 (2006).
28. S. K. Meher and G. R. Rao, Ultralayered Co3O4 for High-performance Supercapacitor Applications. *J. Phys. Chem. C* **115**(31), 15646–15654 (2011).
29. J. P. Zheng, P. J. Cygan and T. R. Jow, Hydrous Ruthenium Oxide as an Electrode Material for Electrochemical Capacitors. *J. Electrochem. Soc.* **142**, 2699–2703 (1995).
30. C. C. Hu, K. H. Chang, M. C. Lin and Y. T. Wu, Design and Tailoring of the Nanotubular Arrayed Architecture of Hydrous RuO2 for Next Generation Supercapacitors. *Nano Lett.* **6**(12), 2690–2695 (2006).
31. V. Sudha and M. V. Sangaranarayanan, Underpotential Deposition of Metals: Structural and Thermodynamic Considerations. *J. Phys. Chem. B* **106**(10), 2699–2707 (2002).
32. E. Herrero, L. J. Buller and H. D. Abruña, Underpotential Deposition at Single Crystal Surfaces of Au, Pt, Ag and Other Materials. *Chem. Rev.* **101**(7), 1897–1930 (2001).
33. R. R. Bi, X. L. Wu, F. F. Cao, L. Y. Jiang, Y. G. Guo and L. J. Wan, Highly Dispersed RuO2 Nanoparticles on Carbon Nanotubes: Facile Synthesis and Enhanced Supercapacitance Performance. *J. Phys. Chem. C* **114**(6), 2448–2451 (2010).
34. X. Lang, A. Hirata, T. Fujita and M. Chen, Nanoporous Metal/oxide Hybrid Electrodes for Electrochemical Supercapacitors. *Nat. Nanotechnol.* **6**, 232–236 (2011).
35. L. Kong, C. Zhang, S. Zhang, J. Wang, R. Cai, C. Lv, W. Qiao, L. Ling and D. Long, High-power and High-energy Asymmetric Supercapacitors based on Li+-Intercalation into a T-Nb2O5/graphene Pseudocapacitive Electrode. *J. Mater. Chem. A* **2**, 17962–17970 (2014).
36. T. Brezesinski, J. Wang, J. Polleux, B. Dunn and S. H. Tolbert, Templated Nanocrystal-based Porous TiO2 Films for Next-generation Electrochemical Capacitors. *J. Am. Chem. Soc.* **131**(5), 1802–1809 (2009).
37. P. Forouzandeh and S. C. Pillai, Two-dimensional (2D) Electrode Materials for Supercapacitors. *Materials Today: Proceedings* **41** 498–505 (2020).

38. X. Peng, L. Peng, C. Wu and Y. Xie, Two Dimensional Nanomaterials for Flexible Supercapacitors. *Chem. Soc. Rev.* 43 3303–3323 (2014).
39. S. V. Morozov, K. S. Novoselov, M. I. Katsnelson, F. Schedin, D. C. Elias, J. A. Jaszczak and A. K. Geim, Giant Intrinsic Carrier Mobilities in Graphene and Its Bilayer. *Phys. Rev. Lett.* **100**, (2007).
40. A. A. Balandin, Thermal Properties of Graphene and Nanostructured Carbon Materials. *Nat. Mater.* **10**, 569–581 (2011).
41. P. Kamedulski, M. Skorupska, P. Binkowski, W. Arendarska, A. Ilnicka and J. P. Lukaszewicz, High Surface Area Micro-mesoporous Graphene for Electrochemical Applications. *Sci. Rep.* **11**, 22054 (2021).
42. C. Lee, X. Wei, J. W. Kysar and J. Hone, Measurement of the Elastic Properties and Intrinsic Strength of Monolayer Graphene. *Science* **321**(5887) 385–388 (2008).
43. J. Chen, C. Li and G. Shi, Graphene Materials for Electrochemical Capacitors. *J. Phys. Chem. Lett.* **4**(8), 1244–1253 (2013).
44. S. R. C. Vivekchand, C. Sekhar, A. Govindaraj and C.N.R. Rao, Graphene-based Electrochemical Supercapacitors, *J. Chem. Sci.* **120**, 9–13(2008).
45. M. D. Stoller, S. Park, Z. Yanwu, J. An and R. S. Ruoff, Graphene-based Ultracapacitors. *Nano Lett.* **8**(10), 3498–3502 (2008).
46. Y. Zhu, S. Murali, M. D. Stoller, K. J. Ganesh, W. Cai, P. J. Ferreira, A. Pirkle, R. M. Wallace, K. A. Cychosz, M. Thommes, D. Su, E. A. Stach and R. S. Ruoff, Carbon-based Supercapacitors Produced by Activation of Graphene. *Science (1979)* **332**(6037), 1537–1541 (2011).
47. C. Liu, Z. Yu, D. Neff, A. Zhamu and B. Z. Jang, Graphene-based Supercapacitor with an Ultrahigh Energy Density. *Nano Lett.* **10**(12), 4863–4868 (2010).
48. W. Lv, D. M. Tang, Y. B. He, C. H. You, Z. Q. Shi, X. C. Chen, C. M. Chen, P. X. Hou, C. Liu and Q. H. Yang, Low-temperature Exfoliated Graphenes: Vacuum-Promoted Exfoliation and Electrochemical Energy Storage. *ACS Nano* **3**(11), 3730–3736 (2009).
49. W. Gao, N. Singh, L. Song, Z. Liu, A. L. M. Reddy, L. Ci, R. Vajtai, Q. Zhang, B. Wei and P. M. Ajayan, Direct Laser Writing of Micro-supercapacitors on Hydrated Graphite Oxide Films. *Nat. Nanotech.* **6**, 496–500 (2011).
50. M. F. El-Kady, V. Strong, S. Dubin and R. B. Kaner, Laser Scribing of High-Performance and Flexible Graphene-based Electrochemical Capacitors. *Science (1979)* **335**, 1326–1330 (2012).
51. Y. Si and E. T. Samulski, Exfoliated Graphene Separated by Platinum Nanoparticles. *Chem. Mater.* **20**(21), 6792–6797 (2008).
52. G. Wang, X. Sun, F. Lu, H. Sun, M. Yu, W. Jiang, C. Liu and J. Lian, Flexible Pillared Graphene-paper Electrodes for High-performance Electrochemical Supercapacitors. *Small* **8**(3), 452–459 (2012).
53. L. Sheng, T. Wei, Y. Liang, L. Jiang, L. Qu, Z. Fan, L. Z. Sheng, T. Wei, L. L. Jiang, Z. J. Fan, Y. Liang and L. T. Qu, Vertically Oriented Graphene Nanoribbon Fibers for

High-Volumetric Energy Density All-Solid-State Asymmetric Supercapacitors. *Small* **13**(22), 1700371 (2017).
54. P. Pachfule, D. Shinde, M. Majumder and Q. Xu, Fabrication of Carbon Nanorods and Graphene Nanoribbons from a Metal–organic Framework. *Nat. Chem.* **8**, 718–724 (2016).
55. C.N.R. Rao, K. Gopalakrishnan and A. Govindaraj, Synthesis, Properties and Applications of Graphene doped with Boron, Nitrogen and Other Elements. *Nano Today* **9**(3), 324–343 (2014).
56. L. Sun, L. Wang, C. Tian, T. Tan, Y. Xie, K. Shi, M. Li and H. Fu, Nitrogen-doped Graphene with High Nitrogen Level *via* a One-step Hydrothermal Reaction of Graphene Oxide with Urea for Superior Capacitive Energy Storage, *RSC Adv.* **2**, 4498–4506, (2012).
57. Z. Wen, X. Wang, S. Mao, Z. Bo, H. Kim, S. Cui, G. Lu, X. Feng and J. Chen, Crumpled Nitrogen-doped Graphene Nanosheets with Ultrahigh Pore Volume for High-performance Supercapacitor. *Adv. Mater.* **24**(41), 5610–5616 (2012).
58. F. M. Hassan, V. Chabot, J. Li, B. K. Kim, L. Ricardez-Sandoval and A. Yu, Pyrrolic-structure Enriched Nitrogen Doped Graphene for Highly Efficient Next Generation Supercapacitors. *J. Mater. Chem. A* **1**, 2904–2912 (2013).
59. W. Zhang, C. Xu, C. Ma, G. Li, Y. Wang, K. Zhang, F. Li, C. Liu, H. M. Cheng, Y. Du, N. Tang and W. Ren, Nitrogen-Superdoped 3D Graphene Networks for High-Performance Supercapacitors. *Adv. Mater.* **29**(36), 1701677 (2017).
60. N. A. Elessawy, J. el Nady, W. Wazeer and A. B. Kashyout, Development of High-Performance Supercapacitor based on a Novel Controllable Green Synthesis for 3D Nitrogen Doped Graphene. *Sci. Rep.* **9**, 1–10 (2019).
61. V. Šedajová, A. Bakandritsos, P. Błoński, M. Medveď, R. Langer, D. Zaoralová, J. Ugolotti, J. Dzíbelová, P. Jakubec, V. Kupka and M. Otyepka, Nitrogen Doped Graphene with Diamond-like Bonds Achieves Unprecedented Energy Density at High Power in a Symmetric Sustainable Supercapacitor. *Energy Environ. Sci.* **15**, 740–748 (2022).
62. D. Y. Yeom, W. Jeon, N. D. K. Tu, S. Y. Yeo, S. S. Lee, B. J. Sung, H. Chang, J. A. Lim and H. Kim, High-concentration Boron Doping of Graphene Nanoplatelets by Simple Thermal Annealing and Their Supercapacitive Properties. *Sci. Rep.* **5**, 136–791 (2015).
63. P. M. Pandian and A. Pandurangan, Copper Nanoparticles Anchored onto Boron-doped Graphene Nanosheets for Use as a High Performance Asymmetric Solid-state Supercapacitor. *RSC Adv.* **9**, 3443–3461 (2019).
64. D. Cui, H. Li, M. Li, C. Li, L. Qian, B. Zhou and B. Yang, Boron-Doped Graphene Directly Grown on Boron-Doped Diamond for High-Voltage Aqueous Supercapacitors. *ACS Appl. Energy Mater.* **2**(2), 1526–1536 (2019).
65. J. Sung and C. Shin, Recent Studies on Supercapacitors with Next-Generation Structures. *Micromachines,* **11**, 1125 (2020).

66. J. Zhang, J. Jiang, H. Li and X. S. Zhao, A High-performance Asymmetric Supercapacitor Fabricated with Graphene-based Electrodes. *Energy Environ. Sci.* **4**, 4009–4015 (2011).
67. Y. Wang, W. Lai, N. Wang, Z. Jiang, X. Wang, P. Zou, Z. Lin, H. J. Fan, F. Kang, C. P. Wong and C. Yang, A Reduced Graphene Oxide/mixed-valence Manganese Oxide Composite Electrode for Tailorable and Surface Mountable Supercapacitors with High Capacitance and Super-long Life. *Energy Environ. Sci.* **10**, 941–949 (2017).
68. J. Chang, M. Jin, F. Yao, T. H. Kim, V. T. Le, H. Yue, F. Gunes, B. Li, A. Ghosh, S. Xie and Y. H. Lee, Asymmetric Supercapacitors Based on Graphene/MnO2 Nanospheres and Graphene/MoO3 Nanosheets with High Energy Density. *Adv. Funct. Mater.* **23**(40), 5074–5083 (2013).
69. M. Cui, S. Tang, Y. Ma, X. Shi, J. A. Syed, X. Meng, M. Cui, S. Tang, Y. Ma, X. Shi, J. A. Syed and X. Meng, Monolayer standing MnO2-Nanosheet Covered Mn3O4 Octahedrons Anchored in 3D N-Doped Graphene Networks as Supercapacitor Electrodes with Remarkable Cycling Stability. *J. Power Sources* **396**(31), 483–490 (2018).
70. M. Sethi, U. S. Shenoy and D. K. Bhat, A Porous Graphene–NiFe2O4 Nanocomposite with High Electrochemical Performance and High Cycling Stability for Energy Storage Applications. *Nanoscale Adv.* **2**, 4229–4241 (2020).
71. S. K. Kaverlavani, S. E. Moosavifard and A. Bakouei, Designing Graphene-wrapped Nanoporous CuCo2O4 Hollow Spheres Electrodes for High-performance Asymmetric Supercapacitors. *J. Mater. Chem. A* **5**, 14301–14309 (2017).
72. A. H. Loo, A. Bonanni, Z. Sofer and M. Pumera, Exfoliated Transition Metal Dichalcogenides (MoS2, MoSe2, WS2, WSe2): An Electrochemical Impedance Spectroscopic Investigation. *Electrochem. Commun.* **50**, 39–42 (2015).
73. Y. Han, Y. Ge, Y. Chao, C. Wang, G. G. Wallace, Y. Han, Y. Chao, Y. Wang and C. & Wallace, Recent Progress in 2D Materials for Flexible Supercapacitors, *Journal of Energy Chemistry*, **27**(1), 52-57 (2018).
74. Y. Chao, R. Jalili, Y. Ge, C. Wang, T. Zheng, K. Shu, G. G. Wallace, Y. Chao, R. Jalili, Y. Ge, C. Wang, T. Zheng, K. Shu and G. G. Wallace, Self-Assembly of Flexible Free-Standing 3D Porous MoS2-Reduced Graphene Oxide Structure for High-Performance Lithium-Ion Batteries. *Adv. Funct. Mater.* **27**(22), 1700234 (2017).
75. M. A. Bissett, I. A. Kinloch and R. A. W. Dryfe, Characterization of MoS2-Graphene Composites for High-Performance Coin Cell Supercapacitors. *ACS Appl. Mater. Interfaces* **7**(31), 17388–17398 (2015).
76. N. Li, T. Lv, Y. Yao, H. Li, K. Liu and T. Chen, Compact Graphene/MoS2 Composite Films for Highly Flexible and Stretchable All-solid-state Supercapacitors. *J. Mater. Chem. A* **5**, 3267–3273 (2017).
77. S. Ratha and C. S. Rout, Supercapacitor Electrodes Based on Layered Tungsten Disulfide-reduced Graphene Oxide Hybrids Synthesized by a Facile Hydrothermal Method. *ACS Appl. Mater. Interfaces* **5**(21), 11427–11433 (2013).

78. M. Naguib, M. Kurtoglu, V. Presser, J. Lu, J. Niu, M. Heon, L. Hultman, Y. Gogotsi and M. W. Barsoum, Two-Dimensional Nanocrystals Produced by Exfoliation of Ti3AlC2. *Adv. Mater.* **23**(37), 4248–4253 (2011).
79. M. Han, C. E. Shuck, R. Rakhmanov, D. Parchment, B. Anasori, C. M. Koo, G. Friedman and Y. Gogotsi, Beyond Ti3C2Tx: MXenes for Electromagnetic Interference Shielding. *ACS Nano* **14**(4), 5008–5016 (2020).
80. M. R. Lukatskaya, S. Kota, Z. Lin, M. Q. Zhao, N. Shpigel, M. D. Levi, J. Halim, P. L. Taberna, M. W. Barsoum, P. Simon and Y. Gogotsi, Ultra-high-rate Pseudocapacitive Energy Storage in Two-dimensional Transition Metal Carbides, *Nat. Energy*, **2**(8), 17105 (2017).
81. M. R. Lukatskaya, O. Mashtalir, C. E. Ren, Y. Dall'Agnese, P. Rozier, P. L. Taberna, M. Naguib, P. Simon, M. W. Barsoum and Y. Gogotsi, Cation Intercalation and High Volumetric Capacitance of Two-dimensional Titanium Carbide. *Science.* **341**(6153), 1502–1505 (2013).
82. Y. Dall'Agnese, M. R. Lukatskaya, K. M. Cook, P. L. Taberna, Y. Gogotsi and P. Simon, High Capacitance of Surface-modified 2D Titanium Carbide in Acidic Electrolyte. *Electrochem. Commun.* **48**, 118–122 (2014).
83. M. R. Lukatskaya, S.-M. Bak, X. Yu, X.-Q. Yang, M. W. Barsoum and Y. Gogotsi, Probing the Mechanism of High Capacitance in 2D Titanium Carbide Using In Situ X-Ray Absorption Spectroscopy. *Adv. Energy Mater.* **5**(15), 1500589 (2015).
84. M. D. Levi, M. R. Lukatskaya, S. Sigalov, M. Beidaghi, N. Shpigel, L. Daikhin, D. Aurbach, M. W. Barsoum and Y. Gogotsi, Solving the Capacitive Paradox of 2D MXene using Electrochemical Quartz-Crystal Admittance and In Situ Electronic Conductance Measurements. *Adv. Energy Mater.* **5**(1), 1400815 (2015).
85. L. Li, M. Zhang, X. Zhang and Z. Zhang, New Ti3C2 Aerogel as Promising Negative Electrode Materials for Asymmetric Supercapacitors. *J. Power Sources* **364**, 234–241 (2017).
86. Y. Deng, T. Shang, Z. Wu, Y. Tao, C. Luo, J. Liang, D. Han, R. Lyu, C. Qi, W. Lv, F. Kang, Q.-H. Yang, Y. Deng, C. Luo, R. Lyu, W. Lv, F. Kang, T. Shang, Z. Wu, Y. Tao, J. Liang, D. Han, C. Qi and Q. Yang, Fast Gelation of Ti3C2Tx MXene Initiated by Metal Ions. *Adv. Mater.* **31**(43), 1902432 (2019).
87. J. Kong, H. Yang, X. Guo, S. Yang, Z. Huang, X. Lu, Z. Bo, J. Yan, K. Cen and K. K. Ostrikov, High-Mass-Loading Porous Ti3C2TxFilms for Ultrahigh-Rate Pseudocapacitors. *ACS Energy Lett.* **5**, 2266–2274 (2020).
88. Y. Xia, T. S. Mathis, M. Q. Zhao, B. Anasori, A. Dang, Z. Zhou, H. Cho, Y. Gogotsi and S. Yang, Thickness-independent Capacitance of Vertically Aligned Liquid-crystalline MXenes. *Nature* **557**, 409–412 (2018).
89. S. Kumar, M. A. Rehman, S. Lee, M. Kim, H. Hong, J. Y. Park and Y. Seo, Supercapacitors Based on Ti3C2Tx MXene Extracted from Supernatant and Current Collectors Passivated by CVD-graphene. *Sci. Rep.* **11**, 1–9 (2021).

90. S. Ren, J.L. Xu, L. Cheng, X. Gao, and S.D. Wang, Amine-Assisted Delaminated 2D Ti3C2Tx MXenes for High Specific Capacitance in Neutral Aqueous Electrolytes, *ACS Appl. Mater. Interfaces* **30**(13), 35878–35888 (2021).

https://doi.org/10.1142/9789811270956_0005

# Chapter 5

## Lead Halide Perovskites

K. V. Saurav and Pratap Vishnoi*

*New Chemistry Unit, International Centre for Materials Science & School of Advanced Materials, Jawaharlal Nehru Centre for Advanced Scientific Research (JNCASR), Jakkur P. O., Bangalore – 560 064*
**pvishnoi@jncasr.ac.in*

Lead halide perovskites (LHPs) have emerged as one of the most exciting classes of materials for solar photovoltaics and other opto-electronic applications. Among different LHPs, cesium lead halide ($CsPbX_3$), methylammonium lead halide ($MAPbX_3$) and formamidinium lead halide ($FAPbX_3$) are the most popular ones, mainly the iodides, due to their narrow bandgaps. Lead iodide perovskite-based PV cells have achieved power conversion efficiency (PCE) of over 25% in single-junction solar cell, and over 29% in perovskite/silicon tandem solar cell within only 13 years since 2009. Their superior photovoltaic properties are attributed to high-absorption coefficient, long carrier diffusion length, narrow and direct bandgap and low-cost solution processability. In this chapter, we discuss the recent developments on lead iodide perovskites in terms of their crystal structures, electronic properties, solar cell applications as well as the challenges for their real-time applications.

## 1. Introduction

With increasing global efforts to cut down the emission of greenhouse gases and transitioning to a green energy landscape, there has been an unprecedented surge of interest in solar photovoltaic cells in the recent past. Typically, photovoltaic solar cells are comprised of semiconducting materials which directly convert the sunlight into electrical energy by making use of the photovoltaic effect. This process generally includes three steps: (i) photon absorption, (ii) formation of photo-generated charge carriers (electrons and holes) and (iii) separation of the charge carriers. The extra energy received by the charge carriers from photons allows the carriers to flow through the material as an electric current which is extracted to power electric devices. So far, the semiconductor-based solar cells use Si, GaAs, CdTe, $CuIn_{1-x}Ga_xSe_2$, and more recently, lead halide perovskites (LHPs).

The 2009 discovery of $(CH_3NH_3)PbX_3$ ($X$ = Br, I) perovskite-sensitized solar cell stimulated an explosion of investigations on perovskites for a variety of properties of both fundamental and practical interests.[1,2] LHPs are a class of solution-processable crystalline semiconductors with a general formula of $APbX_3$ (A = $Cs^+$, $CH_3NH_3^+$ and $HC(NH_2)_2^+$; $X$ = $Cl^-$, $Br^-$ or $I^-$) (Figure 1). Due to narrow, direct and tunable bandgaps as well as high optical absorption of $\sim 10^5$ $cm^{-1}$, these materials absorb a wide range of the electromagnetic spectrum. They exhibit a long carrier diffusion length and a low exciton binding energy. In addition, the excellent defect tolerance of perovskite allows the fabrication of high-quality thin film devices at low-cost through solution processing. These properties have allowed perovskite solar cells (PSCs) to witness rapid improvement in PV power conversion efficiency (PCE) from 3.8% to 25.7% within only 13 years (from 2009 to 2022) (Figure 2). Rapid surge in the PCE is attributed to tuneable chemical composition, film deposition methods as well as cell architectures. When compared to the efficiency of multi-crystalline silicon cell, which is just under 23%, LHPs have a promising future as energy materials. Due to higher ambient stability and more ideal bandgaps for

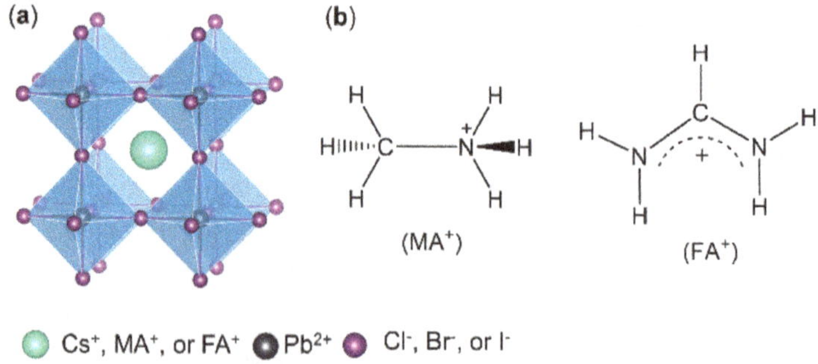


Figure 1. (a) Crystal structure of $APbX_3$ perovskite showing corner-connected $PbX_6$ octahedra and large mono-cation in the $A$-site cavity. (b) Organic amine cations used as the $A$-cation in 3D $APbX_3$ perovskites.

solar cells, almost all high-performance devices are mostly fabricated with the archetypal $CsPbI_3$, $MAPbI_3$, $FAPbI_3$ and their mixed halides. Further, layered perovskites, such as 2D $(PEA)_2PbI_4$ (PEA = phenethyl-ammonium), have been reported to improve ambient stability and reduce charge recombination, thereby increasing efficiency of 2D-3D solar cells.[3] In this chapter, we discuss the archetypical LHPs $CsPbI_3$, $MAPbI_3$ and $FAPbI_3$ including their crystal structures, electronic structures, optical properties and recent develop-ments on LHPs based photovoltaics.

## 2. Renaissance of Lead Halide Perovskites

$CsPbX_3$ ($X$ = Cl, Br or I) perovskites were first synthesized in 1893 by Wells from the reaction of cesium halide and lead halide in aqueous media.[4] In 1957 and 1958, much after the discovery of X-ray crystallography by Bragg in 1912, Møller structurally characterized these perovskites through X-ray diffraction and found that $CsPbCl_3$ and $CsPbBr_3$ have distorted 3-dimensional perovskite structures, whilst $CsPbI_3$ has a 1-dimensional chain structure.[5,6] Møller further noted that at higher temperatures, these compounds undergo phase transition to intensively coloured cubic phases — $CsPbCl_3$ is yellow, $CsPbBr_3$ is orange and $CsPbI_3$ is black — and are sensitive to the electromagnetic spectrum. On light absorption, these materials exhibit photoconductivity due to photo-generated charge carriers. In 1978, Weber had shown that $Cs^+$ can be replaced by methylammonium cation (MA) to form hybrid halide perovskites of general formula $MAPbX_3$ (MA = $CH_3NH_3$; $X$ = Cl, Br or I) and also investigated fundamental optoelectronic properties.[7] These hybrid systems were suggested to be cubic in symmetry. Subsequent solid state $^2H$ and $^{14}N$ NMR spectroscopy,[8] and millimetre-wave spectroscopy studies revealed a series of temperature-dependent phases due to rotation of the MA cation.[9] Since then, for about three decades, the studies on lead halide perovskites progressed almost steadily on phase transition,[10,11] and electronic properties.[12,13] The progress on halide perovskites has been accelerated since 2009 when the group of Miyasaka, at the Toin University of Yokohama, discovered that $MAPbBr_3$ and $MAPbI_3$ can efficiently sensitize mesoporous *n*-type $TiO_2$ films for visible-light conversion with a PCE of 3.13% and 3.81%, respectively.[2] Post this pioneer work, hybrid lead halide perovskites captured a wide attention of many other research groups around the world. Within two years, the PCE was increased beyond 10% for mixed-halide $MAPbI_2Cl$ perovskite coated on insulating $Al_2O_3$.[14] The liquid electrolyte was then replaced with a solid-state hole-conducting material: spiro-MeOTAD (2,2′7,7′-tetrakis(*N*,*N*-di-*p*-methoxyphenyl amine)-9,9′-spirobifluorene. With joint efforts from chemists, materials scientists and physicists, a flurry of follow up activities on materials design, device engineering and device physics have immensely contributed to increasing the PCE of perovskite solar cells (PSCs).[15] Within only 13 years (from 2009 to 2022), the PCE of single junction PSC crossed 25% (certified 25.5%) (Figure 2),[16] which is comparable with the efficiency of Si solar cell. However, the efficiency is still below the Shockley-Queisser (Q-S) limit of 30%[17] and there is room for further improvement. Recently, the PCE crossed 29% for perovskite/ silicon tandem solar cell.[18]

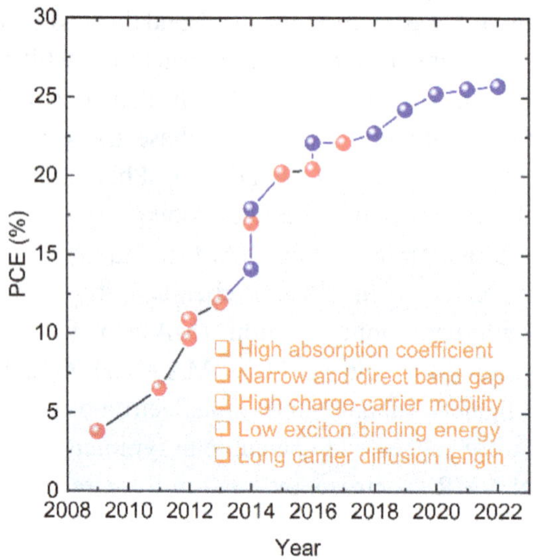


Figure 2. Evolution in PCE of perovskite solar cells from 2009 to 2022. The blue data points are indicative of certified efficiencies published in *Progress in Photovoltaics: Research and Applications*.

## 3. Synthesis

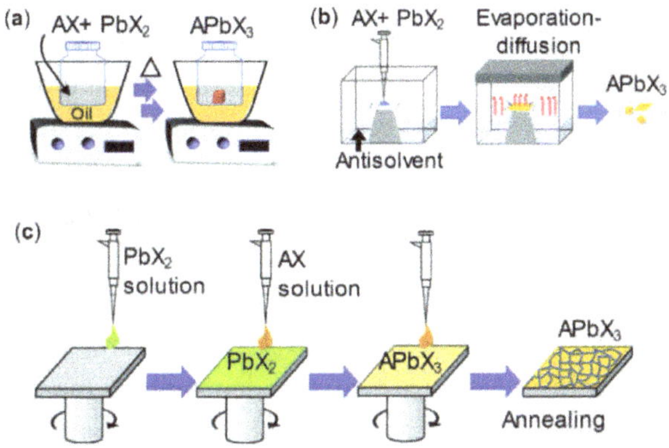


Figure 3. Schematic representation of methods for perovskite synthesis. (a) Single crystal synthesis through inverse temperature crystallization, (b) single crystal synthesis through antisolvent vapour-assisted crystallization and (c) formation of crystalline films through sequential deposition of precursor solutions. Adapted with permission from Ref. 19.

LHPs are typically prepared as single crystals (Figures 3a and 3b) or crystalline thin films (Figure 3c). Single crystals are typically synthesized by solvothermal process which is a very popular method for making these materials. In this process, a solution of the precursors is heated in an autoclave at high temperature, either in high boiling point solvents such as *N,N*-dimethylformamide (DMF),[20] or in aqueous hydrohalic acid.[21] The perovskite crystals form on cooling the saturated solution or evaporation of the solvent from the saturated solution. The crystallization of hybrid perovskites depends greatly on the halide and organic amine cation. Bulk single crystals of $MAPbI_3$ can be synthesized from the precursors in aqueous HI solvent by decreasing the solution temperature to induce the saturation of the solute.[22] Millimeter-sized crack-free $MAPbBr_3$ and $MAPbI_3$ single crystals have been synthesized from DMF or γ-butyrolactone (GBA) solution containing the crystal precursors by slow diffusion of dichloromethane (DCM) as the antisolvent.[23] It has been noted that the solubility of the $PbX_2$ and MAX precursors varies in these solvents; hence $MAPbBr_3$ crystallizes more aptly from DMF, while $MAPbI_3$ crystallizes better from GBA. Generally, a good solvent for crystallization dissolves more precursors on heating, while cooling down induces supersaturation which leads to crystal formation. On the contrary, inverse temperature or retrograde solubility has been shown to rapidly grow high quality single crystals of $MAPbX_3$ in selected solvents: $MAPbBr_3$ in DMF and $MAPbI_3$ in GBA.[24] In this method, the crystallization is induced due to the decrease of solubility of crystal precursors in solvents with increasing temperature. High quality nanocrystals of all-inorganic $CsPbX_3$ ($X$ = Cl, Br, I) perovskites can be solvothermally synthesized from heating precursors in a mixture of oleylamine and oleic acid.[25]

In PSCs and other device studies, perovskite thin films are deposited on suitable substrates by one-step or two-step deposition. One-step deposition involves simultaneous deposition of a solution containing multiple precursors, whereas the precursors are added in a sequential order onto a spinning substrate in two-step deposition. The excess solvent on the substrate is removed by spinning at high speed, followed by annealing. The annealing step helps in further removal of any excess solvent and facilitates the nucleation and growth of perovskite thin film.[26] Although the one-step and two-step depositions have their own advantages, uncontrolled nucleation and rough morphology of thin films remain a concern for fabricating high-efficiency solar cells. Jeon *et al.* introduced an anti-solvent method in which the precursor is deposited onto a substrate and an anti-solvent is employed during spin coating process to enhance the crystallinity of the perovskite films.[27] Chen *et al.* fabricated $MAPbI_3$ films from MAI and $PbI_2$ precursors dissolved in a 9:1 (v/v) mixture of DMF/DMSO.[28] The films were prepared on to PTAA-covered ITO glass substrates (PTAA = 2,4,6-trimethylphenyl)amine; ITO = indium tin oxide) by one-step spin coating with toluene as an antisolvent, followed by annealing at 100°C.

## 4. Crystal Structure

Typically, $APbX_3$ perovskites have infinite 3-dimensional $(PbI_3)_n$ polyanionic framework comprised of corner-shared $PbX_6$ octahedra. The large $A$-site mono-cation resides in the 12-coordinated cuboctahedral cavities created by these corner-shared octahedra. In principle, any combination of ions can form stable perovskites when two empirical parameters, namely

Goldschmidt's tolerance factor $t = (r_A + r_X)/\sqrt{2}(r_B + r_X)$ and Pauling's radius ratio $\mu = r_B/r_X$ are satisfied, where $r_A$, $r_B$, and $r_X$ represent the ionic radii of the *A*-, *B*- and *X*-site ions.[29] A perovskite structure is expected when $t$ is in the 0.8–1.0 range and $\mu$ in the 0.41–0.73 range. The structures with $t$ close to 0.8 are likely to be distorted, while those close to 1.0 are cubic. At lower temperatures, a given perovskite may adopt a low symmetry structure such as tetragonal or orthorhombic phases due to metal halide octahedral tilting.

Among the commonly investigated $Pb^{2+}$ halide perovskites, iodides are the most sought after owing to their suitable bandgaps for PV application. Their ambient stability can be tuned with $Cs^+$, $CH_3NH_3^+$ (MA), or $HC(NH_2)_2^+$ (FA) at the *A*-sites. Each of these three halide perovskites undergoes multiple phase transitions (Table 1) and therefore understanding the structures of different phases is critical since they show different electronic band structures and optoelectronic properties.

Table 1. Structurally known polymorphs of $CsPbI_3$, $MAPbI_3$ and $FAPbI_3$.

| Phases | Unit cell constants (Å) | Ref. |
|---|---|---|
| | **$CsPbI_3$** | |
| α-phase (black) | Cubic *Pm*-3m (No. 221); $a$ = 6.1769(3) | 33 |
| β-phase (black) | Tetragonal *P*4/mbm (No. 127); $a = b$ = 8.82690; $c$ = 6.29902 | 33 |
| γ-phase (black) | Orthorhombic *P*nma (No. 62); $a$ = 8.86370; $b$ = 12.48380; $c$ = 8.57830 | 30 |
| δ-phase (yellow) | Orthorhombic *P*nma (No. 62); $a$ = 10.4342(7); $b$ = 4.7905(3); $c$ = 17.7610(10) | 31 |
| | **$MAPbI_3$** | |
| α-phase (black) | Cubic *Pm*-3m (No. 221); $a$ = 6.31728(27) | 38 |
| β-phase (black) | Tetragonal *I*4/mcm (No. 140); $a = b$ = 8.8660(3); $c$ = 12.6524(9) | 36 |
| γ-phase (black) | Orthorhombic *P*nma (No. 62); $a$ = 8.836(1); $b$ = 12.580(1); $c$ = 8.555(1) | 39 |
| | **$FAPbI_3$** | |
| α-phase (black) | Cubic *Pm-3m* (No. 221); $a$ = 6.3620(8) | 44 |
| β-phase (black) | Tetragonal *P*4/mbm (No. 127); $a = b$ = 8.971650(17); $c$ = 6.348355(24) | 46 |
| γ-phase (black) | Tetragonal *P*4/mbm (No. 127); $a = b$ = 8.887796(22), $c$ = 6.288927(32) | 46 |
| δ-phase (yellow) | Hexagonal *P*6₃mc (No. 194); $a = b$ = 8.6603(14); $c$ = 7.9022(6) | 31 |

### 4.1. *Cesium lead iodide ($CsPbI_3$)*

Figure 4 shows the different polymorphs of $CsPbI_3$ which form at different temperatures. Under an inert atmosphere at ambient temperature and pressure, $CsPbI_3$ crystallizes in the γ-phase which is an orthorhombic black perovskite-phase.[30] On exposure to moist air, this phase changes to a yellow non-perovskite δ-phase.[31] The δ-phase comprises of 1-dimensional infinite chains of $(PbI_3)^-$ formed by edge-sharing $PbI_6$ octahedra and the Cs cations in the interchain spaces. Complete conversion of the γ-phase into the δ-phase occurs in ~48 h.[32] At temperatures > 633 K, both these phases change to black cubic phase (α-phase).[33] On slow cooling to room temperature, the α-phase changes back to the δ-phase, whilst it changes to the γ-phase on fast cooling. This suggests that the δ-phase is the thermodynamic product, and the γ-phase is the kinetic product.[34] Recently, it has been shown that on cooling, the α-phase converts to the black

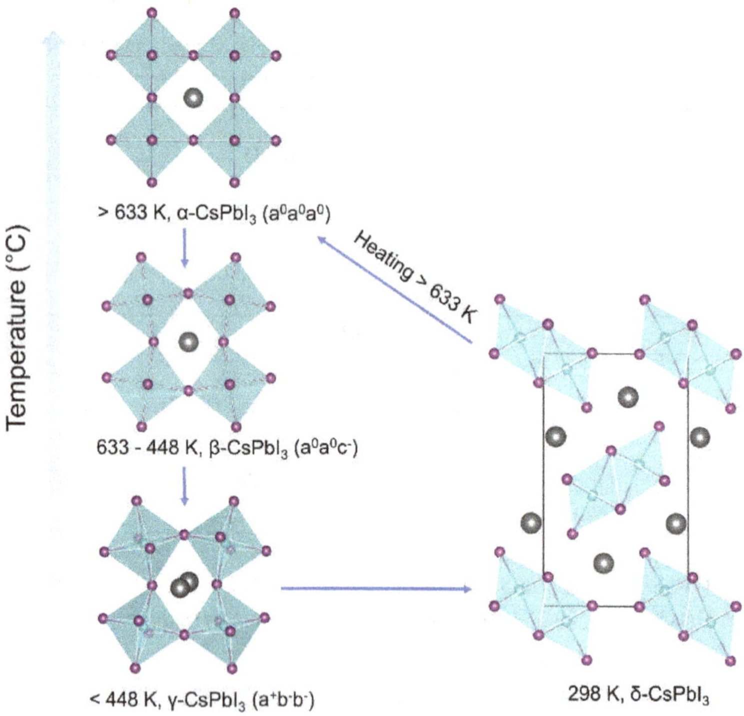


Figure 4. Temperature-dependent phases of $CsPbI_3$. Parentheses show the Galzer tile notation.

tetragonal phase (β-phase) due to cooperative in-plane octahedral tilting which then converts to the γ-phase at 448 K due to in-plane as well as out-of-plane octahedral tilting.[33] Theoretical calculations on the formation energies have suggested that the γ-phase is the most stable among the perovskite phases of $CsPbI_3$, but it is less stable than the non-perovskite δ-phase.[30] By carrying out all processing in an inert atmosphere, $\alpha$-$CsPbI_3$ has been kinetically trapped at room temperature which has been shown to exhibit a much higher PV efficiency of 10% when compared to its other polymorphs.[35]

### 4.2. *Methylammonium lead iodide ($MAPbI_3$)*

At ambient pressure and temperature, $MAPbI_3$ crystallizes in a tetragonal β-phase (*I4/mcm*; No. 140) which undergoes a phase transition to higher symmetry upon heating and lower symmetry upon cooling (Figure 5).[36] Some reports have also found that $MAPbI_3$ crystallizes in a polar tetragonal space group *I4cm* (No. 108), which no longer contains the *c*-glide plane.[31,37] However, the exact assignment of the room temperature phase to either a non-polar *I4/mcm* or a polar *I4cm* space group with potential ferroelectric character is still unclear. Nevertheless, this phase remains symmetrically unchanged in the temperatures range of *ca.* 327 K to *ca.* 162 K. At higher temperatures, it reversibly changes to a cubic $\alpha$-phase.[38] The lattice parameters of the β-phase show a pseudo-cubic relationship with the $\alpha$-phase through $a_\beta = b_\beta = \sqrt{2}a_\alpha$ and $c_\beta = 2a_\alpha$.[39] The $\alpha$-phase ⇌ β-phase transition occurs due to a cooperative tilting of the $PbI_6$ octahedra around the *c*-axis and order-disorder transitions of the *A*-site MA cation.[36] The symmetry lowering leads to rather dense crystal packing within the *ab*-plane. Since the MA cation has a $C_{3v}$ symmetry, it is expected to be disordered in the cubic $\alpha$-phase and tetragonal β-phase to satisfy the *Oh* symmetry. As a result of symmetry lowering, MA partially orders to 8 disordered states in the β-phase from 24 disordered states in the $\alpha$-phase. At temperatures below 162 K, the β-phase changes to orthorhombic γ-phase which is due to both in-plane and out-of-plane octahedral tilting.[39] The MA cation is in a fully ordered state in the γ-phase. Furthermore, high-resolution synchrotron X-ray powder diffraction data have suggested that the γ-phase remains symmetrically unchanged down

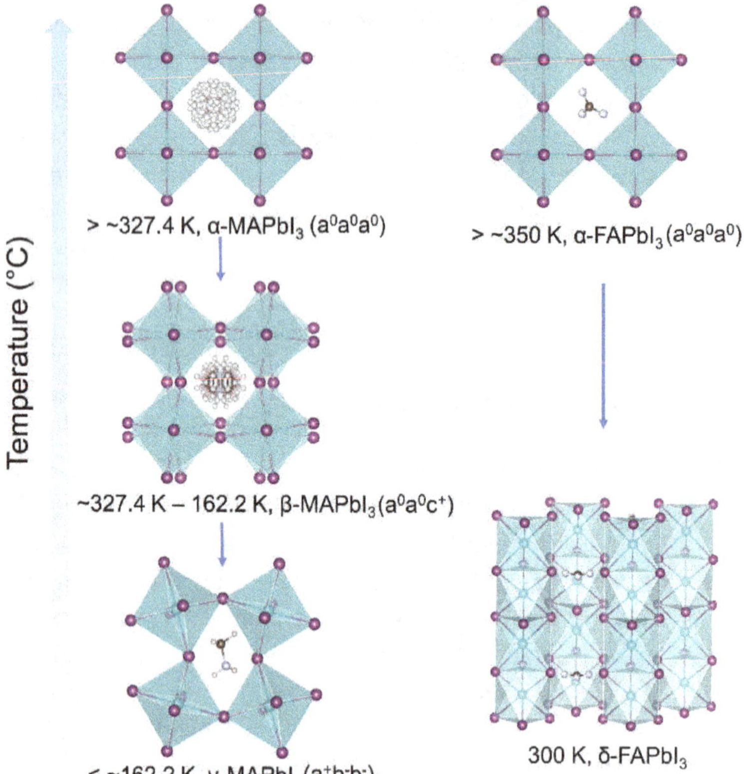


Figure 5. Temperature-dependent phases of $MAPbI_3$ and $FAPbI_3$. Parentheses show the Galzer tile notation.

to 10 K.[40] In addition, there has been reports on pressure-dependent phase transition in this compound. At ~0.4 GPa pressure, a $ReO_3$-type cubic phase (space group *Im-3*; $a$ = 12.3053 Å; No. 204) has been observed between the β-phase and the γ-phase with 8 times larger unit cell than that of the $\alpha$-phase.[31,41,42] In all these phases, the compound retains a corner connected $(PbI_3)_n$ poly-anionic framework with the MA cation in $A$-site voids. In the presence of solvents with hydrogen bonding capability such as $H_2O$, DMF and DMSO, $MAPbI_3$ can form yellow non-perovskite phases $MAPbI_3 \cdot S$ ($S$ = solvent molecule).[43]

### 4.3. *Formamidinium lead iodide ($FAPbI_3$)*

At room temperature and pressure, $FAPbI_3$ crystallizes in a yellow non-perovskite hexagonal $\delta$-phase (Figure 5) which has a wide bandgap of 2.43 eV. The structure is comprised of a 1-dimensional chain structure of face-sharing $PbI_6$ octahedra with the FA cation occupying the interchain spaces. It is worth noting that δ-$FAPbI_3$ is strikingly different in the structure compared to δ-$CsPbI_3$ which is formed by double chain $PbI_6$ polyhedra. The early single-crystal X-ray diffraction studies have proposed that on heating the $\delta$-$FAPbI_3$ at ~623 K, it changes to a trigonal α-phase.[31] The α-$FAPbI_3$ has been directly synthesized at room temperature, the structure of which has been suggested to be cubic.[44] Subsequently, it was found that on heating at 350 K, the δ-$FAPbI_3$ phase undergoes a sharp first-order transition into the cubic α-$FAPbI_3$.[45] Upon cooling, the α-phase changes back to the $\delta$-phase, but at a slightly lower temperature of 290 K. Some studies have found that the phase change on cooling from $\alpha$- to $\delta$-phase was either irreversible or very slow.[31] The FA cation is disordered over 12 crystallographically equivalent positions so that the C–H bond is directed into the face of cubic unit cell, whilst the $-NH_2$ groups interact with the iodide atoms of the $[PbI_3]^-$ framework through hydrogen bonds. Strikingly, the α-$FAPbI_3$ can be kinetically trapped by thermal quenching even at a temperature of 8.2 K.[45] The kinetically trapped phase slowly changes to the $\delta$-phase in a humid atmosphere. Complete conversion of yellow δ-$FAPbI_3$ into the black α-phase has been achieved *in situ* by heating it to 490 K in a Kapton capillary. Density functional theory calculations have shown that the formation enthalpy of the δ-phase is lower by ~ 70 meV per formula unit compared to that of the α-phase, and the α- to δ-phase transition is driven by entropy. It was further suggested that the α- to δ-phase transition occurs through many intermediate structures. A high-resolution synchrotron X-ray powder diffraction study between 90 K and 490 K has revealed that on cooling, the α-$FAPbI_3$ undergoes a continuous phase transition to a tetragonal β-phase at 285 K, and then to a tetragonal γ-phase at 140 K (Figure 6).[46]

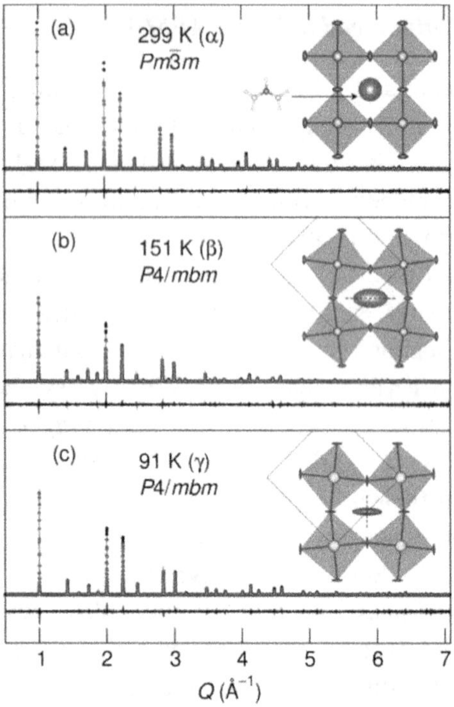


Figure 6. X-ray powder diffraction patterns and resulting crystal structures of the α-, β- and γ-phases of $FAPbI_3$ obtained from Rietveld refinement of synchrotron data. Adapted with permission from Ref. 46.

## 5. Moisture-Induced Degradation

Besides temperature-dependent phase transition due to intrinsic instability, lead iodide perovskites can undergo degradation in atmospheric humidity. Specifically, the effect of humidity on $MAPbI_3$ is of significant interest due to its detrimental effect on the solar cell performance. Depending on relative humidity (RH), $MAPbI_3$ goes through sequential chemical processes including hydration, decomposition, or dissolution. Thus, understanding moisture-induced degradation of perovskite is essential for developing appropriate control strategies to alleviate the degradation. Trace amount of water is found to be helpful for obtaining high-quality perovskite films with improved device performance.[47] At a low RH of 10%, water infiltrates into $MAPbI_3$ on a time scale of seconds which is reversible upon exposure to dry nitrogen or on applying vacuum.[48] Water molecules can infiltrate the $MAPbI_3$ lattice on prolonged exposure to water vapour and form hydrogen bonds with the ammonium group of the MA cation. This leads to the formation of wide bandgap (yellow) non-perovskite monohydrate $MAPbI_3{\cdot}H_2O$ and dihydrate $MA_4PbI_6{\cdot}2H_2O$ as shown in Scheme 1.[49] Figure 7 shows the crystal structures of the mono-hydrate and the dihydrate phases. The monohydrate is an intermediate which consists of 1-dimensional $(PbI_3)^-$ chains interspersed with hydrogen bonded MA cation and $H_2O$ molecules.[43] The $(PbI_3)^-$ chains adopt "double-chain" assembly, just like in the case of δ-$CsPbI_3$ (Figure 4, *vide supra*). This can be easily converted back to $MAPbI_3$ perovskite phase by heating. On prolonged exposure to water vapour, formation of the dihydrate $(MA)_4PbI_6{\cdot}2H_2O$ and amorphous lead iodide ($PbI_2$) occurs. The dihydrate has a zero-dimensional structure composed of isolated $[PbI_6]^{4-}$ octahedra and hydrogen bonded $[(MA{\cdots}H_2O{\cdots}MA)_2]^{4+}$ dimers. Although both the hydrated species are found to be co-existing on prolonged exposure to humid air, their formation is not simultaneous but a two-step sequential process.[49] Exposure to high RH or liquid water usually gives rise to irreversible dissolution of $MAPI_3$ forming water-soluble MAI and $PbI_2$,

$$4MAPbI_3 + 4H_2O \xrightleftharpoons{H_2O} 4[MAPbI_3.H_2O] \rightleftharpoons (MA)_4PbI_6.2H_2O + 3PbI_2 + 2H_2O \quad (\mathbf{1})$$

$$(MA)_4PbI_6.2H_2O \longrightarrow 4MAI + PbI_2 + 2H_2O \quad (\mathbf{2})$$

Scheme 1. Reversible structural transition of $MAPbI_3$ perovskite phase to non-perovskite monohydrate and dihydrate phases on exposure to water/humidity.

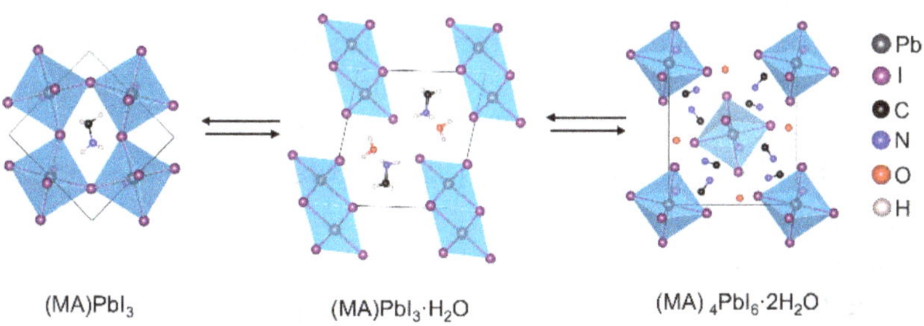


Figure 7. Crystal structures of $MAPbI_3$ perovskite and non-perovskite monohydrate and dihydrate phases which form on exposure to water/humidity.

which are then dissolved into excess water. Unlike $MAPbI_3$, $FAPbI_3$ changes to its δ-phase or decomposes into constituents rather than forming hydrated products. On exposure to low RH of 30% over 90 days, $FAPbI_3$ showed no signs of decomposition, but at 50 % RH it changes to the δ-$FAPbI_3$, and completely decomposes into yellow $PbI_2$ on dipping in water.[50]

## 6. Electronic Structure and Optical Property

Figure 8a shows the Pb-I molecular orbitals (MOs) diagram for the $APbI_3$ class of perovskites. The valence band maxima (VBM) are mainly the antibonding MOs composed of Pb $6s^2$ and I $5p$ orbitals, with the latter being a major component. The conduction band minima (CBM) are mainly the MOs composed of a mixture of Pb $6p$ and I $5p$ with less antibonding and more nonbonding Pb $6p$ character.[51,52] As a result, the bandgaps of $MAPbI_3$ and the related perovskites are strongly direct and determined at the same point (*R* point) in the first Brillouin zone boundary (Figure 8b). The total density of states (DOS) near the band edges is comprised mainly of Pb $6s$, Pb $6p$ and I $5p$ orbitals and the electronic states of the *A*-cation appear only several electron volts below the VBM as an essentially non-dispersive state (Figure 8c). This suggests that the *A*-cation does not contribute electronically to the frontier orbitals. However, the interaction of the *A*-cation with the Pb-I framework can indirectly affect the band structure and bandgap through two ways:[53] (i) smaller *A*-cation can cause octahedral tilting-led symmetry lowering, or (ii) by contracting the lattice without symmetry change. Generally, the former effect increases the bandgap, whilst the latter effect decreases the bandgap. Since the VBM is composed of antibonding hybrid states of the Pb *s* and I *p* orbitals, any change in perovskite lattice that increases the extent of Pb-I orbital overlap will raise the energy level of the VBM. The opposite happens when the extent of Pb-I orbital overlap decreases. Compared to the VBM, the CBM energy level is less affected by the changes in the extent of Pb-I orbital overlap due to non-bonding nature. As a result, structural changes that increase Pb-I overlap will decrease the bandgap and *vice versa*. Under applied pressure, bandgap narrowing has been experimentally observed in $MAPbI_3$[57] and $FAPbI_3$.[58] Bandgap widening has been observed due to

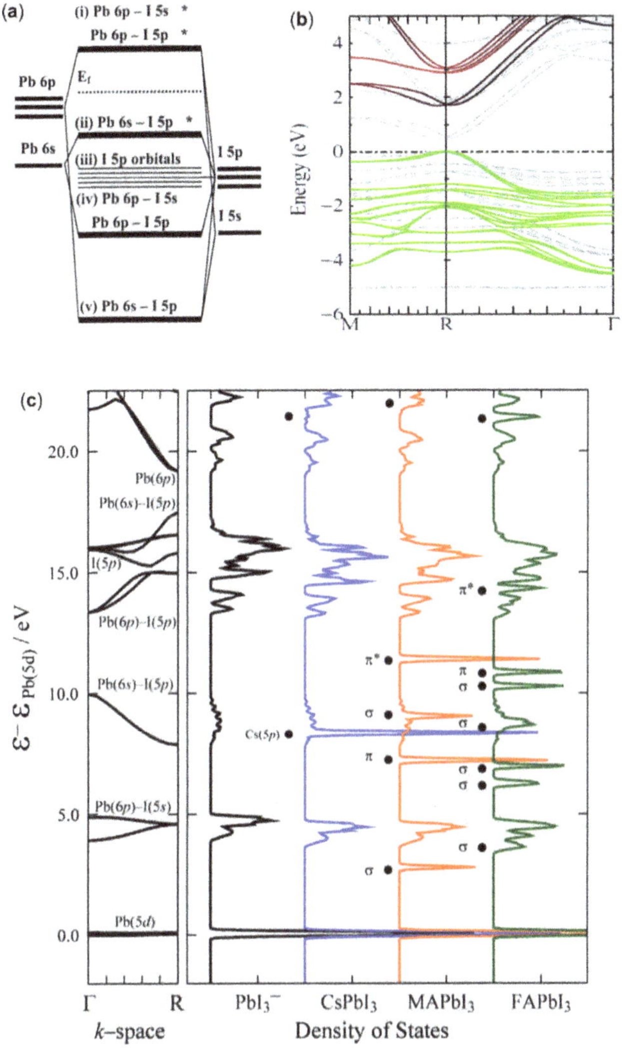


Figure 8. (a) Molecular orbital diagram of $(PbI_3)^-$ framework, showing interaction between the orbitals of Pb and I. (b) Band structure of $MAPbI_3$ indicating direct bandgap at the *R* point in the reciprocal space (*k*-space). (c) Band structure of $(PbI_3)^-$ framework with the density of states (DOS) of $PbI_3^-$, $CsPbI_3$, $MAPbI_3$ and $FAPbI_3$. Panel (a) adapted with permission from Ref. 54; panel (b) adapted with permission from Ref. 55 and panel (c) adapted with permission from Ref. 56.

temperature related symmetry lowering.[59] However, there exists some inconsistencies; for example, the bandgaps reduce in order of $CsPbI_3$

($E_g$ = 1.73 eV), $MAPbI_3$ ($E_g$ = 1.57 eV) and $FAPbI_3$ ($E_g$ = 1.47 eV) (Figure 9a), although their lattice size increases without phase transition.[60] Similarly, a steady increase in the bandgaps has been observed in mixed *A*-cation $Cs_xFA_{1-x}PbI_3$ (Figure 9b),[53] and $Cs_xMA_{1-x}PbI_3$[61] on Cs substitution of FA and MA, respectively. The increase in the bandgaps is attributed to the tilting of Pb-I-Pb bond angle which dominates over the lattice contraction.[53] Some calculations suggest that MA or FA cation electronically interact with the non-boding Pb *p* orbitals of the CBM, lowering their energy level, whilst they do not change the VBM.[56] As a result, the bandgaps of $MAPbI_3$ and $FAPbI_3$ are lower compared to that of $CsPbI_3$. Additionally, the bandgaps can be tuned by using a wide range of

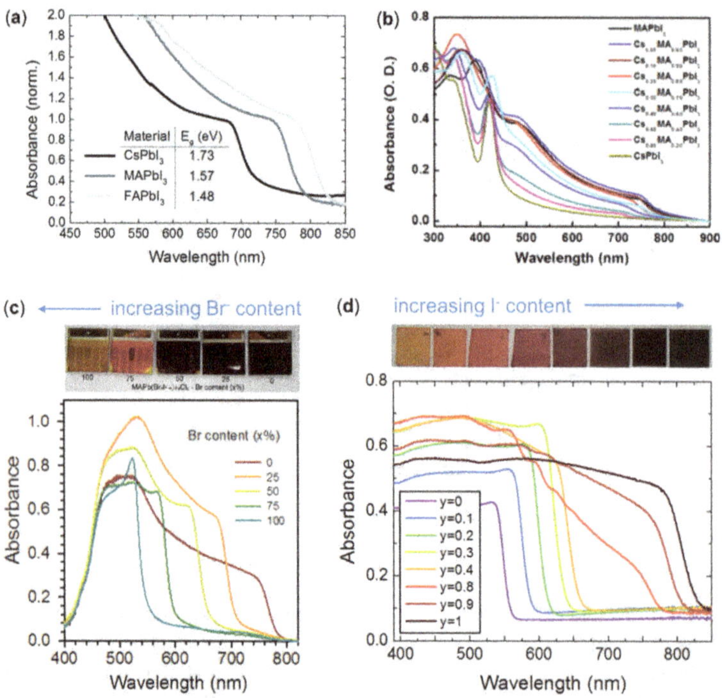


Figure 9. UV-Vis absorption spectra of (a) $APbI_3$ (*A* = Cs, MA, FA), (b) $Cs_xMA_{1-x}PbI_3$, (c) $MAPb(Br_xI_{1-x})_{3-y}Cl_y$ ($0 \leq x \leq 1$), and (d) $FAPbI_yBr_{3-y}$. The optical images of the $MAPb(Br_xI_{1-x})_{3-y}Cl_y$ and $FAPbI_yBr_{3-y}$ perovskite films are also provided in panels (c) and (d), respectively. Panel (a) adapted with permission from Ref. 60; panel (b) adapted with permission from Ref. 61; panel (c) adapted with permission from Ref. 64 and panel (d) adapted with permission from Ref. 60.

halide substitutions; decreasing the halide size, increases the bandgap (Figures 9c and 9d).

Broad and flat absorption covering complete visible light wavelengths make halide perovskites an excellent class of materials for solar cell. The absorption edge of prototype $MAPbI_3$ can reach the near- infrared (NIR) region, exhibiting the absorption coefficient of $10^5$ $cm^{-1}$, comparable with typical PV materials, such as crystalline-silicon, CsTe and GaAs (Figure 10). The absorption coefficient of $MAPbI_3$ is such that its 300 nm thick layer can capture almost all the absorbable incident light.[62] It should be noted that $FAPbI_3$ is more attractive than $MAPbI_3$ and $CsPbI_3$ because of its bandgap of 1.48 eV, which is close to the theoretically optimal bandgap of 1.4 eV for photovoltaic cells. Furthermore, the hybrid lead halide perovskite lattices are highly anharmonic due to large fluctuations caused by a dynamic disorder of the organic cations as well as stereo-chemically active Pb $6s^2$ lone-pair. The dynamic nature of the perovskite lattice influences the emission spectra and charge carrier dynamics.[63]

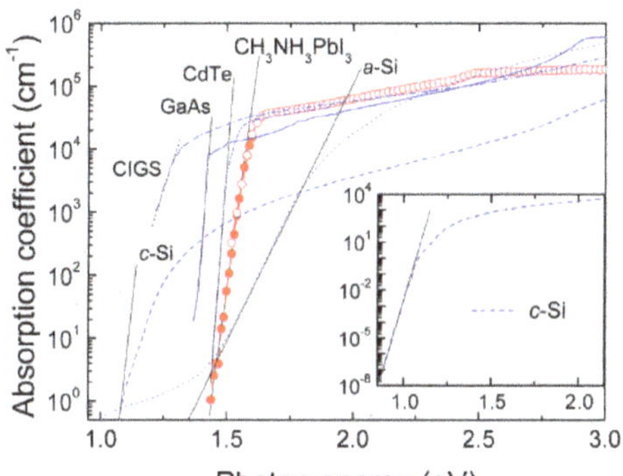


Figure 10. Room temperature absorption coefficients of a $MAPbI_3$ perovskite thin film compared with other materials, typically used in photovoltaic cells. Adapted with permission from Ref. 65.

## 7. Perovskite Solar Cell (PSC)

### 7.1. *Structure and working mechanism*

A perovskite solar cell (PSC) differs from the typical crystalline-silicon solar cell in the layer stacking pattern and the working mechanism.[66,67] The PSCs are found primarily in three forms as shown in Figure 11. In the regular planar cell, a few hundred nanometers thick perovskite absorbing layer is sandwiched between the hole-transporting layer (HTL) and electron-transporting layer (ETL). The regular cell can be further classified into two; n-i-p and p-i-n type heterojunction depending on the sequence of ETL and HTL. In the mesoporous cell, the perovskite layer is infiltrated in mesoporous oxide (Figure 11a), whilst in the regular planar cell, the perovskite layer is deposited on compact oxide layer (Figure 11b). In these types of PSCs, the ETL encounters the sunlight first, just like in the case of DSSCs. In inverted planar PSCs (Figure 11c), the HTL encounters the light before the ETL which is reminiscent of organic solar cells but has perovskite as the absorber layer instead of organic material. The regular structure has been extensively investigated among these three types of device structures, which typically has the following components stacked vertically:

(i) *Glass coated transparent conducting oxide (TCO)*: Glass coated fluorine doped tin oxide (FTO) or indium doped tin oxide (ITO) which acts as the transparent electrode of the perovskite solar cell.

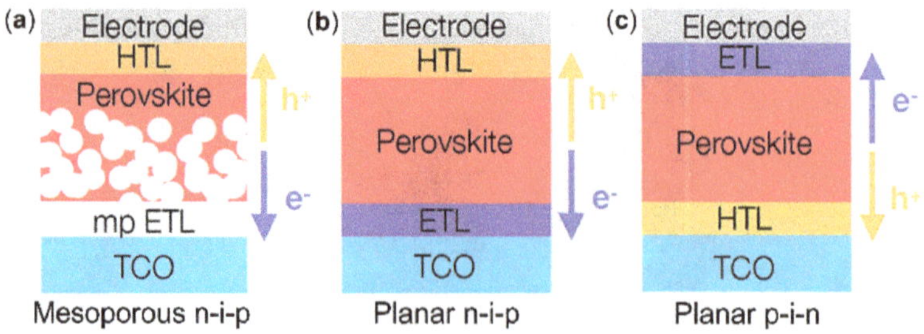


Figue 11. Schematics of three typical PSC device structures. Adapted with permission from Ref. 68.

(ii) *Electron transporting layer (ETL)*: ETL facilitates the diffusion of the photoexcited electrons from the perovskite layer into the FTO/ ITO electrode and then to the external circuit. $TiO_2$ is commonly used as the electron transporting material owing to its wide bandgap, appropriate alignment of conduction band with the perovskite for efficient electron injection, and good electron mobility.

(iii) *Light absorbing layer*: A few hundred nanometres thick halide perovskite layer, spin-coated over the ETL, serves as a light sensitizer or absorber.

(iv) *Hole transporting layer (HTL)*: HTL is adjacent to the perovskite layer which facilitates diffusion of the photoexcited holes from perovskite layer into the metallic counter electrode for extraction. Spiro-OMeTAD (2,2',7,7'-tetrakis (N,N-di-p-methoxy phenylamine)-9,9'-spirobifluorene) is the commonly used hole transporter.

(v) *Back contact/metal contact*: Typically, metals such as Au, Ag or Al are deposited on the HTL by thermal vaporization which act as the counter electrode, also known as back contact.

When light falls on the solar cell, the perovskite layer absorbs photons to produce electron-hole charge carriers (also called excitons). The electron in the valance band (VB) of the perovskite layer is excited to its conduction band leaving the perovskite layer oxidized, which is then compensated by the movement of the electron from the HOMO of the adjacent hole transporting layer. The recombination of electrons and holes within the perovskite is limited due to the crystallinity and low trap densities in perovskite. Since perovskite materials have different exciton binding energies, these excitons can form free carriers. The electrons in the CB then move to the lower lying $TiO_2$ layer and subsequently transported to the FTO by diffusion. Simultaneously, the holes from the VB move towards the HTL and then to back electrode (Au/Ag). As shown in Figue 12, the hopping of electrons and holes in opposite directions is essentially due to the difference in the work function of ETL and HTL, generating electric current connected via external circuit.[69]

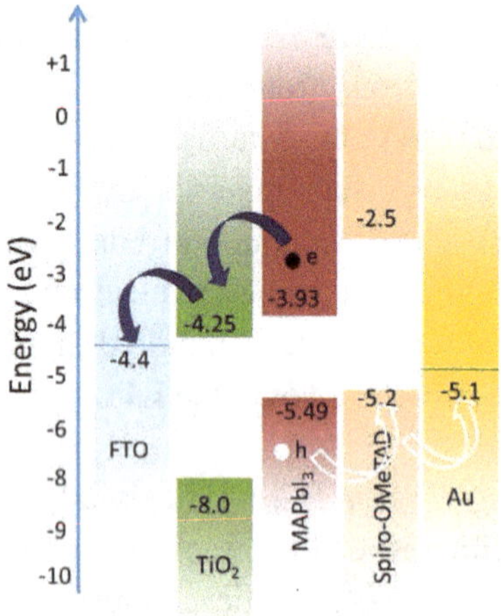


Figure 12. Energy level schematic of a planar n-i-p PSC comprised of $MAPbI_3$ as the perovskite layer, $TiO_2$ as the ETL, spiro-OMeTAD as the HTL, FTO as the front contact and Au as the back contact. Adapted with permission from Ref. 69.

## 7.2. *Single-junction Perovskite Solar Cell*

The area of single-junction PSC has seen a flurry of research activities since 2009, resulting in a rapid increase in the efficiencies over 25.7% from 3.8% (Figure 2, *vide supra*) and the notable PCE listed in Table 2. In 2009, Kojima *et al.* fabricated the first perovskite solar cells based on a DSSC structure by depositing $MAPbI_3$ onto a mesoporous $TiO_2$ film as an electron collector with a liquid electrolyte containing a mixture of LiI and $I_2$.[2] This cell showed an efficiency of 3.81%. Subsequently, the efficiency has been enhanced to 6.54% by using $MAPbI_3$ quantum dots of *ca.* 2–3 nm size on a nanocrystalline $TiO_2$ surface.[67] By optimizing the device structure with a thicker perovskite QD layer and thinner $TiO_2$ film, optimal absorption coefficient and photocurrent density have been achieved which led to enhancement in efficiency. However, in these cells, the performance rapidly degrades due to the dissolution of the perovskite layer in the liquid electrolyte. This shortcoming has been addressed by replacing the liquid electrolyte with a solid-state hole-conducting material *spiro*-MeOTAD (2,2′7,7′-tetrakis(*N*,*N*-di-*p*-methoxyphenyl amine)-9,9′-spirobifluorene.[90] A PCE of 9.7% was achieved with this all-solid-state solar cell comprising $MAPbI_3$ coated $TiO_2$ as a light absorber and *spiro*-MeOTAD as a

Table 2. Efficiencies, perovskite composition and device structure of different halide perovskite solar cells.

| S. No. | Perovskites | Device type | PCE (%) | Year | Ref |
|---|---|---|---|---|---|
| 1 | $MAPbI_3$ | liquid junction | 3.81 | 2009 | 2 |
| 2 | $MAPbI_3$ | mesoporous $TiO_2$ | 6.54 | 2011 | 67 |
| 3 | $MAPbI_3$ | mesoscopic $TiO_2$ | 9.7 | 2012 | 70 |
| 4 | $MAPbI_2Cl$ | mesoscopic $Al_2O_3$ | 10.9 | 2012 | 14 |
| 5 | $MAPbI_3$ | mesoscopic $TiO_2$ | 12.0 | 2013 | 71 |
| 7 | $MAPbI_{3-x}Cl_x$ | mesoscopic $TiO_2$ | 15.4 | 2013 | 72 |
| 6 | $MAPbI_3$ | mesoscopic $TiO_2$ | 14.1[c] | 2014 | 73 |
| 8 | $FAPbI_3$ | mesoscopic $TiO_2$ | 16.01 | 2014 | 74 |
| 8 | $MAPbI_3$ | mesoscopic $TiO_2$ | 17.01 | 2014 | 75 |
| 9 | $MAPb(I_{1-x}Br_x)_3$ | mesoscopic $TiO_2$ | 17.9[c] | 2014 | 76 |
| 10 | $MAPb(I_{1-x}Br_x)_3$ | mesoscopic $TiO_2$ | 20.1[c] | 2015 | 77 |
| 11 | $(FAPbI_3)_{0.95}(MAPbBr_3)_{0.05}$ | mesoscopic $TiO_2$ | 20.2 | 2015 | 78 |
| 12 | $MAPbI_3$ | mesoscopic $TiO_2$ | 20.4 | 2016 | 79 |
| 13 | $(FAPbI_3)_{0.85}(MAPbBr_3)_{0.15}$ | mesoscopic $TiO_2$ | 22.1[c] | 2016 | 80 |
| 14 | $Cs_x(MA_{0.17}FA_{0.83})_{(1-x)}Pb(I_{0.83}Br_{0.17})_3$ | mesoporous $TiO_2$ | 21.1 | 2016 | 81 |
| 15 | $FAPbI_3$ | mesoscopic $TiO_2$ | 22.1 | 2017 | 82 |
| 16 | $(FAPbI_3)_{0.85}(MAPbBr_3)_{0.15}$ | mesoscopic $TiO_2$ | 22.7[c] | 2018 | 83 |
| 17 | $(FAPbI_3)_{1-x}(MAPbBr_3)_x$ | planar-structure $SnO_2$ | 23.7[c] | 2019 | 84 |
| 18 | $HTAB_{0.3}(FAPbI_3)_{0.95}(MAPbBr_3)_{0.05}$ | mesoscopic $TiO_2$ | 24.2[c] | 2019 | 85 |
| 16 | $Cs_{0.05}FA_{0.85}MA_{0.10}Pb(I_{0.97}Br_{0.03})_3$ | mesoscopic $TiO_2$ | 23.5 | 2020 | 86 |
| 19 | $HTAB_{0.3}(FAPbI_3)_{0.95}(MAPbBr_3)_{0.05}$ | mesoscopic $TiO_2$ | 25.2[c] | 2020 | 87 |
| 20 | $FAPbI_3$ | mesoscopic $TiO_2$ | 25.5[c] | 2021 | 88 |
| 21 | $FAPbI_3$ | mesoscopic $TiO_2$ | 25.7[c] | 2022 | 89 |

[c]Certified values published in *Progress in Photovoltaics: Research and Applications*

MA = methylammonium; FA = formamidinium; HTAB = *n*-hexyl trimethylammonium bromide

hole-transporting layer. The device showed excellent ambient stability even without encapsulation in protecting layer. In these devices, the perovskite layer acts as the sensitizer, just like the dye in DSSCs.

Snaith and co-worker postulated that a mixed halide perovskite, $CH_3NH_3PbI_{3-x}Cl_x$ can not only act as a light absorber but also support electron transport when coated with inert scaffold such as insulating mesoporous alumina (mp-$Al_2O_3$) in a "meso-superstructured solar cell (MSSC)".[14] This meso-superstructured cell can generate open-circuit photovoltages ($V_{OC}$) of > 1.1 volts and PCE of 10.9% under simulated full sunlight, while the analogous sensitized-$TiO_2$ cell exhibits $V_{oc}$ of 0.80 V and overall PCE of 7.6%. In the case of $TiO_2$ based cell, the photoexcited electrons move to the ETL layer comprised of *n*-type $TiO_2$ (Figure 13a). On the other hand, the wide bandgap insulator (7 to 9 eV) mp-$Al_2O_3$ and perovskites layers form a type-I heterojunction. The CBM and VBM of $Al_2O_3$ straddle the perovskite band edges (Figure 13b). As a result, the photoexcited electrons remain in the perovskite layer until they are collected at the $TiO_2$-coated FTO electrode and thus must be carried through the perovskite layer. The insulating scaffold of mp-$Al_2O_3$ also contributes to the enhanced performance by minimizing energy losses at the interfaces with the perovskite layers. It was observed that replacing the $TiO_2$ with the insulating $Al_2O_3$ led to an increase in the charge transport through the perovskite layer, whilst overall photocurrent remained unaffected. Ball *et al.* further enhanced the efficiency of MSSC over 12.3% by optimizing the thickness of alumina film, processed at a much lower temperature of 150°C (Figures 13c and 13d).[91]

So far, the perovskite layer has been deposited onto conducting metal oxide layer via a single-step synthesis in which a solution of the required perovskite is drop casted. However, single-step synthesis leads to large morphological variations in the perovskite film, thereby causing variation in the solar cell efficiencies. Burschka *et al.* introduced a two-step sequential deposition of perovskite layer on mesoporous metal oxide.[92] The $MAPbI_3$ film has been fabricated via deposition of $PbI_2$ into a mesoporous $TiO_2$ substrate followed by exposing it to a solution of MAI solution. The $TiO_2$ pores are filled with $PbI_2$ in the first step which converts into perovskite during the second step. The formation of perovskite occurs quickly with much better control over morphology and high reproducibility of the photovoltaic performance. The device yielded a PCE of 15% (certified 14.1%) under standard AM1.5G sunlight. Im *et al.* showed that spin-coating of $PbI_2$ solution (20 μl in DMF) on the mesoporous $TiO_2$ film (~40 nm in diameter) followed by MAI solution of different concentrations led to the formation of $MAPbI_3$ cuboids with the average size between ~720 nm and ~90 nm depending on the MAI concentration and reaction time. The highest PCE of 17.01% was achieved with the cuboids of ~300–400 nm size as the perovskite layer with

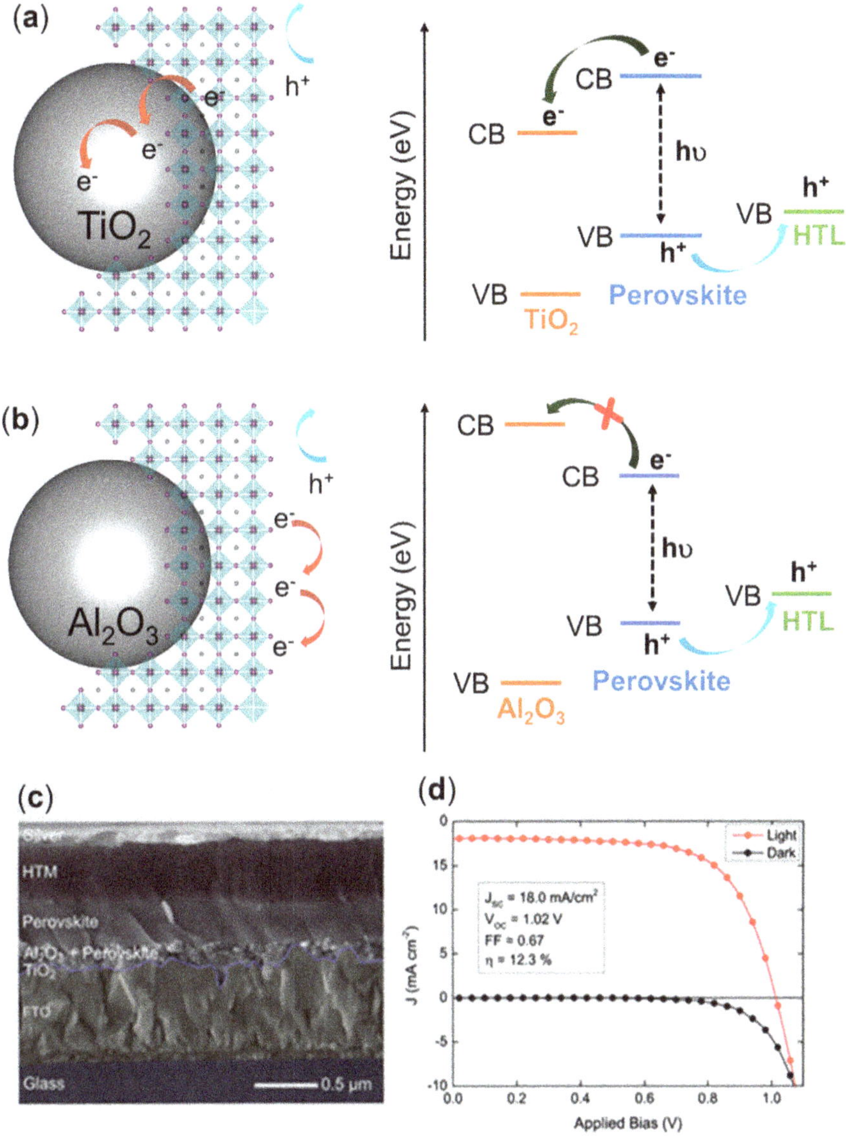


Figure 13. Schematic illustration for the charge transfer and charge transport in (a) a perovskite sensitized $TiO_2$ solar cell, (b) $Al_2O_3$-based MSSC, (c) color-enhanced and annotated cross-section SEM image showing schematic of the $Al_2O_3$-based MSSC. (d) I-V curve for the MSSC with the highest PCE of 12.3%, measured under simulated AM1.5, 100 mW $cm^{-2}$ sunlight. Panels (c) and (d) adapted with permission from Ref. 91.

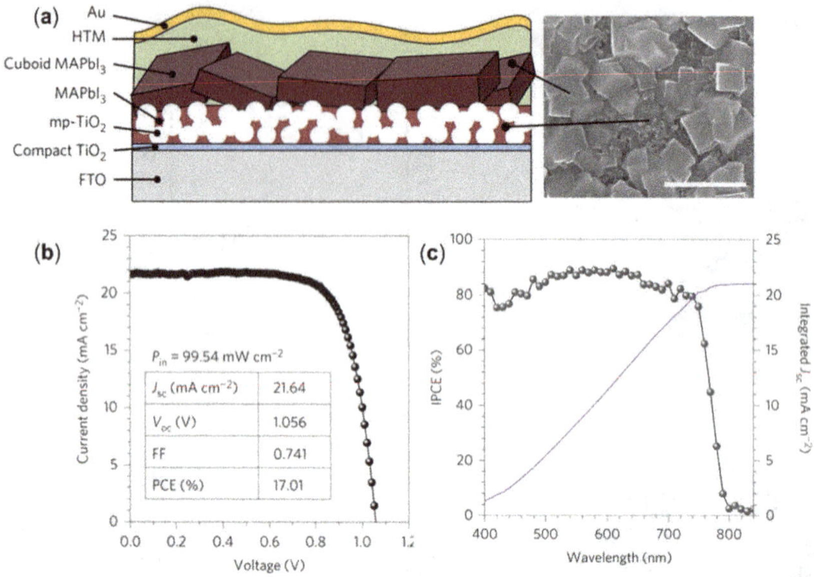


Figure 14. (a) Device schematic of $MAPbI_3$ PSC with SEM images of $MAPbI_3$ cuboid. (b) J-V curve showing PCE of 17.01% and fill factor of 74%. (c) Incident photon to electron conversion efficiency (IPCE) spectrum (black circle and line) and integrated photocurrent density curve (purple line) of $MAPbI_3$ PSC. Adapted with permission from Ref. 75.

FTO/b-$Al_2O_3$/mp-$TiO_2$/$MAPbI_3$/spiro-MeOTAD/Au device configuration (Figure 14).[75]

Although $MAPbI_3$ is a promising PV material due to the rapid growth of power conversion efficiency in recent years, ambient instability in the presence of water and oxygen molecules is still a major concern for the commercialization of perovskite solar cells. In particular, $MAPbI_3$ decomposes faster in moist air compared to $MAPbBr_3$. Studies on mixed-halide $MAPb(I_{1-x}Br_x)_3$ ($0 \leq x \leq 1$) perovskite have shown that the stability of solar cells can be greatly improved via the substitution of larger $I^-$ ions with smaller $Br^-$ ions.[93] The device fabricated with $MAPb(I_{1-x}Br_x)_3$ ($x \geq 0.2$) as light harvesters on mesoporous (mp) $TiO_2$ and polytriarylamine (PTAA) as hole-transporting materials (HTMs) exhibits a power conversion efficiency of 12.3%. The enhancement in the efficiency is attributed to better device stability and bandgap tuning over the entire visible spectrum. It is observed that the solar cell performance reduces in the case of mixed halide perovskites due to light-induced trap-state formation and reversible halide segregation upon illumination.[94]

Designing perovskite with different cations has become important because it enhances the reproducibility, stability, and efficiency of the solar cell. There have been many reports on the enhancement in efficiency and stability of PSC by introducing a small quantity of either Cs, MA or FA in the lattice. Pellet *et al.* fabricated the PSC by sequential deposition of a solution containing MA and FA cations onto $PbI_2$ infiltered mesoporous $TiO_2$ films. The device with the black phase of $MA_{0.6}FA_{0.4}PbI_3$ yielded a PCE of 14.9%, superior to their single-cation analogues.[95] Similarly, Lee *et al.* fabricated $FA_{0.9}Cs_{0.1}PbI_3$ cell via partial substitution of Cs for FA in $FAPbI_3$, giving rise to a PCE to 16.5%.[96] The enhanced PCE has been attributed to suppression of charge recombination as well as better ambient stability. Saliba *et al.* fabricated Cs/MA/FA containing triple cation perovskite $Cs_x(MA_{0.17}FA_{0.83})_{(1-x)}Pb(I_{0.83}Br_{0.17})_3$. The PSC fabricated with this perovskite exhibited an efficiency of 21.1% which slightly decreased to ~18% after operational conditions of 250 hours.[81]

As shown in Table 2, among a large number of compositions of lead iodide perovskites investigated, the higher efficiency perovskite solar cells are predominantly based on black $\alpha$-$FAPbI_3$ and its mixed compositions at the *A*-site and the *X*-site. The higher PCE in the $FAPbI_3$ based solar cells has been attributed to its smaller bandgap, long-range electron and hole diffusion lengths and higher ambient stability compared to $CsPbI_3$ and $MAPbI_3$ based devices. Furthermore, the bandgap of $FAPbI_3$ (1.48 eV) is closer to the optimal single-junction solar cell bandgap between 1.1 eV–1.4 eV. Hence, there has been a great interest in exploring $FAPbI_3$ for photovoltaic application.

Eperon *et al.* reported $FAPbI_3$ based PSC by the spin-coating technique of a single precursor solution to fabricate film for planar heterojunction PSC and achieved an efficiency of up to 14.2% with a high short-circuit current density of > 23 mA cm$^{-2}$ under simulated full sunlight.[60] The current density decreases on increasing the Br content in mixed halide $FAPbI_yBr_{3-y}$ planar heterojunction solar cells. Although the photoactive black $FAPbI_3$ is more stable than the $MAPbI_3$ analogue under ambient conditions, it is unstable in the solar cell operating temperature. To enhance the stability, Lee *et al.* fabricated a $FAPbI_3$ based photostable cell by the immersion of $PbI_2$ coated electrode in $HC(NH_2)_2I$ solution followed by annealing at different temperatures.[74] The annealing temperature was found to have a critical role in photovoltaic performance. The best efficiency of 16% was obtained with the device annealed at 150°C mainly due to the high photocurrent density ($J_{sc}$) and open-circuit voltage ($V_{oc}$). The solar cell was found to be photostable in the temperature range of 296 K–423 K.

Yang *et al.* employed an intramolecular exchange process for the fabrication of $FAPbI_3$ solar cell by introducing a triiodide ion ($I_3^-$) and a small amount of $MAPbBr_3$.[82] The presence of triiodide ions in the $FAPbI_3$ layers with small amounts of $MAPbBr_3$ suppresses the formation of deep-level defects responsible for limiting the performance of PSCs by promoting nonradiative recombination centres in perovskite layers.[82]

Lu *et al.* demonstrated a method to completely convert photoinactive yellow δ-$FAPbI_3$ to black α-$FAPbI_3$ phase by exposing it to methylammonium thiocyanate (MASCN) or formamidinium thiocyanate (FASCN) vapours at 100°C.[97] Once the α-$FAPbI_3$ is formed in pure phase, its conversion back to the δ-phase is inhibited by a high activation energy barrier. This makes the photoactive α-$FAPbI_3$ stable for a longer time, retaining the structure even after 500 hours at 85°C. The thiocyanate ions ($SCN^-$) are mostly located near the surface of the $FAPbI_3$ films and therefore facilitate stabilization of the α-$FAPbI_3$ phase below the phase transition temperature. The low-defect density device fabricated with $SnO_2$ as ETL and spiro-MeOTAD as HTL yielded a short-circuit current density of 24.4 mA/cm$^2$, open-circuit voltage of 1.165 V, fill factor of 81.3% and PCE of 23.1%.[97] The α-$FAPbI_3$ phase can also be stabilized by doping with methylenediammonium dichloride ($MDACl_2$), leading to a further increase in the PCE.[99] The open-circuit voltage ($V_{oc}$) in the case of α-$FAPbI_3$ based solar cells, still lags behind the radiative limit which suggests that the non-radiative charge recombination due to iodide vacancies need to be further suppressed. Jeong *et al.* recently fabricated high stability PSC by treating α-$FAPbI_3$ with 2% formate ($HCOO^-$), which resulted in a fivefold reduction in the non-radiative recombination rate (Figure 15).[98] Due to the small size of the formate ion, it fits into the iodide vacancies that are present at the grain boundaries and at the surface of the perovskite layer, thereby passivating iodide vacancies. The formate treated $FAPbI_3$ solar cell has shown an efficiency of 25.6% (certified 25.2%)

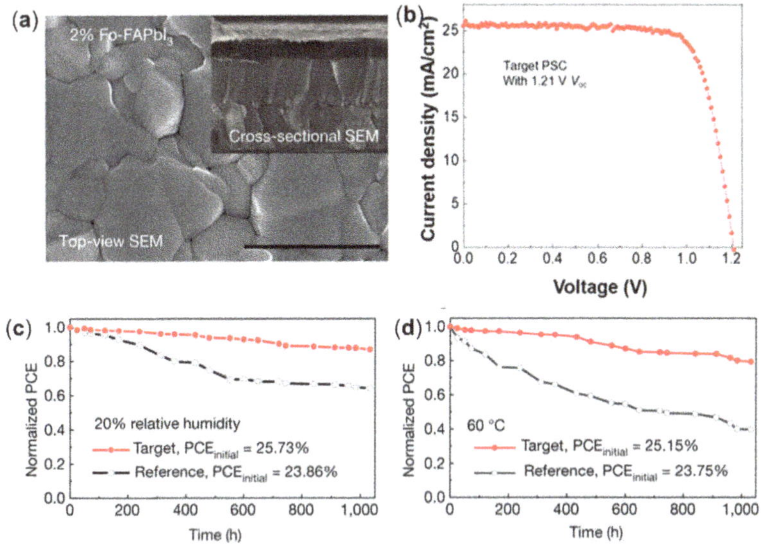


Figure 15. (a) Cross-sectional SEM image of 2% formate-doped-$FAPbI_3$, (b) corresponding J-V curve. Stability studies of $FAPbI_3$ PSC at (c) 20% relative humidity, (d) at temperature 60°C. Adapted with permission from Ref. 98.

which is the highest value among single-junction PSCs to date and close to that of the Shockley–Queisser (S–Q) theoretical limit.[98] The $V_{oc}$ of 1.21 V has been obtained, which is 96% of the S–Q limit of 1.25 V.

## 8. Perovskite/Si Tandem Solar Cell

Photovoltaics efficiency of a typical single-junction is limited by S–Q limit of 33.7% for a bandgap of 1.34 eV under 1 sun illumination.[17] For an efficient solar cell, the light absorption spectrum is expected to cover most of the solar radiation incident on the earth's surface.[100] When the photon energy is higher than the bandgap of the active layer, the excess energy is lost by the thermalisation process when the electrons relax back to the bottom of conduction band. This process also causes heating to the system. Attaining a balance between capturing more photons with less thermalisation loss limits the theoretical PSC of solar cell.[101] A tandem solar cell provides an opportunity to minimize these losses and break the S-Q efficiency limit for single-junction solar cells. A perovskite/Si tandem solar cell consists of conventionally used silicon as the low bandgap bottom cell and perovskite layer as the wide bandgap top absorber cell. Schulze *et al.* fabricated a tandem solar cell with a planar p-i-n PSCs on a textured heterojunction silicon solar cell with a perovskite composition of $FA_{0.75}Cs_{0.25}Pb(I_{0.8}Br_{0.2})_3$ having a bandgap of 1.68 eV (Figure 16).[102] ITO recombination layer connects the top perovskite layer with the bottom silicon layer. The top perovskite layer is fabricated with PTAA as HTL coated with poly(9,9-bis(30-(N,N-dimethyl)-N-ethylammoinium-propyl-2,7-fluorene)-alt-2,7-(9,9-dioctylfluorbene))dibromide (PFN-Br) as wetting agent and $C_{60}$ as ETL. To avoid damage caused during ITO sputtering, a buffer layer of $SnO_x$ is deposited on top of ETL. The bottom silicon layer has boron doped amorphous silicon and phosphorous as contacts. Ag electrode connected with ITO acts as contact pads for external electrical connections and a layer $MgF_2$ reduces the front side reflection. With this tandem configuration, the cell successfully yielded a PCE of 25.1% with a fill factor of 80%.

Al-Ashouri *et al.* have recently fabricated a monolithic perovskite/Si tandem solar using a high bandgap (1.68 eV) triple-cation top perovskite

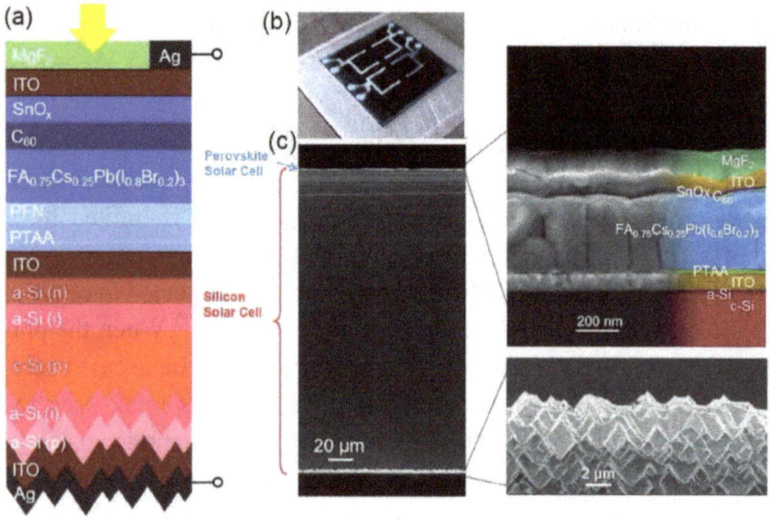


Figure 16. (a) Schematic diagram of a monolithic perovskite silicon tandem solar cell. (b) Optical image of tandem device. (c) Cross-sectional (SEM) image of tandem device with top and bottom cell magnified. Adapted with permission from Ref. 102.

monolayer of $Cs_{0.05}(FA_{0.77}MA_{0.23})_{0.95}Pb(I_{0.77}Br_{0.23})_3$.[89] In the air without encapsulation, the tandem cell retained 95% of its initial performance of 29% after 300 hours of operation.

## 9. Concluding Remarks and Future Directions

High-absorption coefficient (~$10^5$ $cm^{-1}$), long carrier diffusion length and desirable direct bandgaps (~1.5 eV) make lead halide perovskites potential candidates for next-generation thin-film PV materials. We have discussed that these properties arise from the nature of chemical bonding, crystal structure, and electronic band structures of metal halide perovskites. Due to these properties, within only 13 years (from 2009 to 2022), the power conversion efficiency of perovskite photovoltaic cell of single junction PSC crossed 25% which is comparable with the efficiency of Si solar cell. However, the efficiency is still below the Shockley-Queisser (Q-S) limit of 30%, and there is room for further improvement. Although there are challenges for lead halide perovskites due to environmental toxicity and lack of long-term ambient stability, the field is still thriving, and we anticipate many more exciting developments on solar cell devices and other optoelectronics.

## Acknowledgments

KVS and PV thank the Science & Engineering Research Board (SERB) of the Govt. of India for the Ramanujan Fellowship to Pratap Vishnoi (Award No. RJF/2020/000106).

## References

1. J. J. Yoo, S. S. Shin and J. Seo, *ACS Energy Lett.* (2022), **7**, 2084–2091.
2. A. Kojima, K. Teshima, Y. Shirai and T. Miyasaka, *J. Am. Chem. Soc.* (2009), **131**, 6050–6051.
3. P. Chen, Y. Bai, S. Wang, M. Lyu, J.-H. Yun and L. Wang, *Adv. Funct. Mater.* (2018), **28**, 1706923.
4. H. L. Wells, *Z. Anorg. Allg. Chem.* (1893), **3**, 195–210.
5. C. K. Møller, *Nature* (1958), **182**, 1436.

6. C. K. Møller, *Nature* (1957), **180**, 981–982.
7. D. Weber, *Zeitschrift für Naturforsch. B* (1978), **33**, 1443–1445.
8. R. E. Wasylishen, O. Knop and J. B. Macdonald, *Solid State Commun.* (1985), **56**, 581–582.
9. A. Poglitsch and D. Weber, *J. Chem. Phys.* (1987), **87**, 6373–6378.
10. A. Maalej, Y. Abid, A. Kallel, A. Daoud, A. Lautié and F. Romain, *Solid State Commun.* (1997), **103**, 279–284.
11. R. J. Worhatch, H. Kim, I. P. Swainson, A. L. Yonkeu and S. J. L. Billinge, *Chem. Mater.* (2008), **20**, 1272–1277.
12. I. B. Koutselas, L. Ducasse and G. C. Papavassiliou, *J. Phys. Condens. Matter* (1996), **8**, 1217–1227.
13. K. Yamada, H. Kawaguchi, T. Matsui, T. Okuda and S. Ichiba, *Bull. Chem. Soc. Jpn.* (1990), **63**, 2521–2525.
14. M. M. Lee, J. Teuscher, T. Miyasaka, T. N. Murakami and H. J. Snaith, *Science* (2012), **338**, 643–647.
15. J. Y. Kim, J.-W. Lee, H. S. Jung, H. Shin and N.-G. Park, *Chem. Rev.* (2020), **120**, 7867–7918.
16. H. Min, D. Y. Lee, J. Kim, G. Kim, K. S. Lee, J. Kim, M. J. Paik, Y. K. Kim, K. S. Kim, M. G. Kim, T. J. Shin and S. Il Seok, *Nature* (2021), **598**, 444–450.
17. W. Shockley and H. J. Queisser, *J. Appl. Phys.* (1961), **32**, 510–519.
18. A. Al-Ashouri, E. Köhnen, B. Li, A. Magomedov, H. Hempel, P. Caprioglio, J. A. Márquez, A. B. Morales Vilches, E. Kasparavicius, J. A. Smith, N. Phung, D. Menzel, M. Grischek, L. Kegelmann, D. Skroblin, C. Gollwitzer, T. Malinauskas, M. Jošt, G. Matič, B. Rech, R. Schlatmann, M. Topič, L. Korte, A. Abate, B. Stannowski, D. Neher, M. Stolterfoht, T. Unold, V. Getautis and S. Albrecht, *Science* (2020), **370**, 1300–1309.
19. L. Chouhan, S. Ghimire, C. Subrahmanyam, T. Miyasaka and V. Biju, *Chem. Soc. Rev.* (2020), **49**, 2869–2885.
20. T. Zhang, M. Yang, E. E. Benson, Z. Li, J. van de Lagemaat, J. M. Luther, Y. Yan, K. Zhu and Y. Zhao, *Chem. Commun.* (2015), **51**, 7820–7823.
21. H.-R. Xia, W.-T. Sun and L.-M. Peng, *Chem. Commun.* (2015), **51**, 13787–13790.
22. Y. Dang, Y. Liu, Y. Sun, D. Yuan, X. Liu, W. Lu, G. Liu, H. Xia and X. Tao, *Cryst. Eng. Comm.* (2015), **17**, 665–670.
23. D. Shi, V. Adinolfi, R. Comin, M. Yuan, E. Alarousu, A. Buin, Y. Chen, S. Hoogland, A. Rothenberger, K. Katsiev, Y. Losovyj, X. Zhang, P. A. Dowben, O. F. Mohammed, E. H. Sargent and O. M. Bakr, *Science* (2015), **347**, 519–522.
24. M. I. Saidaminov, A. L. Abdelhady, B. Murali, E. Alarousu, V. M. Burlakov, W. Peng, I. Dursun, L. Wang, Y. He, G. Maculan, A. Goriely, T. Wu, O. F. Mohammed and O. M. Bakr, *Nat. Commun.* (2015), **6**, 7586.
25. M. Chen, Y. Zou, L. Wu, Q. Pan, D. Yang, H. Hu, Y. Tan, Q. Zhong, Y. Xu, H. Liu, B. Sun and Q. Zhang, *Adv. Funct. Mater.* (2017), **27**, 1701121.

26. W. A. Dunlap-Shohl, Y. Zhou, N. P. Padture and D. B. Mitzi, *Chem. Rev.* (2019), **119**, 3193–3295.
27. N. J. Jeon, J. H. Noh, Y. C. Kim, W. S. Yang, S. Ryu and S. Il Seok, *Nat. Mater.* (2014), **13**, 897–903.
28. S. Chen, X. Xiao, B. Chen, L. L. Kelly, J. Zhao, Y. Lin, M. F. Toney and J. Huang, *Sci. Adv.* (2022), **7**, eabb2412.
29. C. Li, X. Lu, W. Ding, L. Feng, Y. Gao and Z. Guo, *Acta Crystallogr. Sect. B* (2008), **64**, 702–707.
30. R. J. Sutton, M. R. Filip, A. A. Haghighirad, N. Sakai, B. Wenger, F. Giustino and H. J. Snaith, *ACS Energy Lett.* (2018), **3**, 1787–1794.
31. C. C. Stoumpos, C. D. Malliakas and M. G. Kanatzidis, *Inorg. Chem.* (2013), **52**, 9019–9038.
32. C. C. Stoumpos and M. G. Kanatzidis, *Acc. Chem. Res.* (2015), **48**, 2791–2802.
33. A. Marronnier, G. Roma, S. Boyer-Richard, L. Pedesseau, J.-M. Jancu, Y. Bonnassieux, C. Katan, C. C. Stoumpos, M. G. Kanatzidis and J. Even, *ACS Nano* (2018), **12**, 3477–3486.
34. D. B. Straus, S. Guo and R. J. Cava, *J. Am. Chem. Soc.* (2019), **141**, 11435–11439.
35. G. E. Eperon, G. M. Paternò, R. J. Sutton, A. Zampetti, A. A. Haghighirad, F. Cacialli and H. J. Snaith, *J. Mater. Chem. A* (2015), **3**, 19688–19695.
36. M. Szafrański and A. Katrusiak, *J. Phys. Chem. Lett.* (2016), **7**, 3458–3466.
37. J. Breternitz, F. Lehmann, S. A. Barnett, H. Nowell and S. Schorr, *Angew. Chem. Int. Ed.* (2020), **59**, 424–428.
38. M. T. Weller, O. J. Weber, P. F. Henry, A. M. Di Pumpo and T. C. Hansen, *Chem. Commun.* (2015), **51**, 4180–4183.
39. T. Baikie, Y. Fang, J. M. Kadro, M. Schreyer, F. Wei, S. G. Mhaisalkar, M. Graetzel and T. J. White, *J. Mater. Chem. A* (2013), **1**, 5628–5641.
40. P. S. Whitfield, N. Herron, W. E. Guise, K. Page, Y. Q. Cheng, I. Milas and M. K. Crawford, *Sci. Rep.* (2016), **6**, 35685.
41. A. Jaffe, Y. Lin, C. M. Beavers, J. Voss, W. L. Mao and H. I. Karunadasa, *ACS Cent. Sci.* (2016), **2**, 201–209.
42. S. Jiang, Y. Fang, R. Li, H. Xiao, J. Crowley, C. Wang, T. J. White, W. A. Goddard III, Z. Wang, T. Baikie and J. Fang, *Angew. Chem. Int. Ed.* (2016), **55**, 6540–6544.
43. F. Hao, C. C. Stoumpos, Z. Liu, R. P. H. Chang and M. G. Kanatzidis, *J. Am. Chem. Soc.* (2014), **136**, 16411–16419.
44. M. T. Weller, O. J. Weber, J. M. Frost and A. Walsh, *J. Phys. Chem. Lett.* (2015), **6**, 3209–3212.
45. T. Chen, B. J. Foley, C. Park, C. M. Brown, L. W. Harriger, J. Lee, J. Ruff, M. Yoon, J. J. Choi and S.-H. Lee, *Sci. Adv.* (2022), **2**, e1601650.
46. D. H. Fabini, C. C. Stoumpos, G. Laurita, A. Kaltzoglou, A. G. Kontos, P. Falaras, M. G. Kanatzidis and R. Seshadri, *Angew. Chem. Int. Ed.* (2016), **55**, 15392–15396.

47. G. E. Eperon, S. N. Habisreutinger, T. Leijtens, B. J. Bruijnaers, J. J. van Franeker, D. W. deQuilettes, S. Pathak, R. J. Sutton, G. Grancini, D. S. Ginger, R. A. J. Janssen, A. Petrozza and H. J. Snaith, *ACS Nano* (2015), **9**, 9380–9393.
48. C. Müller, T. Glaser, M. Plogmeyer, M. Sendner, S. Döring, A. A. Bakulin, C. Brzuska, R. Scheer, M. S. Pshenichnikov, W. Kowalsky, A. Pucci and R. Lovrinčić, *Chem. Mater.* (2015), **27**, 7835–7841.
49. A. M. A. Leguy, Y. Hu, M. Campoy-Quiles, M. I. Alonso, O. J. Weber, P. Azarhoosh, M. van Schilfgaarde, M. T. Weller, T. Bein, J. Nelson, P. Docampo and P. R. F. Barnes, *Chem. Mater.* (2015), **27**, 3397–3407.
50. J. S. Yun, J. Kim, T. Young, R. J. Patterson, D. Kim, J. Seidel, S. Lim, M. A. Green, S. Huang and A. Ho-Baillie, *Adv. Funct. Mater.* (2018), **28**, 1705363.
51. T. Miyasaka, *Bull. Chem. Soc. Jpn.* (2018), **91**, 1058–1068.
52. J.-H. Lee, N. C. Bristowe, J. H. Lee, S.-H. Lee, P. D. Bristowe, A. K. Cheetham and H. M. Jang, *Chem. Mater.* (2016), **28**, 4259–4266.
53. R. Prasanna, A. Gold-Parker, T. Leijtens, B. Conings, A. Babayigit, H.-G. Boyen, M. F. Toney and M. D. McGehee, *J. Am. Chem. Soc.* (2017), **139**, 11117–11124.
54. T. Umebayashi, K. Asai, T. Kondo and A. Nakao, *Phys. Rev. B* (2003), **67**, 155405.
55. F. Brivio, K. T. Butler, A. Walsh and M. van Schilfgaarde, *Phys. Rev. B* (2014), **89**, 155204.
56. R. Jono and H. Segawa, *Chem. Lett.* (2019), **48**, 877–880.
57. L. Kong, G. Liu, J. Gong, Q. Hu, R. D. Schaller, P. Dera, D. Zhang, Z. Liu, W. Yang, K. Zhu, Y. Tang, C. Wang, S.-H. Wei, T. Xu and H. Mao, *Proc. Natl. Acad. Sci.* (2016), **113**, 8910–8915.
58. G. Liu, L. Kong, J. Gong, W. Yang, H. Mao, Q. Hu, Z. Liu, R. D. Schaller, D. Zhang and T. Xu, *Adv. Funct. Mater.* (2017), **27**, 1604208.
59. J. Tian, Q. Xue, Q. Yao, N. Li, C. J. Brabec and H.-L. Yip, *Adv. Energy Mater.* (2020), **10**, 2000183.
60. G. E. Eperon, S. D. Stranks, C. Menelaou, M. B. Johnston, L. M. Herz and H. J. Snaith, *Energy Environ. Sci.* (2014), **7**, 982–988.
61. H. Choi, J. Jeong, H.-B. Kim, S. Kim, B. Walker, G.-H. Kim and J. Y. Kim, *Nano Energy* (2014), 7, 80–85.
62. T. M. Brenner, D. A. Egger, L. Kronik, G. Hodes and D. Cahen, *Nat. Rev. Mater.* (2016), **1**, 15007.
63. X. Huang, X. Li, Y. Tao, S. Guo, J. Gu, H. Hong, Y. Yao, Y. Guan, Y. Gao, C. Li, X. Lü and Y. Fu, *J. Am. Chem. Soc.* (2022), **144**, 12247–12260.
64. B. Suarez, V. Gonzalez-Pedro, T. S. Ripolles, R. S. Sanchez, L. Otero and I. Mora-Sero, *J. Phys. Chem. Lett.* (2014), **5**, 1628–1635.
65. S. De Wolf, J. Holovsky, S.-J. Moon, P. Löper, B. Niesen, M. Ledinsky, F.-J. Haug, J.-H. Yum and C. Ballif, *J. Phys. Chem. Lett.* (2014), **5**, 1035–1039.
66. M. Petrović, V. Chellappan and S. Ramakrishna, *Sol. Energy* (2015), **122**, 678–699.
67. J.-H. Im, C.-R. Lee, J.-W. Lee, S.-W. Park and N.-G. Park, *Nanoscale* (2011), **3**, 4088–4093.

68. C. C. Boyd, R. Cheacharoen, T. Leijtens and M. D. McGehee, *Chem. Rev.* (2019), **119**, 3418–3451.
69. A. K. Jena, A. Kulkarni and T. Miyasaka, *Chem. Rev.* (2019), **119**, 3036–3103.
70. H.-S. Kim, C.-R. Lee, J.-H. Im, K.-B. Lee, T. Moehl, A. Marchioro, S.-J. Moon, R. Humphry-Baker, J.-H. Yum, J. E. Moser, M. Grätzel and N.-G. Park, *Sci. Rep.* (2012), **2**, 591.
71. J. H. Heo, S. H. Im, J. H. Noh, T. N. Mandal, C.-S. Lim, J. A. Chang, Y. H. Lee, H. Kim, A. Sarkar, M. K. Nazeeruddin, M. Grätzel and S. Il Seok, *Nat. Photonics* (2013), **7**, 486–491.
72. M. Liu, M. B. Johnston and H. J. Snaith, *Nature* (2013), **501**, 395–398.
73. M. A. Green, K. Emery, Y. Hishikawa, W. Warta and E. D. Dunlop, *Prog. Photovoltaics Res. Appl.* (2014), **22**, 1–9.
74. J.-W. Lee, D.-J. Seol, A.-N. Cho and N.-G. Park, *Adv. Mater.* (2014), **26**, 4991–4998.
75. J.-H. Im, I.-H. Jang, N. Pellet, M. Grätzel and N.-G. Park, *Nat. Nanotechnol.* (2014), **9**, 927–932.
76. M. A. Green, K. Emery, Y. Hishikawa, W. Warta and E. D. Dunlop, *Prog. Photovoltaics Res. Appl.* (2014), **22**, 701–710.
77. M. A. Green, K. Emery, Y. Hishikawa, W. Warta and E. D. Dunlop, *Prog. Photovoltaics Res. Appl.* (2015), **23**, 1–9.
78. W. S. Yang, J. H. Noh, N. J. Jeon, Y. C. Kim, S. Ryu, J. Seo and S. Il Seok, *Science* (2015), **348**, 1234–1237.
79. D.-Y. Son, J.-W. Lee, Y. J. Choi, I.-H. Jang, S. Lee, P. J. Yoo, H. Shin, N. Ahn, M. Choi, D. Kim and N.-G. Park, *Nat. Energy* (2016), **1**, 16081.
80. M. A. Green, K. Emery, Y. Hishikawa, W. Warta and E. D. Dunlop, *Prog. Photovoltaics Res. Appl.* (2016), **24**, 905–913.
81. M. Saliba, T. Matsui, J.-Y. Seo, K. Domanski, J.-P. Correa-Baena, M. K. Nazeeruddin, S. M. Zakeeruddin, W. Tress, A. Abate, A. Hagfeldt and M. Grätzel, *Energy Environ. Sci.* (2016), **9**, 1989–1997.
82. W. S. Yang, B.-W. Park, E. H. Jung, N. J. Jeon, Y. C. Kim, D. U. Lee, S. S. Shin, J. Seo, E. K. Kim, J. H. Noh and S. Il Seok, *Science (80-. )* (2017), **356**, 1376–1379.
83. M. A. Green, Y. Hishikawa, E. D. Dunlop, D. H. Levi, J. Hohl-Ebinger and A. W. Y. Ho-Baillie, *Prog. Photovoltaics Res. Appl.* (2018), **26**, 3–12.
84. M. A. Green, Y. Hishikawa, E. D. Dunlop, D. H. Levi, J. Hohl-Ebinger, M. Yoshita and A. W. Y. Ho-Baillie, *Prog. Photovoltaics Res. Appl.* (2019), **27**, 3–12.
85. A. Green, MA, Dunlop, ED, Levi, DH, Hohl-Ebinger, J, Yoshita, M, Ho-Baillie, *Prog. Photovoltaics Res. Appl.* (2019), **27**, 565– 575.
86. H. Zhu, Y. Liu, F. T. Eickemeyer, L. Pan, D. Ren, M. A. Ruiz-Preciado, B. Carlsen, B. Yang, X. Dong, Z. Wang, H. Liu, S. Wang, S. M. Zakeeruddin, A. Hagfeldt, M. I. Dar, X. Li and M. Grätzel, *Adv. Mater.* (2020), **32**, 1907757.
87. M. A. Green, E. D. Dunlop, J. Hohl-Ebinger, M. Yoshita, N. Kopidakis and A. W. Y. Ho-Baillie, *Prog. Photovoltaics Res. Appl.* (2020), **28**, 3–15.

88. M. Green, E. Dunlop, J. Hohl-Ebinger, M. Yoshita, N. Kopidakis and X. Hao, *Prog. Photovoltaics Res. Appl.* (2021), **29**, 3-15.
89. M. A. Green, E. D. Dunlop, J. Hohl-Ebinger, M. Yoshita, N. Kopidakis, K. Bothe, D. Hinken, M. Rauer and X. Hao, *Prog. Photovoltaics Res. Appl.* (2022), **30**, 687–701.
90. H.-S. Kim, C.-R. Lee, J.-H. Im, K.-B. Lee, T. Moehl, A. Marchioro, S.-J. Moon, R. Humphry-Baker, J.-H. Yum, J. E. Moser, M. Grätzel and N.-G. Park, *Sci. Rep.* (2012), **2** , 591.
91. J. M. Ball, M. M. Lee, A. Hey and H. J. Snaith, *Energy Environ. Sci.* (2013), **6**, 1739–1743.
92. J. Burschka, N. Pellet, S.-J. Moon, R. Humphry-Baker, P. Gao, M. K. Nazeeruddin and M. Grätzel, *Nature* (2013), **499**, 316–319.
93. J. H. Noh, S. H. Im, J. H. Heo, T. N. Mandal and S. Il Seok, *Nano Lett.* (2013), **13**, 1764–1769.
94. E. T. Hoke, D. J. Slotcavage, E. R. Dohner, A. R. Bowring, H. I. Karunadasa and M. D. McGehee, *Chem. Sci.* (2015), **6**, 613–617.
95. N. Pellet, P. Gao, G. Gregori, T.-Y. Yang, M. K. Nazeeruddin, J. Maier and M. Grätzel, *Angew. Chem. Int. Ed.* (2014), **53**, 3151–3157.
96. J.-W. Lee, D.-H. Kim, H.-S. Kim, S.-W. Seo, S. M. Cho and N.-G. Park, *Adv. Energy Mater.* (2015), **5**, 1501310.
97. H. Lu, Y. Liu, P. Ahlawat, A. Mishra, W. R. Tress, F. T. Eickemeyer, Y. Yang, F. Fu, Z. Wang, C. E. Avalos, B. I. Carlsen, A. Agarwalla, X. Zhang, X. Li, Y. Zhan, S. M. Zakeeruddin, L. Emsley, U. Rothlisberger, L. Zheng, A. Hagfeldt and M. Grätzel, *Science* (2020), **370**, eabb8985.
98. J. Jeong, M. Kim, J. Seo, H. Lu, P. Ahlawat, A. Mishra, Y. Yang, M. A. Hope, F. T. Eickemeyer, M. Kim, Y. J. Yoon, I. W. Choi, B. P. Darwich, S. J. Choi, Y. Jo, J. H. Lee, B. Walker, S. M. Zakeeruddin, L. Emsley, U. Rothlisberger, A. Hagfeldt, D. S. Kim, M. Grätzel and J. Y. Kim, *Nature* (2021), **592**, 381–385.
99. H. Min, M. Kim, S.-U. Lee, H. Kim, G. Kim, K. Choi, J. H. Lee and S. Il Seok, *Science* (2019), **366**, 749–753.
100. Z. Wang, Z. Song, Y. Yan, S. (Frank) Liu and D. Yang, *Adv. Sci.* (2019), **6**, 1801704.
101. G. E. Eperon, M. T. Hörantner and H. J. Snaith, *Nat. Rev. Chem.* (2017), **1**, 95.
102. P. S. C. Schulze, A. J. Bett, M. Bivour, P. Caprioglio, F. M. Gerspacher, Ö. Ş. Kabaklı, A. Richter, M. Stolterfoht, Q. Zhang, D. Neher, M. Hermle, H. Hillebrecht, S. W. Glunz and J. C. Goldschmidt, *Sol. RRL* (2020), **4**, 2000152.

https://doi.org/10.1142/9789811270956_0006

# Chapter 6

## Plasmon and Phonon-Polaritons in Refractory Transition Metal Nitrides

Krishna Chand Maurya[1,2,3] and Bivas Saha[1,2,3,*]

[1]*Chemistry and Physics of Materials Unit, Jawaharlal Nehru Centre for Advanced Scientific Research, Bangalore 560064, India*
[2]*International Centre for Materials Science, Jawaharlal Nehru Centre for Advanced Scientific Research, Bangalore 560064, India*
[3]*School of Advanced Materials and Sheikh Saqr Laboratory, Jawaharlal Nehru Centre for Advanced Scientific Research, Bangalore 560064, India*
[*]*bsaha@jncasr.ac.in, bivas.mat@gmail.com*

Performance and efficiencies in modern electronic and optoelectronic devices are fundamentally governed by the resistance-capacitance (RC) delay and diffraction limit. Polaritons, the hybrid quasiparticles of the photon with the electric dipole in materials, can overcome these limitations. Light coupling with collective oscillation of free carriers known as plasmon-polaritons and with polar lattice vibrations known as phonon-polaritons has attracted significant interest. However, to harness the full potential of polaritons, it is necessary and important to design material platforms that host such optical resonances with low losses and high *figures-of-merit*. Ag and Au (noble metals) are well-known plasmonic materials because of their low optical loss in the ultraviolet (UV)-to-visible spectral range. But due to several limitations such as complementary metal-oxide-semiconductor (CMOS) incompatibility, difficulty in fabrication and integration, softness, and high cost, practical applications of noble metals are limited. Similarly, phonon polariton is an alternative method for achieving mid-to-far infrared optical resonance but requires stable, scalable, CMOS compatible, and cost-effective materials for practical applications. In this chapter, we present transition metal and metal nitride-based different classes of plasmonic and phonon-polaritonic materials such as refractory transition metals (Ta, W, Mo, etc.) and transition metal mononitrides (TiN, HfN, ZrN, etc.), and polar nitride dielectrics such as AlN, GaN and bi-polaritonic ScN. Polariton formation's basic working principle, properties, and applications in several energy transport and conversion devices are highlighted.

## 1. Introduction

Materials play an important role for humanity by enabling the new scientific concept into next-generation technology. Lifestyle has completely changed from the stone era to the nanomaterial era. Semiconductor technologies and optical communication are ubiquitous and have revolutionized our information technology-based society. The volume of information processing and data storage have risen exponentially, mainly due to the reduction in the size of semiconductor components, and transistor counts have increased generally following Moore's law.[1] In semiconductor manufacturing, scaling of the electronic component from 10 μm (in 1971) to 3 nm (in 2022) semiconductor process technology has commercially reached to increase density volume density. For continuous demand for more energy efficient and faster devices, further size reduction will face several hurdles such as gate leakage, Ohmic loss (RC delay) and increment in power consumption. On the other hand, in photonic devices such as optical fiber, electromagnetic (EM) wave (or light) is the carrier, and it is well known that nothing can move faster than light. But the problem with the photonic devices is their size since the EM wave cannot be confined directly below the diffraction limit. So, to overcome these limitations and to improve the existing technology, a methodology is required where the information can be processed at the speed of light in small-scale dimensions to achieve high volume density. Also, the host materials should be compatible with the existing electronic and optoelectronic device technologies. Polaritons, half-light-half-matter quasiparticles, are promising candidates to overcome the fundamental limitations. When light couples with collective oscillation of free carriers in metals, they are termed as plasmon-polariton (or simply called 'plasmon'), while light coupled to polar lattice vibrations are known as the phonon-polaritons.

Polaritonics has recently been the subject of significant attention, but some of its phenomena have been used for decades. For example, in the ancient Roman era, the Lycurgus Cup (see Figure 1) shows vibrant colors due to the optical plasmonic resonances induced in the metal nanoparticles embedded in the glass. In the 19th century, Michael Faraday systematically studied the synthesis and optical response of colloidal gold. But researchers could not identify their underlying color-controlling

Figure 1. Ancient Roman Lycurgus cup. It appears green in reflected light while red in transmitted light. These glasses are mixed with gold nanoparticles (typically 10–100 nm in size).[14]

surface plasmonic principles until the 1950s. In 1970, after the discovery of surface-enhanced Raman scattering research on plasmonics was boosted.[2] Light interacts with free electrons in metals, consequently, a surface wave of charge density propagates on the metal's surface. This surface wave was named as surface plasmon wave that enabled the light confinement below the diffraction limit. Light can strongly couple with the metals via surface plasmon, and it has two important consequences: first, light can be confined below the diffraction limit, and second, the local electromagnetic field enhances by many orders of magnitudes due to the presence of plasmon excitation. However, in recent years, plasmonic materials are essential[3,4] for subwavelength imaging, plasmon-induced hot carrier,[5] electro-thermo-plasmonic nanotweezer,[6] efficient absorbers,[7] hot-electron generation,[8] nanoparticle trapping,[9,10] photo-catalysis,[11] solar-thermophotovoltaics,[12,13] etc.

Metals are also known as plasmonic materials because of free electron oscillation (bulk plasmon) interaction with light. Dielectric permittivity ($\varepsilon$) is a complex parameter which governs by the optical response of the materials. Its real part ($\varepsilon_1$) signifies the polarization strength while the imaginary part ($\varepsilon_2$) is related to the optical losses, originating due to the interband or intraband transition in the materials. Plasmonic property of the metal can be explained by the Drude model[15]

$$\varepsilon = \varepsilon_1 + i\varepsilon_2 = \varepsilon_\infty - \frac{\omega_p^2}{\omega^2 + i\gamma\omega} \quad (1)$$

Here $\varepsilon_\infty$ is the dielectric constant at high frequency, $\omega_p$ and $\gamma$ are the plasma frequency ($\omega_p = \sqrt{\frac{ne^2}{m^*\epsilon_0}}$, $n$ is carrier concentration, $m^*$ is effective mass) and damping factor (describing the scattering and ohmic losses), respectively. Below the plasma frequency (or above the corresponding wavelength), metals exhibit a negative real permittivity ($\varepsilon_1$) and the optical losses ($\varepsilon_2$) that arise from the electronic transitions (mostly intraband transition). Optical loss is the crucial factor limiting the performance of optical devices.

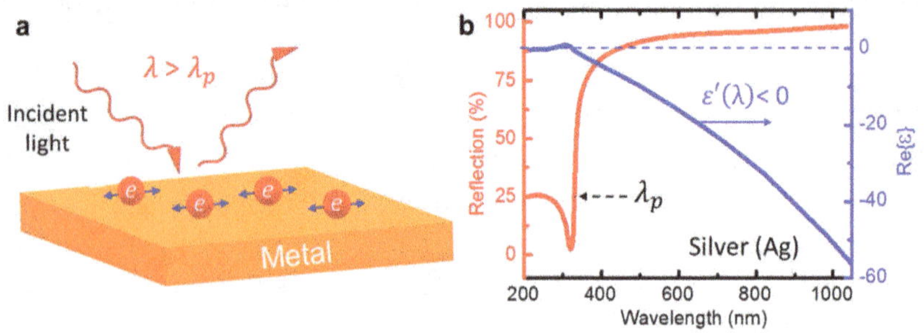


Figure 2. (a) Schematics for interaction of light with free electrons in metal (bulk plasmon). (b) Reflection curve of silver along with its real part of dielectric permittivity ($\varepsilon_1$). Beyond 330 nm, Ag is highly reflective, and $\varepsilon_1$ turns from positive-to-negative, this crossover wavelength is known as plasma wavelength ($\lambda_p$) and corresponding frequency as plasma frequency.

Due to the discrepancy in momentum between free space light and the free electrons, the surface polariton mode cannot be driven directly on the surface of the metal with light. However, there are several methods to probe them by overcoming the momentum mismatch problem. Surface polariton modes can be excited through a prism (Kretschmann configuration is very common (see Figure 3(a))) or through the fabrication of the nanostructure (see Figure 3(b)) of the polaritonic medium. In general, two types of surface waves are observed, first, propagating

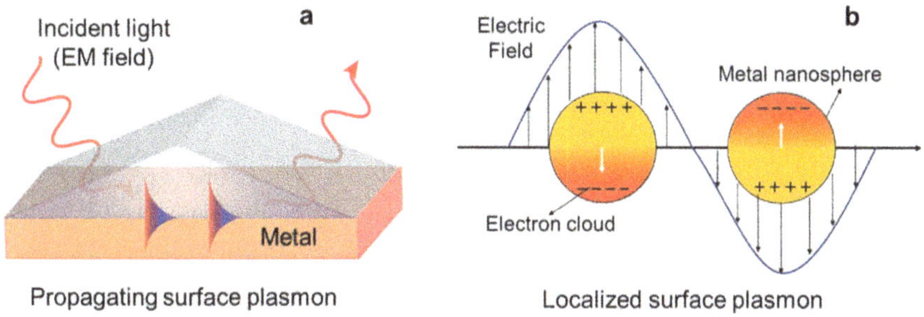


Figure 3. Schematics diagram for the types of surface modes excitation: (a) excitation of surface-plasmon-plasmon in the Kretschmann configuration, (b) excitation of localized surface-plasmons in the nanostructures.

surface polariton wave at metal/dielectric interface, and second, localized surface wave in nanostructures.

In general, the plasma frequency of metals lies between the UV-visible range, and by doping, it can be tuned up to near-infrared region (NIR). Metal and doped semiconductors can be used for plasmonic application only in the UV-NIR range (depending on the materials), but they are not very efficient for mid-to-far infrared range because of high optical losses. So, apart from the plasmon-polariton, phonon polariton is used for mid-to-far infrared nanophotonic applications. Phonon-polariton is the hybridization of light and ionic oscillations (optical phonon) in polar dielectric materials such as AlN, SiC, GaN, 2D van der Waals materials etc. The polar lattice behaves like a dipole and oscillates collectively to oppose the incident light of a certain wavelength range (see Figure 4(a)). Consequently, a screening of incident photons produces a spectral region of high reflectivity between the longitudinal optical (LO) and transverse optical (TO) phonon modes (see Figure 4(b)), the gap between the LO and TO phonon is referred to as the Reststrahlen band. The permittivity contributions of optical phonons (LO and TO) can be expressed as:[16]

$$\varepsilon_{PhP} = \varepsilon_b\left(1 + \frac{{\omega_{LO}}^2 - {\omega_{TO}}^2}{{\omega_{TO}}^2 - \omega^2 - i\omega\gamma}\right)$$

Here $\omega_{LO}$ and $\omega_{TO}$ are the LO and TO phonon frequencies, respectively, and $\gamma$ is the damping of the phonons, $\varepsilon_b$ is the background permittivity due to bound charges.

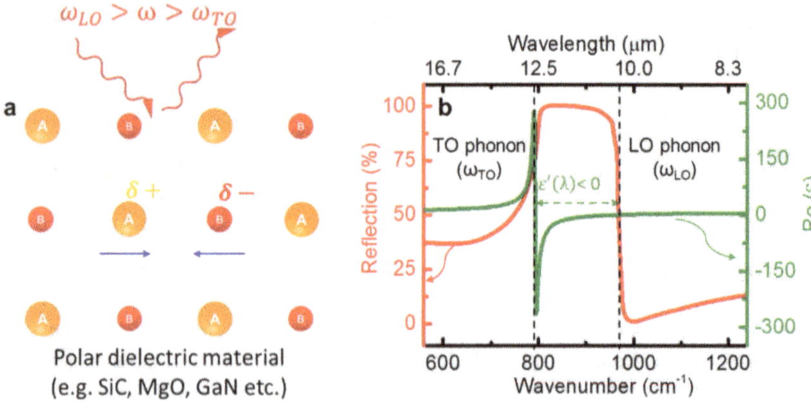


Figure 4. (a) Schematics for interaction of light with ionic lattice of polar dielectric material. (b) Reflection curve of a polar dielectric material along with its real part of the dielectric permittivity. Polar dielectric material shows high reflectivity between the LO and TO phonon frequencies, also known as the Reststrahlen band.

Also, light can be confined below the diffraction limit via excitation of surface-phonon-polariton within the Reststrahlen band, where the permittivity becomes negative. Along with that, the lifetimes of phonon modes (~ picosecond) are longer than the free carriers (~ femtosecond). So, phonon-polariton is another approach to employing low-loss alternative to plasmonic for mid-infrared nanophotonic applications.

## 2. Plasmon Polariton in Refractory Metals

### 2.1. *Plasmon polariton in refractory transition metals*

Au and Ag (Noble metals) are one of the most common plasmonic material[17] in the visible to the near-UV spectral range due to low optical loss and high carrier concentrations ($\geq 10^{21}$ cm$^{-3}$), also exhibit a sharp dip in the reflection curve at plasma frequency (see Figure 2b). But due to softness,[18] low melting temperature of nanostructures, and CMOS incompatibility, the noble metals are not suitable for practical refractory applications.[19,20] Refractory transition metals (RTMs) such as Ta, W and Mo are widely used for hard coating, adhesive layer and high-temperature applications.[21] Also, due to high conductivity and compatibility with Si technology, RTMs have the potential to be promising alternative plasmonic materials.[22]

In Figure 5(a), the dielectric permittivity of sputter grown RTMs (Ta, W, Mo) is shown along with noble metals (Ag and Au). The permittivity is obtained by ellipsometer measurements (see Ref. 22 for details). Compared to the noble metals, where $\varepsilon_1$ decreased monotonically and sharply at the plasma frequency, RTMs exhibit comparatively flat $\varepsilon_1$ in the UV-to-visible spectral range, and a subsequent monotonic and sharp decrease in the near-IR regions. However, compared to the Au and Ag, $\varepsilon_1$ of RTMs are smaller which is expected due to their lower carrier densities, and $\varepsilon_2$ are relatively larger. Traditionally, the plasmonic material with negative real component of the permittivity ($\varepsilon_1$) and very small optical losses ($\varepsilon_2$) were regarded as a good plasmonic material. But recently, plasmonic materials with high optical loss are utilized for for the confinement and impedance effects,[23] such as loss-induced heating,[24] plasmon-induced hot carrier,[5] thermo-plasmonic nanotweezer[6] etc.

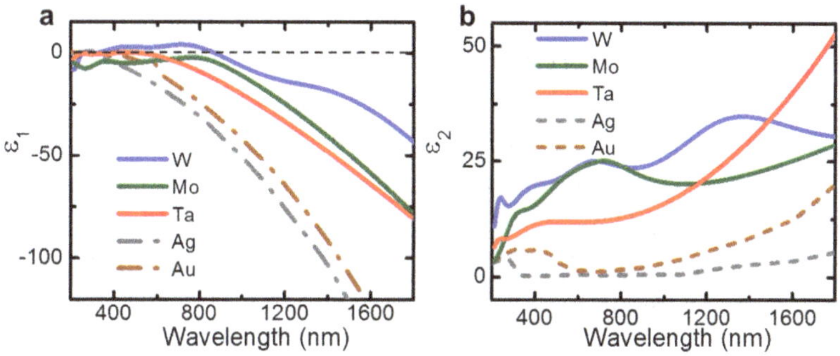


Figure 5. (a) The real ($\varepsilon_1$) and imaginary ($\varepsilon_2$) components of the dielectric permittivity sputter-deposited Ta, W and Mo thin films are compared with the optical parameters of noble metals (such as Au[17] and Ag[17]). Reproduced with permission from Maurya *et al.*, *Appl. Phys. Lett.* **118**, 041902 (2021).[22]

In order to demonstrate the quality of metal RTM films grown by sputtering reflection, high-resolution X-ray diffraction (HRXRD) and atomic force microscopy (AFM) are presented (see Figure 6 for Ta that is representative for W and Mo). The peak of HRXRD at 38.26° representing (110) oriented of Ta growth on (002) MgO substrates (see Figure 6(a)). The full-width-half-maxima (FWHM) of the rocking curve is found $2.4^0$

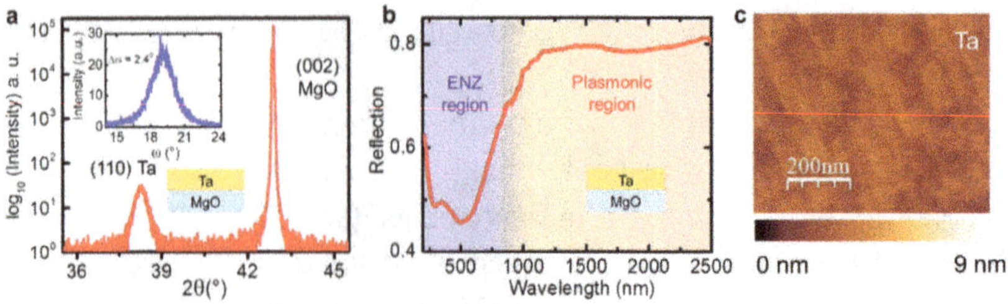


Figure 6. (a) Symmetric 2θ-ω X-ray diffractogram of Ta thin film on MgO substrate and its rocking curve (ω-scan) with 2.4$^0$ FWHM in the inset. The growth orientation is with (110) on (002) MgO substrate. (b) Reflectivity of Ta film exhibits a sharp dip close to ENZ regions. (c) Atomic force micrographs (AFM) of exhibit rms. surface roughness of 0.5 nm in Ta film. Reproduced with permission from Maurya *et al.*, *Appl. Phys. Lett.* **118**, 041902 (2021).[22]

in the Ta/MgO film that representing its highly crystalline nature. XRD data showed that Ta thin film grown at high temperature crystallizes in body-centred cubic α-Ta phase while β-Ta is usually obtained at low temperatures. High reflectivity (~ 80%) is obtained in the near-IR spectral range with a broad dip from the Ta thin film.

In conclusion, highly crystalline transition metals (Ta, W, and Mo) deposited from the sputtering show the plasmonic nature from the visible-to-near IR spectral range. Since RTMs exhibit, higher optical losses compared to noble metals, but they can still be useful for confinement and impedance effects. The improvement in the optical response of refractory transition metals (Ta, W, Mo) marked an important step in the field of plasmonic and nanophotonic materials research.

## 2.2. *Plasmon polariton in refractory transition metal nitrides*

In the previous section, we saw the plasmonic property of transition metals (Ta, Mo, W) with high optical loss. But for many applications such as optical nano-circuits,[25] photon tunneling,[26,27] super-coupling,[28] extreme nonlinear interactions,[29] ultrafast switching,[30] etc., plasmonic materials with low optical loss are required. Metals from the transition metal nitrides family such as titanium nitride (TiN), hafnium nitride, and zirconium nitride (ZrN) emerged as promising alternatives to noble metals.[31–35] Since these nitrides have similar plasmonic properties, only TiN properties are discussed further.

TiN has high melting temperature, high electrical conductivity, and chemical and mechanical stability, which make it an excellent plasmonic material, especially for high-temperature applications.[36,37] In addition, TiN exhibits a low surface energy of 24–39 mJ/m$^2$,[38] which allows uniform, smooth surfaces and ultrathin film deposition. Also, TiN is CMOS compatible and already has been used as a buffer layer in the integrated circuit (IC) fabrication process.

In this section, the plasmonic response of MBE-grown single-crystalline TiN thin film is discussed. In Figure 7, the permittivity of TiN film grown on MgO and $Al_2O_3$ substrates exhibit a crossover wavelength (transition from positive-to-negative $\varepsilon_1$ representative of the metallic nature) near 470 nm. The optical properties depend on the growth parameter, growth condition and substrate as well. In the present samples, films grown on the MgO substrate have higher carrier concentrations in comparison to the films deposited on the Al2O3 substrate, which results in high $\varepsilon_1$ (magnitude) in TiN film on the MgO substrate. Intraband transition dominant above the visible spectral range, results in a monotonic increment in the optical loss ($\varepsilon_2$) (see Figure 7(b)). The Drude damping is found smaller in TiN/$Al_2O_3$ in comparison to TiN/MgO, resulting in a lower optical loss ($\varepsilon_2$). The lower damping appeared in the TiN/$Al_2O_3$ film because of the better crystal quality, which was verified by the lower FWHM of the rocking curve in HRXRD (see Figure 8). The optical loss ($\varepsilon_2$) in TiN/$Al_2O_3$, in most parts of the visible to the near-IR spectral range, is lower than Au.

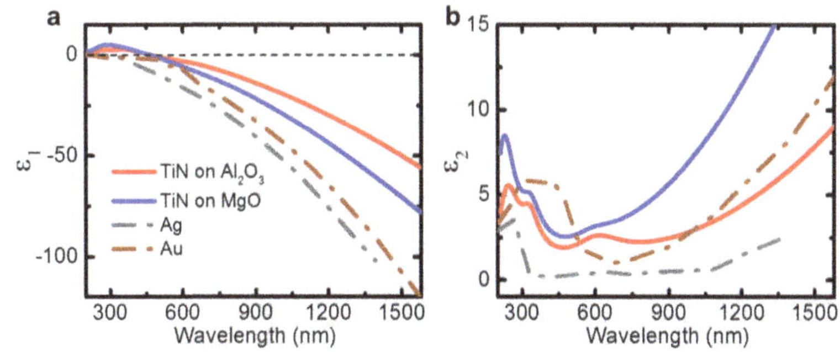


Figure 7. The complex dielectric permittivity of MBE-grown TiN thin film on MgO and Al2O3 substrates. (a) The real ($\varepsilon_1$) and (b) imaginary ($\varepsilon_2$) parts are presented. Reproduced with permission from Maurya *et al.*, *Optical Materials Express* 2679, (2020).[34]

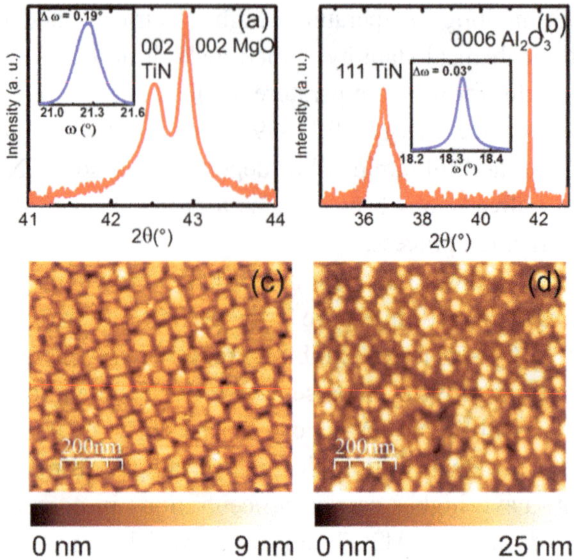


Figure 8. HRXRD curve of TiN/MgO (a) and TiN/$Al_2O_3$ (b) thin films and their rocking curve ($\omega$-scan) in the inset. Atomic force micrographs of TiN/MgO (c) and TiN/$Al_2O_3$ (d). Reproduced with permission from Maurya *et al.*, *Optical Materials Express* 2679, (2020).[34]

HRXRD spectra of TiN films are shown in Figure 8, TiN thin film grow along 002 orientation on the (001) MgO substrate, while 111 orientations on the (0001) Al2O3 substrates. The FWHM of the rocking curve was found at $0.19^0$ for TiN/MgO and $0.03^0$ for TiN/$Al_2O_3$ sample. That indicates that TiN/$Al_2O_3$ has superior crystalline nature to TiN/MgO. The lattice parameter (*c*-axis) was found to be 4.25Å and 4.24 Å for TiN/MgO and TiN/$Al_2O_3$, respectively. Plan-view atomic force microscopy (AFM) images showed square-shaped features (see Figure 8(c)) for TiN/MgO film while pyramidal shapes (see Figure 8(d)) for TiN/Al2O3 film. The surface RMS-roughness were 1.2 nm and 3.8 nm for TiN/MgO film and TiN/$Al_2O_3$ film, respectively.

In Figure 9, the Reflection (R) and transmission (T) curves of TiN film, as a function of the angle of incidence, are presented. Near plasma frequency (~430 nm), TiN showed a peak in the transmission curve (see Figure 9(a)) while a dip in the reflection curve (see Figure 9(b)).

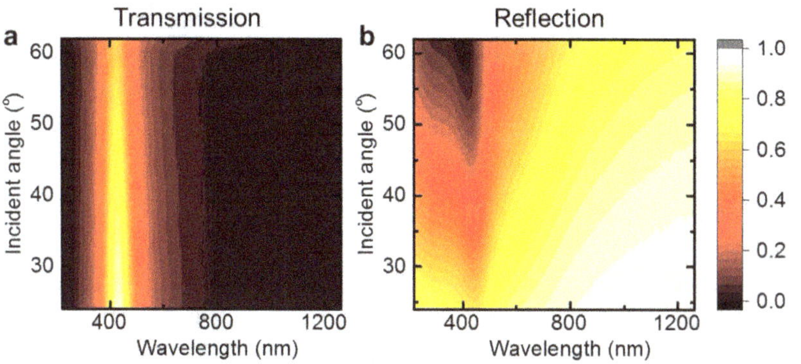


Figure 9. Angle and wavelength dependent (a) transmission and (b) reflection spectrum of TiN film. Reproduced with permission from Maurya *et al.*, *Optical Materials Express* 2679, (2020).[34]

In Figure 10, the plasmonic property of transition metal (W, Ta, Mo), transition metal nitride (TiN, ZrN, HfN) and noble metals (Ag, Au, Al) are compared. Optical loss at their crossover wavelength is compared. The optical property strongly depends on the growth parameters. For example, RTMs grown at room temperature (RT-RTMs) exhibit lower optical loss but their optical response (real part of permittivity or $\varepsilon_1$) is not very strong.

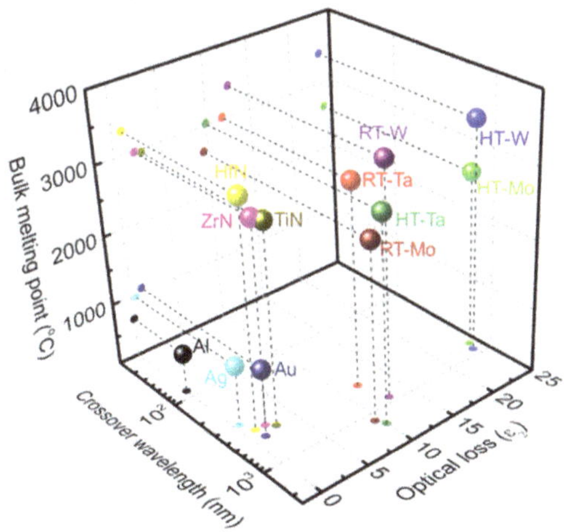


Figure 10. Comparison of metals with their optical loss at crossover wavelength and their bulk meting point. Reproduced with permission from Maurya *et al.*, *Appl. Phys. Lett.* **118**, 041902 (2021).[22]

While RTMs grown at high temperature (700°C) (HT-RTMs) have strong optical response (large negative $\varepsilon_1$) and their optical loss is also high. So, there is a trade-off between $\varepsilon_1$ and $\varepsilon_2$ that can be easily controlled by the growth condition according to the required application.

## 3. Phonon Polaritons in Dielectrics

As discussed earlier, metal plasmonic can be only applicable for the UV-to-visible range and doped semiconductor (lower carrier concentration) can push such limits to the near-infrared region. Infrared light confinement and electric field enhancement via plasmonics is challenging. Optical phonons are the alternative carrier to overcome this problem. In polar dielectric materials such as AlN,[39] SiC,[40] GaN,[41] MgO[42] etc., infrared light couples with the ionic oscillation (optical phonon) to form phonon-polariton. Since, GaN thin film has wurtzite crystal structure which belongs to $C_{6V}^4$ ($P6_3mc$) space group and according to group theory only A1(Z), E1(X, Y), and E2 phonon modes are Raman active.[43,44] So, in the Raman spectra, peaks appear at 563 cm$^{-1}$ and 739 cm$^{-1}$ which correspond to the (transverse optical) TO ($E_1$) and (longitudinal optical) LO ($A_1$) phonon respectively (see Figure 11(a)). GaN film shows high reflective band in the region between LO and TO phonon frequency which corresponds to the Reststrahlen band, the region where light couples with the optical phonon to form the phonon polaritons.[16] Like plasmonic material, the real part dielectric permittivity of the GaN film shows

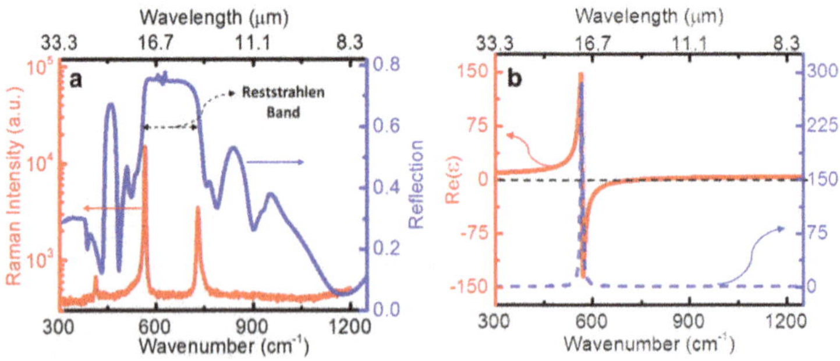


Figure 11. (a) FTIR reflection (blue curve) and Raman (red curve) spectra obtained from a GaN epitaxial layer. (b) Dielectric permittivity (ε) of GaN thin film.

negative value in the Reststrahlen band (see Figure 11(b)) which is the necessary condition to excite the surface phonon polariton.
Reststrahlen band of polar dielectric material open the window for infrared photonics and emerging applications such as radiative cooling,[45,46] telecommunication,[47] sensing,[48] thermophotovoltaics,[13,49,50] surveillance[51] etc.[52]

## 4. Bi-Polariton in Degenerate Semiconductor

As we saw in previous sections, metals are used to achieve plasmon-polaritons while polar dielectrics (or insulator and wide bandgap semi-conductor) are utilized for phonon-polaritons. However, it is difficult to demonstrate low-loss plasmon and phonon polariton in one host material due to mutually contradictory electronic configuration requirements. In this section, we shall show that ScN a degenerate semiconductor, can have both types of plasmon and phonon polaritons. There are several applications such as radiative cooling,[45,46] thermo-photovoltaics,[13,49,50] where multiple resonances in the different electro-magnetic spectrums are required. Since plasmon resonance occurs in the UV-NIR range primarily, phonon polariton excitation takes shape in the mid-far infrared, harnessing both the polaritons simultaneously which can be useful for these types of multifunctional applications.
ScN is a degenerate semiconductor with carrier concentration varying from $10^{18}$ to $10^{21}$ $cm^{-3}$.[53–55] Since scandium has a high affinity for oxygen, ScN films usually contain oxygen as impurities. Oxygen impurities incorporated into ScN films influence the electronic as well as plasmonic properties. In Figure 12, the plasmonic property of ScN film is shown where ScN film exhibits plasma frequency (where $\varepsilon_1 \rightarrow 0$) at 1.83 μm and beyond this wavelength, ScN behaves like a Drude metal ($\varepsilon_1 < 0$). At the plasma frequency, optical loss ($\varepsilon_2$) in ScN is 1.2, which is much lower than the conventional plasmonic material such as Ag and Au. The value of plasma frequency is further confirmed by reflectivity measurement (plotted as a ratio between p-polarized and s-polarized intensity), which showed a dip at $\lambda_p$ (see Figure 12(b)).

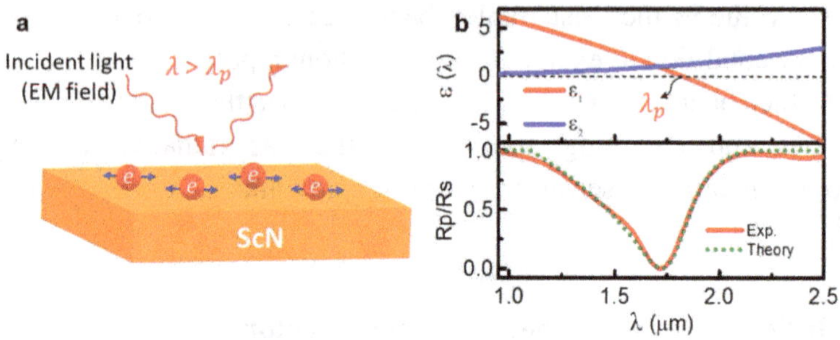


Figure 12. (a) Schematic diagram for the plasmon response in the ScN thin film. (b) The complex dielectric permittivity (the real ($\varepsilon_1$) and imaginary ($\varepsilon_2$) components) of ScN film with the carrier concentration of $1.6 \times 10^{21}$ cm$^{-3}$ is presented. ScN showed a plasmon response at 1.83 μm which corresponds to the plasma frequency. Further, the plasma frequency is verified by the reflectivity measurement (plotted as the ratio between the *p*-intensity to *s*-intensity) which shows a dip near $\lambda_p$. Reproduced with permission from Maurya *et al.*, *Nano Lett.* 2022, **22** (13), 5182–5190.[56]

Apart from the plasmon-polariton, ScN film also exhibits the phonon polariton. ScN film shows the high reflective band between LO (677 cm$^{-1}$) and TO (340 cm$^{-1}$) phonon, the region where a light interacts with the optical phonon to form the phonon polaritons (PhP). The dielectric permittivity of the ScN film shows that the real part of dielectric permittivity has a negative value in the Reststrahlen band, which is the necessary condition to excite the surface phonon polariton.

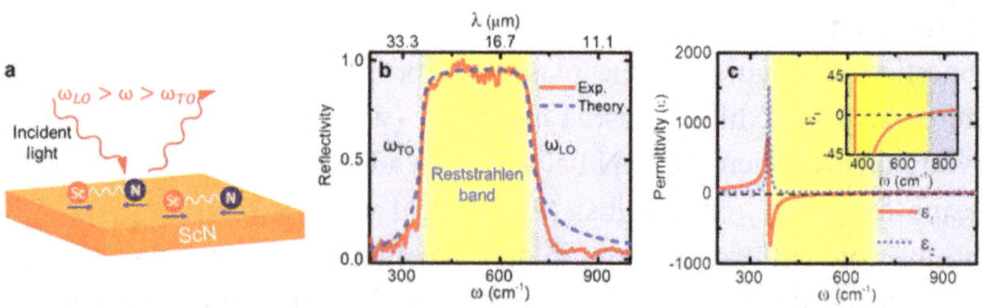


Figure 13. (a) Schematic diagram for the phonon response in the ScN thin film. (b) The normalized FTIR reflection (red) spectrum of ScN film is shown along with the calculated reflection spectrum (in blue dotted line). (c) Calculated dielectric permittivity (ε) of ScN thin film showing negative value within the Reststrahlen band. Reproduced with permission from Maurya *et al.*, *Nano Lett.* 2022, **22** (13), 5182–5190.[56]

Therefore, ScN can host both the plasmon-polariton in the short-wavelength infrared and the phonon-polariton in the mid-far infrared spectrum.

## 5. Applications

The present-day energy crisis stems from the reckonable end of fossil fuels. In addition, the emission of greenhouse gases from conventional fuel, such as carbon dioxide (CO2), causes global warming and damage to the environment and biodiversity. Also, as global energy consumption increases, the shortage of fossil fuels will be felt in the coming decades. So, renewable energy harvesting from the sun as a source has generated a significant interest over the past decades. Several applications such as solar thermophotovoltaics, passive radiative cooling, solar steam generation etc. are promising technologies for sustainable production of renewable energies. In these technologies, optical control is very important to achieve high efficiency in two complementary systems, based on absorption/emission of light in a particular wavelength range. Polariton based artificial structures (or metamaterials) have great potential for spectral control for energy harvesting devices.

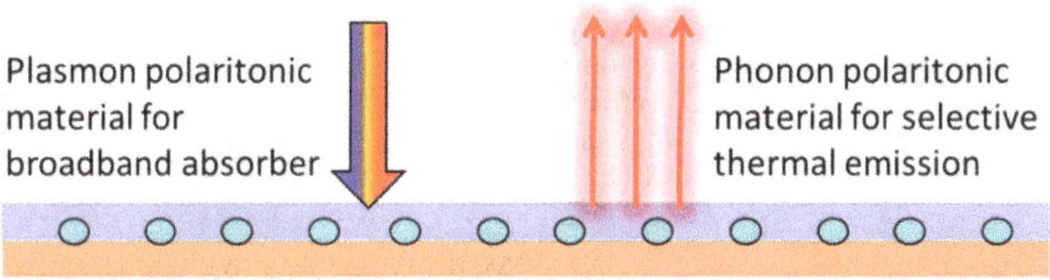


Figure 14. Schematic diagram for plasmon and phonon polaritonic metamaterial for multifunctional application.

### 5.1. *Solar-thermo-photovoltaics*

Photovoltaics (PV) are the most prevalent way for solar energy-harvesting technology using semiconductor *pn* junction. Despite recent advances, the efficiency of single-junction PV cells is still below the basic limit predicted by Shockley and Queisser.[57,58] This low efficiency arises due to energy losses below-bandgap photons and hot-carrier thermalization. Solar thermophotovoltaics (STPV) is a promising alternative to traditional photovoltaics, where an intermediate structure of the absorber/emitter first absorbs the entire spectrum of light, it heats, and then thermal radiation is emitted narrowly (equal to the bandgap of PV cell) to PV cells to stimulate carriers for electricity generation. The theoretical efficiency of STPV is 85.4% higher than that of traditional photovoltaics.[58] However, experimental efficiency is still less than 7% due to various thermal and optical losses.[59]

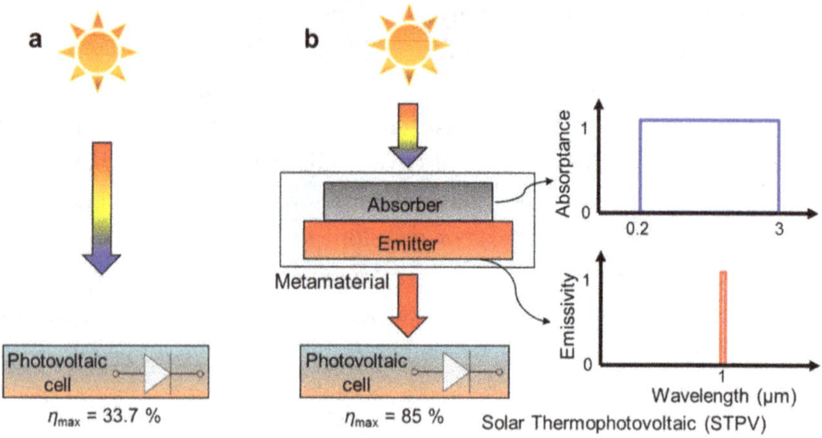


Figure 15. Schematic diagram for (a) photovoltaic and (b) solar-thermo-photovoltaic (STPV) cell. Maximum theoretical efficiency for a single PV cell is 33.7% while, for STPV is 85%. The basic principle of STPV is to absorb whole solar spectrum from 200 nm to 3 μm and emit selective wavelength equal to the wavelength of the bandgap of the PV cell.

Transition metals (Ta, W and Mo)-based metasurface will be useful for solar absorbers design to exhibit high absorption[13,22] where the solar spectral radiation is significant. While for wavelength-selective emitter, the same refractory transition metal and metal nitride can be used. So, hybrid structure of plasmon and phonon polaritonic material can be utilized for broadband absorber and selective emitter for STPV.

### 5.2. *Radiative cooling*

Though the sun is a most abundant heat source at a temperature close to 6,000K, if we just look away from the Sun, we will find that the dark universe is extremely cold which is at a temperature of ~ 4K. From a basic thermodynamic point of view, something that is cold is also extremely important. If we think about just a Carnot efficiency conversion limit, it certainly will be great if all the energy conversion process that is currently going on, does not use the earth as the heat sink, rather use the universe as a heat sink. The molecules in the atmosphere absorb some of that heat and send it back. That is the greenhouse effect which is responsible for the climate change. But the critical fact is, our atmosphere does not absorb all that heat. If it did, we would be on a much warmer planet. At certain wavelengths, between 8–13 microns, our atmosphere has, what is known as, a transmission window.[60] This window allows some of the heat that goes up as infrared light to effectively escape.

During peak summer, the sun heats up the things and around 17% of global electricity (for air conditioning) is consumed just to cool the surrounding that also cost equivalent to the 8% of total global greenhouse gas emission. Radiation cooling is a technique in which the surface emits heat as infrared radiation in the atmospheric window but is inefficient under strong sunlight during the daytime. Due to tunable optical properties of metamaterials,[61] daytime radiative cooling is possible. The key function of metamaterial-based radiative cooling is: The main function of radiative cooling based on metamaterials is as follows: (a) solar spectrum reflection (0.3–3 μm) and (b) thermal emission in the atmospheric window (8–13 μm).[60]

Noble metals such as Ag, Au, Al, etc., are well known for their reflective property but they are not practically applicable due to their softness, low melting temperature, are unstable in harsh conditions and expensive. On the other hand, refractory metals and metal nitrides (W, Ta, TiN, ZrN, etc.) are mechanically hard, exhibit high melting temperature, are stable in harsh conditions and are relatively inexpensive. As a result, they are suitable to be the reflective component of metamaterial. Semiconducting and/or dielectric (such as GaN, ScN, MgO, $SiO_2$, etc.) nano-spheres with variable diameters embedded inside transparent polymer matrix can be engineered for emission in the atmospheric window.

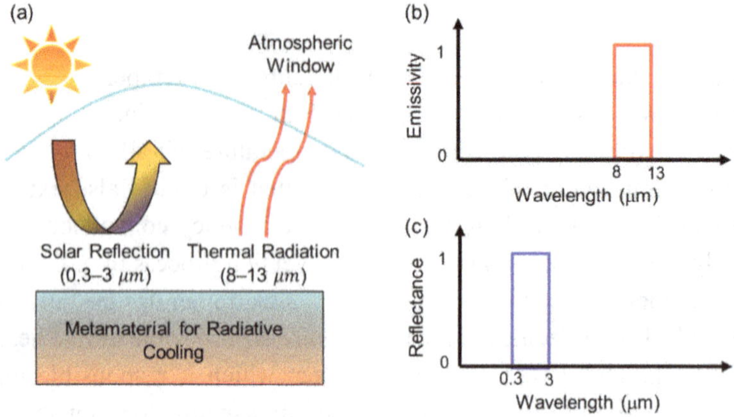


Figure 16. (a) The basic principles of radiative cooling. (b) Emission and (c) reflection properties for radiative cooling. The metamaterial must reflect the incoming radiation from Sun (0.3–3 $\mu$m) and radiate within the atmospheric transparency window (8–13 $\mu$m).

### 5.3. *Solar steam generation*

Water boiling is important not only to make tea and rice dishes but also for applications such as distillation, sterilization, and electricity generation. The use of solar energy to boil water usually requires a costly set of motorized solar tracking mirrors to focus the sunlight. Polaritons can change the way of boiling the water to avoid this cost. The heat generated by the surface of the polaritonic nanoparticles is due to the strong interaction between the electronic oscillator of the nanoparticle and incident solar light. This phenomenon is also known as polaritonic local heating.[3] The heat generated by the polaritonic nanoparticle surface is transferred to the medium. This generated heat is controlled by the localized surface plasmon or phonon coupling of the nanoparticle with the light. The metals (plasmon) with high optical loss (such as W, Ta and Mo) can be useful for local heating.

## 6. Conclusion

In conclusion, material research plays an important role in human civilization enabling novel science and technology. The progress towards the next generation of technology depends on the development of the

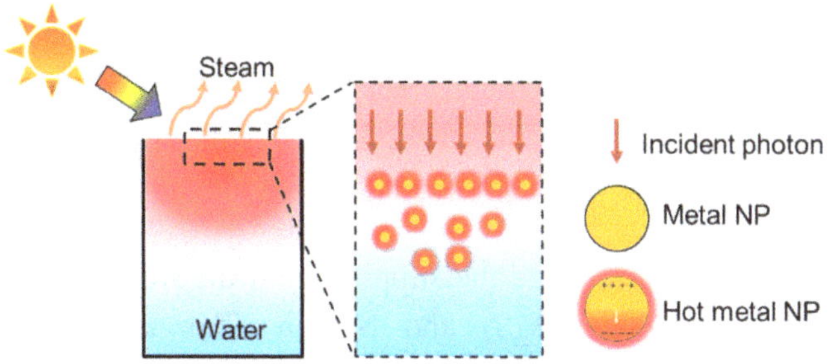


Figure 17. Schematic of the solar steam generation by localized heating via plasmonic nanoparticles (NPs).

materials, for example, the age of stone to the age of nanomaterials is an important example of this progress. At the present time, semiconductor technologies and optical communication are the most valuable technology for information processing devices. Over the last couple of decades, the speed of the devices increased exponentially, because of the miniaturization and integration of the semiconducting electronic components (for example, billions of transistors are embedded on a fingernail-size chip). But the continuous demand for faster devices in smaller dimensions is limited by the existing technology. The major problem with the existing technology is the ohmic loss in the electronic devices and the diffraction limit in the optical (or photonic) devices. For the next-generation technology, polariton is a promising candidate because it has the speed of light and the dimension of charge carriers. Polariton is a quasi-particle, which is the coupled-state of a photon (quanta of light) and charge carrier of material (such as free electron, lattice vibration, exciton, etc.). Polaritons, not only confine the light below the diffraction limit but also enhance the local electromagnetic field by many orders of magnitudes in the materials. This chapter covers the recent development of the plasmon and phonon polaritonic properties in transition metals and nitrides. Transition metals (i.e., Ta, W, Mo, etc.) and transition metal nitrides (i.e., TiN, HfN, ZrN) exhibit plasmon polaritons in the visible range. Wide-bandgap semiconducting GaN exhibits phonon polariton in the mid-IR range. Degenerate semiconducting ScN exhibits both the plasmon (in the near IR range) and phonon (in the mid IR range) polaritons. These materials are refractory, chemically stable, CMOS compatible, stable at high melting temperature making them suitable for optoelectronic and nano-photonic devices.

## Acknowledgment

K.C.M. and B.S. acknowledge the support from the International Centre for Materials Science (ICMS) and Sheikh Saqr Laboratory (SSL) JNCASR. B.S. acknowledges the financial support from the Science and Engineering Research Board (SERB) of the Government of India, Start-Up Research Grant No. SRG/2019/000613. K.C.M. is grateful to the Council of Scientific & Industrial Research (CSIR) for the fellowship.

## References

1. Moore, G. E. Cramming More Components onto Integrated Circuits. *Proc. IEEE* **1998**, *86* (1), 82–85. https://doi.org/10.1109/JPROC.1998.658762.
2. Fleischmann, M., Hendra, P. J. and McQuillan A. J. Raman Spectra of Pyridine Adsorbed at a Silver Electrode. *Chem. Phys. Lett.* **1974**, *26* (2), 163–166. https://doi.org/10.1016/0009-2614(74)85388-1.
3. Ndukaife, J. C., Shalaev, V. M. and Boltasseva, A. Plasmonics--Turning Loss into Gain. *Science (80-. ).* **2016**, *351* (6271), 334–335. https://doi.org/10.1126/science.aad9864.
4. Boriskina, S. V., Cooper, T. A., Zeng, L., Ni, G., Tong, J. K., Tsurimaki, Y., Huang, Y., Meroueh, L., Mahan, G. and Chen, G. Losses in Plasmonics: From Mitigating Energy Dissipation to Embracing Loss-Enabled Functionalities. *Adv. Opt. Photonics* **2017**, *9* (4), 775. https://doi.org/10.1364/aop.9.000775.
5. Brongersma, M. L., Halas, N. J. and Nordlander, P. Plasmon-Induced Hot Carrier Science and Technology. *Nature Nanotechnology*. 2015, pp. 25–34. https://doi.org/10.1038/nnano.2014.311.
6. Ndukaife, J. C., Kildishev, A. V., Nnanna, A. G. A., Shalaev, V. M., Wereley, S. T. and Boltasseva, A. Long-Range and Rapid Transport of Individual Nano-Objects by a Hybrid Electrothermoplasmonic Nanotweezer. *Nat. Nanotechnol.* **2016**, *11* (1), 53–59. https://doi.org/10.1038/nnano.2015.248.
7. Yoon, J., Zhou, M., Badsha, M. A., Kim, T. Y., Jun, Y. C. and Hwangbo, C. K. Broadband Epsilon-Near-Zero Perfect Absorption in the Near-Infrared. *Sci. Rep.* **2015**, *5*, 1–8. https://doi.org/10.1038/srep12788.
8. Tian, Y., García de Arquer, F. P., Dinh, C. T., Favraud, G., Bonifazi, M., Li, J., Liu, M., Zhang, X., Zheng, X., Kibria, M. G., Hoogland, S., Sinton, D., Sargent, E. H. and Fratalocchi, A. Enhanced Solar-to-Hydrogen Generation with Broadband Epsilon-Near-

Zero Nanostructured Photocatalysts. *Adv. Mater.* **2017**. https://doi.org/10.1002/adma.201701165.

9. Juan, M. L., Righini, M. and Quidant, R. Plasmon Nano-Optical Tweezers. *Nat. Photonics* **2011**, *5* (6), 349–356. https://doi.org/10.1038/nphoton.2011.56.
10. Juan, M. L., Gordon, R., Pang, Y., Eftekhari, F. and Quidant, R. Self-Induced Back-Action Optical Trapping of Dielectric Nanoparticles. *Nat. Phys.* **2009**, *5* (12), 915–919. https://doi.org/10.1038/nphys1422.
11. Hou, B., Shen, L., Shi, H., Kapadia, R. and Cronin, S. B. Hot Electron-Driven Photocatalytic Water Splitting. *Phys. Chem. Chem. Phys.* **2017**, *19* (4), 2877–2881. https://doi.org/10.1039/c6cp07542h.
12. Dyachenko, P. N., Molesky, S., Petrov, A. Y., Störmer, M., Krekeler, T., Lang, S., Ritter, M., Jacob, Z. and Eich, M. Controlling Thermal Emission with Refractory Epsilon-near-Zero Metamaterials via Topological Transitions. *Nat. Commun.* **2016**, *7* (May). https://doi.org/10.1038/ncomms11809.
13. Chang, C. C., Kort-Kamp, W. J. M., Nogan, J., Luk, T. S., Azad, A. K., Taylor, A. J., Dalvit, D. A. R., Sykora, M. and Chen, H. T. High-Temperature Refractory Metasurfaces for Solar Thermophotovoltaic Energy Harvesting. *Nano Lett.* **2018**, *18* (12), 7665–7673. https://doi.org/10.1021/acs.nanolett.8b03322.
14. Leonhardt, U. Optical Metamaterials: Invisibility Cup. *Nat. Photonics* **2007**, *1* (4), 207–208. https://doi.org/10.1038/nphoton.2007.38.
15. Maier, S. A. Excitation of Surface Plasmon Polaritons at Planar Interfaces. In *Plasmonics: Fundamentals and Applications*; Springer US: New York, NY, 2007; pp. 39–52. https://doi.org/10.1007/0-387-37825-1_3.
16. Stroscio, M. A. and Dutta, M. Phonons in Bulk Würtzite Crystal. In *Phonons in Nanostructures*; Cambridge University Press, 2001; pp. 16–25. https://doi.org/10.1017/CBO9780511534898.004.
17. Johnson, P. B. and Christy, R. W. Optical Constant of the Nobel Metals. *Phys. Rev. B* **1972**, *6* (12), 4370–4379.
18. Li, W., Guler, U., Kinsey, N., Naik, G. V., Boltasseva, A., Guan, J., Shalaev, V. M. and Kildishev, A. V. Refractory Plasmonics with Titanium Nitride: Broadband Metamaterial Absorber. *Adv. Mater.* **2014**, *26*, 7959–7965. https://doi.org/10.1002/adma.201401874.
19. Logeeswaran, V. J., Kobayashi, N. P., Islam, M. S., Wu, W., Chaturvedi, P., Fang, N. X., Wang, S. Y. and Williams, R. S. Ultrasmooth Silver Thin Films Deposited with a Germanium Nucleation Layer. *Nano Lett.* **2009**, *9* (1), 178–182. https://doi.org/10.1021/nl8027476.
20. Chen, W., Thoreson, M. D., Ishii, S., Kildishev, A. V. and Shalaev, V. M. Ultra-Thin Ultra-Smooth and Low-Loss Silver Films on a Germanium Wetting Layer. *Opt. Express* **2010**, *18* (5), 5124. https://doi.org/10.1364/oe.18.005124.
21. Kalpakjian, S. and Schmid, S. R. *Manufacturing Engineering and Technology*, 6th ed.; Stark, H., Opaluch, W., Eds.; Prentice Hall, 2010.

22. Maurya, K. C., Biswas, B., Rao, D. and Saha, B. Giant Enhancement of Plasmonic Response and Epsilon-near-Zero Signature in Refractory Transition Metals (Ta, W, and Mo) Deposited at High-Temperature. *Appl. Phys. Lett.* **2021**, *118* (4), 041902. https://doi.org/10.1063/5.0027497.
23. Kinsey, N., DeVault, C., Boltasseva, A. and Shalaev, V. M. Near-Zero-Index Materials for Photonics. *Nat. Rev. Mater.* **2019**, *4*, 742–760. https://doi.org/10.1038/s41578-019-0133-0.
24. Baffou, G. and Quidant, R. Thermo-Plasmonics: Using Metallic Nanostructures as Nano-Sources of Heat. *Laser and Photonics Reviews*. 2013. https://doi.org/10.1002/lpor.201200003.
25. Engheta, N. Circuits with Light at Nanoscales: Optical Nanocircuits Inspired by Metamaterials. *Science (80-. )* **2007**, *317* (5845), 1698–1702. https://doi.org/10.1126/science.1133268.
26. Silveirinha, M. and Engheta, N. Tunneling of Electromagnetic Energy through Subwavelength Channels and Bends Using ε-near-Zero Materials. *Phys. Rev. Lett.* **2006**, *97*, 157403. https://doi.org/10.1103/PhysRevLett.97.157403.
27. Adams, D. C., Inampudi, S., Ribaudo, T., Slocum, D., Vangala, S., Kuhta, N. A., Goodhue, W. D., Podolskiy, V. A. and Wasserman, D. Funneling Light through a Subwavelength Aperture with Epsilon-near-Zero Materials. *Phys. Rev. Lett.* **2011**, *107* (13), 133901. https://doi.org/10.1103/PhysRevLett.107.133901.
28. Campione, S., Wendt, J. R., Keeler, G. A. and Luk, T. S. Near-Infrared Strong Coupling between Metamaterials and Epsilon-near-Zero Modes in Degenerately Doped Semiconductor Nanolayers. *ACS Photonics* **2016**, *3* (2), 293–297. https://doi.org/10.1021/acsphotonics.5b00663.
29. Reshef, O., De Leon, I., Alam, M. Z. and Boyd, R. W. Nonlinear Optical Effects in Epsilon-near-Zero Media. *Nat. Rev. Mater.* **2019**, *4* (8), 535–551. https://doi.org/10.1038/s41578-019-0120-5.
30. Kinsey, N., DeVault, C., Kim, J., Ferrera, M., Shalaev, V. M. and Boltasseva, A. Epsilon-near-Zero Al-Doped ZnO for Ultrafast Switching at Telecom Wavelengths. *Optica* **2015**, *2* (7), 616. https://doi.org/10.1364/optica.2.000616.
31. Yu, I.-S., Cheng, H.-E., Chang, C.-C., Lin, Y.-W., Chen, H.-T., Wang, Y.-C. and Yang, Z.-P. Substrate-Insensitive Atomic Layer Deposition of Plasmonic Titanium Nitride Films. *Opt. Mater. Express* **2017**, *7* (3), 777. https://doi.org/10.1364/ome.7.000777.
32. Naik, G. V., Kim, J. and Boltasseva, A. Oxides and Nitrides as Alternative Plasmonic Materials in the Optical Range [Invited]. *Opt. Mater. Express* **2011**, *1* (6), 1090–1099. https://doi.org/10.1103/PhysRevLett.107.133901.
33. Patsalas, P., Kalfagiannis, N., Kassavetis, S., Abadias, G., Bellas, D. V., Lekka, C. and Lidorikis, E. Conductive Nitrides: Growth Principles, Optical and Electronic Properties, and Their Perspectives in Photonics and Plasmonics. *Mater. Sci. Eng. R Reports* **2018**, *123*, 1–55. https://doi.org/10.1016/j.mser.2017.11.001.
34. Maurya, K. C., Shalaev, V. M., Boltasseva, A. and Saha, B. Reduced Optical Losses

in Refractory Plasmonic Titanium Nitride Thin Films Deposited with Molecular Beam Epitaxy. *Opt. Mater. Express* **2020**, *10* (10), 2679. https://doi.org/10.1364/ome.405259.

35. Das, P., Maurya, K. C., Schroeder, J. L., Garbrecht, M. and Saha, B. Near-UV-to-Near-IR Hyperbolic Photonic Dispersion in Epitaxial (Hf,Zr)N/ScN Metal/Dielectric Superlattices. *ACS Appl. Energy Mater.* **2022**, *5* (4), 3898–3904. https://doi.org/10.1021/acsaem.1c03467.
36. Babinova, R. V., Smirnov, V. V., Useenov, A. S., Kravchuk, K. S., Gladkikh, E. V., Shapovalov, V. I. and Mylnikov, I. L. Mechanical Properties of Titanium Nitride Films Obtained by Reactively Sputtering with Hot Target. *J. Phys. Conf. Ser.* **2017**, *872*, 012035.
37. Datta, S., Das, M., Balla, V. K., Bodhak, S. and Murugesan, V. K. Mechanical, Wear, Corrosion and Biological Properties of Arc Deposited Titanium Nitride Coatings. *Surf. Coat. Technol.* **2018**, *344*, 214–222. https://doi.org/10.1016/j.surfcoat.2018.03.019.
38. Pang, X., Zhang, L., Yang, H., Gao, K. and Volinsky, A. A. Residual Stress and Surface Energy of Sputtered TiN Films. *J. Mater. Eng. Perform.* **2015**, *24*, 1185–1191. https://doi.org/10.1007/s11665-015-1393-5.
39. Ng, S. S., Hassan, Z. and Abu Hassan, H. Experimental and Theoretical Studies of Surface Phonon Polariton of AlN Thin Film. *Appl. Phys. Lett.* **2007**, *90* (8), 2005–2008. https://doi.org/10.1063/1.2645970.
40. Hillenbrand, R., Taubner, T. and Keilmann, F. Phonon-Enhanced Light-Matter Interaction at the Nanometre Scale. *Nature* **2002**, *418* (6894), 159–162. https://doi.org/10.1038/nature00899.
41. Torii, K., Koga, T., Sota, T., Azuhata, T., Chichibu, S. F. and Nakamura, S. An Attenuated-Total-Reflection Study on the Surface Phonon-Polariton in GaN. *J. Phys. Condens. Matter* **2000**, *12* (31), 7041–7044. https://doi.org/10.1088/0953-8984/12/31/305.
42. Turner, A. F., Chang, L. and Martin, T. P. Enhanced Reflectance of Reststrahlen Reflection Filters. *Appl. Opt.* **1965**, *4* (8), 927. https://doi.org/10.1364/AO.4.000927.
43. Azuhata, T., Matsunaga, T., Shimada, K., Yoshida, K., Sota, T., Suzuki, K. and Nakamura, S. Optical Phonons in GaN. *Phys. B Condens. Matter* **1996**, *219–220* (1–4), 493–495. https://doi.org/10.1016/0921-4526(95)00789-X.
44. Davydov, V. Y., Kitaev, Y. E., Goncharuk, I., Smirnov, A., Graul, J., Semchinova, O. and Uffmann, D. Phonon Dispersion and Raman Scattering in Hexagonal GaN and AlN. *Phys. Rev. B - Condens. Matter Mater. Phys.* **1998**, *58* (19), 12899–12907. https://doi.org/10.1103/PhysRevB.58.12899.
45. Hossain, M. M., Jia, B. and Gu, M. A Metamaterial Emitter for Highly Efficient Radiative Cooling. *Adv. Opt. Mater.* **2015**, *3* (8), 1047–1051. https://doi.org/10.1002/adom.201500119.
46. Zhai, Y., Ma, Y., David, S. N., Zhao, D., Lou, R., Tan, G., Yang, R. and Yin, X. Scalable-Manufactured Randomized Glass-Polymer Hybrid Metamaterial for

Daytime Radiative Cooling. *Science (80-. )* **2017**, *355* (6329), 1062–1066. https://doi.org/10.1126/science.aai7899.

47. Nikolajsen, T., Leosson, K. and Bozhevolnyi, S. I. Surface Plasmon Polariton Based Modulators and Switches Operating at Telecom Wavelengths. *Appl. Phys. Lett.* **2004**, *85* (24), 5833–5835. https://doi.org/10.1063/1.1835997.
48. Lewis, R. A. A Review of Terahertz Detectors. *J. Phys. D. Appl. Phys.* **2019**, *52* (43), 433001. https://doi.org/10.1088/1361-6463/ab31d5.
49. Chirumamilla, M., Chirumamilla, A., Yang, Y., Roberts, A. S., Kristensen, P. K., Chaudhuri, K., Boltasseva, A., Sutherland, D. S., Bozhevolnyi, S. I. and Pedersen, K. Large-Area Ultrabroadband Absorber for Solar Thermophotovoltaics Based on 3D Titanium Nitride Nanopillars. *Adv. Opt. Mater.* **2017**, *5* (22). https://doi.org/10.1002/adom.201700552.
50. Molesky, S., Dewalt, C. J. and Jacob, Z. High Temperature Epsilon-near-Zero and Epsilon-near-Pole Metamaterial Emitters for Thermophotovoltaics. *Opt. Express* **2013**, *21* (January), 549–554. https://doi.org/10.1364/cleo_qels.2013.qtu1a.6.
51. Corsi, C. Infrared: A Key Technology for Security Systems. *Adv. Opt. Technol.* **2012**, *2012* (2). https://doi.org/10.1155/2012/838752.
52. Sanvitto, D. and Kéna-Cohen, S. The Road towards Polaritonic Devices. *Nat. Mater.* **2016**, *15* (10), 1061–1073. https://doi.org/10.1038/nmat4668.
53. Eklund, P., Kerdsongpanya, S. and Alling, B. Transition-Metal-Nitride-Based Thin Films as Novel Energy Harvesting Materials. *J. Mater. Chem. C* **2016**, *4* (18), 3905–3914. https://doi.org/10.1039/c5tc03891j.
54. Gall, D., Petrov, I., Madsen, L. D., Sundgren, J.-E. and Greene, J. E. Microstructure and Electronic Properties of the Refractory Semiconductor ScN Grown on MgO(001) by Ultra-High-Vacuum Reactive Magnetron Sputter Deposition. *J. Vac. Sci. Technol. A Vacuum, Surfaces, Film.* **1998**, *16* (4), 2411–2417. https://doi.org/10.1116/1.581360.
55. Maurya, K. C., Biswas, B., Garbrecht, M. and Saha, B. Wave-Vector-Dependent Raman Scattering from Coupled Plasmon–Longitudinal Optical Phonon Modes and Fano Resonance in n -type Scandium Nitride. *Phys. status solidi – Rapid Res. Lett.* **2019**, *13* (9), 1900196. https://doi.org/10.1002/pssr.201900196.
56. Maurya, K. C., Rao, D., Acharya, S., Rao, P., Pillai, A. I. K., Selvaraja, S. K., Garbrecht, M. and Saha, B. Polar Semiconducting Scandium Nitride as an Infrared Plasmon and Phonon–Polaritonic Material. *Nano Lett.* **2022**, *22* (13), 5182–5190. https://doi.org/10.1021/acs.nanolett.2c00912.
57. Shockley, W. and Queisser, H. J. Detailed Balance Limit of Efficiency of P-n Junction Solar Cells. *J. Appl. Phys.* **1961**, *32* (3), 510–519. https://doi.org/10.1063/1.1736034.
58. Wang, Y., Liu, H. and Zhu, J. Solar Thermophotovoltaics: Progress, Challenges, and Opportunities. *APL Materials*. AIP Publishing, LLC 2019, p 080906. https://doi.org/10.1063/1.5114829.
59. Bierman, D. M., Lenert, A., Chan, W. R., Bhatia, B., Celanović, I., Soljačić, M. and

Wang, E. N. Enhanced Photovoltaic Energy Conversion Using Thermally Based Spectral Shaping. *Nat. Energy* **2016**, *1* (6), 16068. https://doi.org/10.1038/nenergy.2016.68.

60. Hossain, M. M. and Gu, M. Radiative Cooling: Principles, Progress, and Potentials. *Adv. Sci.* **2016**, *3* (7), 1–10. https://doi.org/10.1002/advs.201500360.
61. Maurya, K. C., Bhui, A., Biswas, K. and Saha, B. Anisotropic Epsilon-near-Pole (ENP) Resonance Leads to Hyperbolic Photonic Dispersion in Homologous (Bi2)m(Bi2Se3)n Topological Quantum Materials. *Appl. Phys. Lett.* **2021**, *119*, 011902. https://doi.org/10.1063/5.0053587.

https://doi.org/10.1142/9789811270956_0007

# Chapter 7

# Electrochemical and Photochemical Hydrogen Evolution

Mohd Monis Ayyub and C.N.R. Rao*

*New Chemistry Unit, International Centre for Materials Science and School of Advanced Materials, Jawaharlal Nehru Centre for Advanced Scientific Research, Jakkur, Bangalore 560064, India*

**cnrrao@jncasr.ac.in*

Considering the worsening of climatic conditions on earth, there is an urgent need to develop clean and renewable sources of energy. Water splitting reaction to produce hydrogen and oxygen is one of the most studied catalytic processes that has the potential to cater to the energy needs of humanity. Water splitting produces hydrogen of very high purity in a non-polluting process. Hydrogen is a high energy density fuel. However, technological challenges exist in its sustainable production. Various approaches like electrochemical, photochemical, photoelectrochemical and solar thermochemical processes have been explored to accomplish water-splitting reaction. This chapter discusses the fundamentals of photochemical and electrochemical hydrogen evolution.

## 1. Introduction

Fossil fuels are advantageous as they store energy in concentrated form which can be easily extracted via combustion. However, these complex molecules require millions of years to form and produce harmful gases upon combustion. A rather simple and efficient way to store energy is in hydrogen which has a high energy density (142 MJ/kg) and burns clean. Currently, reforming natural gas is the method for producing hydrogen which requires high temperature and high pressure, and evolves hazardous gaseous products. Therefore, it is pertinent to move towards generating hydrogen from renewable energy sources such as solar or wind energy. Hydrogen can be produced by the splitting of water, and this reaction can be carried out by using electricity from renewable energy sources (wind and solar) or by directly utilizing solar energy in photocatalysis.[1,2] The process can be further enhanced by utilizing electricity and solar energy in a process called photoelectrochemical water splitting. A substantial amount of electricity produced from intermittent renewable sources (wind turbines and photovoltaics) can be stored in the form of hydrogen produced from the electrolysis of water. Splitting of water can also be achieved by directly utilizing seawater which makes the process more sustainable.[3] In addition to the above methods, solar thermochemical splitting of water to produce hydrogen is also an upcoming field that differs in terms of technology;[4] therefore, we shall not deal with this aspect. This chapter discusses the fundamentals of electrochemical and photochemical production of hydrogen, along with discussion of current material design strategies.

## 2. Electrochemical hydrogen evolution

Water electrolysis is generally carried out by immersing two electrodes; cathode and anode, in an electrolyte, and electricity is passed through the electrodes via an external power supply to carry out electrolysis (Figure 1a). Water electrolysis consists of two half reactions; hydrogen evolution reaction (HER) involving reduction of water to form hydrogen gas at cathode and oxygen evolution reaction (OER) involving oxidation of water to produce oxygen at anode.

$$2H^+ + 2e^- \rightarrow H_2 \quad \text{Reduction (Cathode)} \quad (1)$$

$$2H_2O + 4h^+ \rightarrow O_2(g) + 4H^+(l) \quad \text{Oxidation (Anode)} \quad (2)$$

$$2H_2O \rightarrow 2H_2(g) + O_2(g) \quad \text{Full reaction} \quad (3)$$

The standard potentials for reactions (1) and (2) are $E^0_{H_2} = 0$ V and $E^0_{O_2} = 1.229$ V (*vs.* normal hydrogen electrode (NHE) at 25°C and pH = 0). This implies that the whole reaction (3) has $E^0_{H_2O} = 1.229$ V, which is equivalent to $\Delta G_o = 237.2$ kJ mol$^{-1}$ at 25°C.[5]

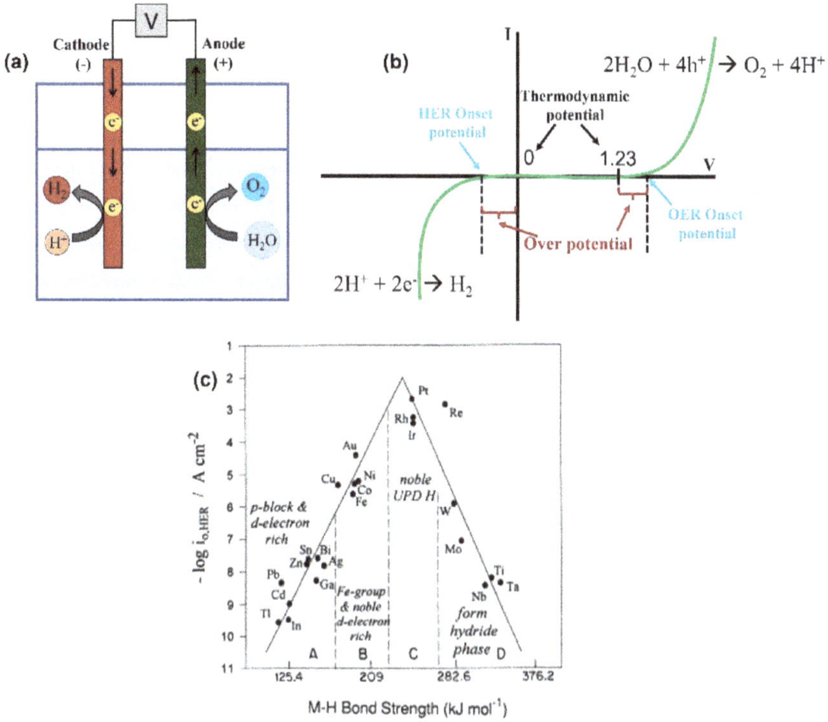


Figure 1. (a) Schematic representation of an electrochemical cell, (b) current (I)-potential (V) characteristic of water splitting reaction (vs NHE at pH = 0), and (c) volcano plot for log $i_o$ value for hydrogen evolution reaction as a function of metal-hydrogen (M-H) bond strength. Reprinted with permission from Ref. 8. Copyright 2000, with permission from Elsevier.

## 2.1. *Mechanism*

Mechanism of hydrogen evolution reaction on an electrode surface (M) proceeds by an initial adsorption of hydrogen on the catalyst surface called the Volmer step.[6]

$$H^+ + e^- + M \rightarrow M\text{-}H_{ads} \quad \text{(Acidic conditions)} \quad (4)$$

$$H_2O + e^- + M \rightarrow M\text{-}H_{ads} + OH^- \quad \text{(Alkaline conditions)} \quad (5)$$

This is followed by either an electrochemical Heyrovsky step involving transfer of another proton-electron pair or chemical Tafel step involving recombination of two adsorbed protons.

$$H^+ + M\text{-}H_{ads} + e^- \rightarrow M + H_2 \quad \text{(Heyrovsky step – Acidic Conditions)} \qquad (6)$$

$$H_2O + M\text{-}H_{ads} + e^- \rightarrow M + H_2 + OH^- \quad \text{(Heyrovsky step – Alkaline Conditions)} \qquad (7)$$

$$2M\text{-}H_{ads} \rightarrow H_2 + 2M \quad \text{(Tafel step)} \qquad (8)$$

Theoretically, the hydrogen evolution reaction becomes thermodynamically downhill at 0 V (vs NHE at pH = 0). However, in actual scenario, HER commences at cathodic potentials than 0 V which is due to the extra energy required to surpass kinetic barriers. The potential at which the thermodynamic and kinetic barriers for the hydrogen evolution reactions become exergonic is defined as the onset potential and the additional potential required to overcome the kinetic barriers is defined as the overpotential. Figure 1b depicts a typical current-potential characteristic of water splitting reaction depicting thermodynamic potential, onset potential and overpotential for HER and OER (vs NHE at pH = 0).

Electrocatalyst materials are employed to reduce the overpotential for HER and a good catalyst should possess low overpotential, high current density and stability.[7] The kinetics of HER on a catalyst surface depends upon the nature of electrode, where the adsorption of H on the electrode surface to form $M\text{-}H_{ads}$ (Volmer step) is a good descriptor of HER rate. A plot of binding energy of hydrogen on catalyst surface as a function of reaction rate forms a volcano type relationship and predicts that catalysts that adsorb hydrogen neither too strongly nor too weakly will possess highest activity (Figure 1c).[8] Apart from hydrogen binding energy, there are other parameters, such as nature of proton source and surface spectator species, that play a role in determining HER rates. In alkaline medium, where $H_2O$ is the proton donor, the catalyst surface will have $M\text{-}OH_{ad}$ species in addition to $M\text{-}H_{ad}$ species. A good catalyst should have a balance between adsorption energies of both the species; neither too strong $M\text{-}H_{ads}$ nor too oxophilic for adsorption of -OH.[9]

When studying the performance of a catalyst, current-voltage measurements are fundamental towards understanding HER, where the kinetic control regime of the current-voltage relation gives information about the intrinsic rate of the reaction at the interface. Tafel in 1905[10] summarised that under this kinetic control region the current is related to overpotential by the following relation,

$$\eta = a + b \log i \qquad \text{Tafel equation} \tag{8}$$

A plot of log $i$ vs $\eta$ yields a straight line with intercept "a" as the exchange current density and slope "b" known as the Tafel slope. Tafel slope essentially gives the potential required to produce a logarithmic increase in current and is an important parameter for judging performance of a catalyst since it contains information about the mechanism of HER, where the elementary kinetic steps possess a fixed value of Tafel slope.[6]

Volmer step; Tafel slope = 120 mV/dec

Heyrovsky step; Tafel slope = 40 mV/dec

Tafel step; Tafel slope = 30 mV/dec

## 2.2. *Materials*

Electrocatalysts are being studied to achieve high water reduction current densities at low overpotentials. The properties discussed above: onset potential, overpotential required for high current densities, Tafel slope and hydrogen binding energies, are studied for electrocatalyst for identifying and designing better catalyst materials. The goal is to design a catalyst with onset potential close to zero and high current densities at low overpotentials. In accordance with the volcano plot for hydrogen binding energy, the platinum group metals occupy the top of the plot (Figure 1c) and hence are the best and most studied metals for HER. However, the high cost of these noble metals has made it imperative to look for alternative inexpensive and earth abundant electrocatalysts. In this regard, there has been much progress in designing non-noble metal electrocatalysts.

Graphene due to its high conductivity and chemical stability is a good choice for electrocatalysis, but its inert surface shows very poor catalytic activity. Catalytic activity of graphene can be enhanced by various methods such as structural engineering, doping and creation of defects

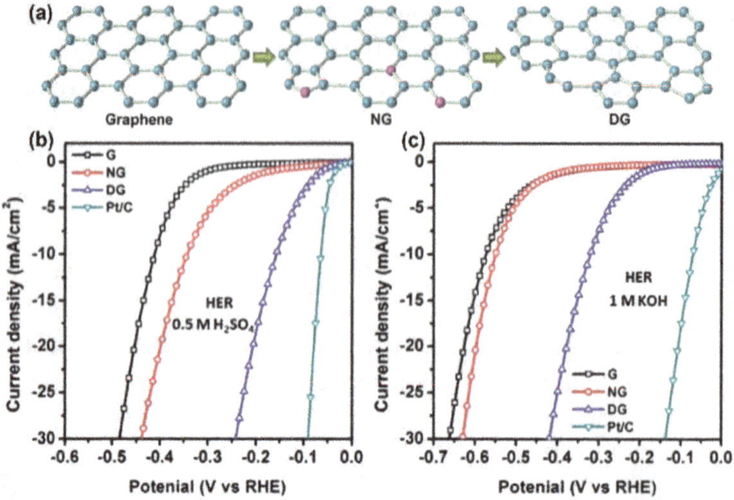


Figure 2. Schematic of the formation of graphene with defects (DG). Linear sweep voltammetry curves for hydrogen evolution reaction of the pristine graphene, N-doped graphene (NG), and DG in (a) acid and (c) alkaline solution respectively. Reprinted with permission from Ref. 12. Copyright 2016, John Wiley and Sons.

(Figure 2).[11,12] Non-metal doping in graphene can enhance the catalytic activity which is due to the introduction of new defect sites, altered electronic structure, and increased local curvature. Doping elements such as nitrogen, phosphorus, boron and sulfur in graphene lattice alter the density of carriers and break the electroneutrality, thereby enhancing the catalytic activity.[13] Dopant atoms affect the valence orbital energy levels and reduce the $\Delta G_H^*$ which promote H* adsorption. Doping two atoms in carbon matrices such as nitrogen and phosphorus, or nitrogen and sulphur, introduces a synergistic effect in tuning the $\Delta G_H^*$ on the catalyst surface and therefore can further enhance the catalytic activity.[14,15] Nitrogen and phosphorus co-doped graphene exhibits a low onset potential of 0.12 V (*vs.* RHE) and a Tafel slope of 79mV/dec.[14] N and S codoped graphene shows onset potential of 0.13 V (*vs.* RHE) and a Tafel slope of 80.5 mV/dec.[15] Doping of boron and nitrogen in graphene lattice forms a codoped structure termed as borocarbonitride which exhibits a low onset potential of −0.056 V (*vs.* RHE) and Tafel slope of 100 mV/dec, where the introduction of B and N in the lattice alters the electronic structure by opening a band gap which is suitable for water reduction.[16]

Transition metal dichalcogenides (TMDs) ($MX_2$; M = Transition metal, X = chalcogens S, Se, Te) are another class of earth-abundant non-noble metal-based materials that have emerged as promising catalyst — materials for HER. $MoS_2$ is the most widely studied electrocatalyst among TMDs and its activity is derived from the edge sites. The activity of $MoS_2$ can be enhanced by tuning the electronic structure of edges to increase M-$H_{ads}$ bond strength.[17] The surface structure of $MoS_2$ can be engineered to preferentially expose edge sites. For example, Jaramillo *et al.*[18] demonstrated the synthesis of thin films of highly ordered double-gyroid $MoS_2$ with nanoscale pores with exposed edge sites which exhibit excellent HER activity (Figure 3).

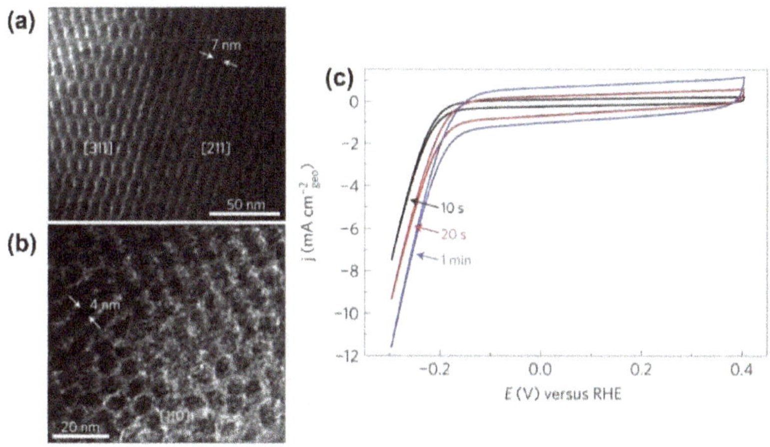


Figure 3. (a) TEM image of double-gyroid $MoS_2$ showing the [311] and [211] projections of the double-gyroid structure. The white arrow indicates the pore-to-pore distance of ~ 7 nm (b) Augmented image of the [211] and [110] projections illustrating the thickness of the double-gyroid $MoS_2$ networks. (c) Cyclic voltammograms of 10 s, 20 s and 1 min samples at 50 mV $s^{-1}$ in 0.5 M $H_2SO_4$ with graphite foil counter electrode and a Hg|$Hg_2SO_4$ (in saturated $K_2SO_4$) reference electrode (Hach). Reprinted by permission from Ref. 18 from Springer Nature, Nature Materials Copyright 2012.

TMDs and carbon-based materials can form heterostructures, which are a composite of two different materials that can be physically or chemically bonded. Formation of heterostructures increases the number of exposed active sites, enhances the electrical conductivity, enhances diffusion of protons to the active sites, and the differences in the electro-negativity of the two components can induce better charge separation. Combining $MoS_2$ with reduced graphene oxide ($MoS_2$/RGO) exposes the active edge sites and thereby enhances HER activity with a Tafel slope of 41 mv/dec, where RGO acts as a supporting substrate, increasing the rate of electron transfer.[19] The interaction of $MoS_2$ with RGO in this case is physical and not chemical. Covalent cross-linking of g-$C_3N_4$ with $MoS_2$ and nitrogenated-RGO (NRGO), and $MoS_2$ with BCN has been shown to significantly enhance the HER catalytic activity, where the covalent linkers between layers acts as electron channels (Figure 4).[20,21] Theoretical calculations predict that the transfer of charge between the cross-linked layers occurs from highest occupied molecular orbital (HOMO) of the carbon/conducting layer ($C_3N_4$, NRGO or BCN) to the lowest unoccupied molecular orbital (LUMO) of the active layer ($MoS_2$) and can occur simultaneously through space and linker bonds. This enhanced charge transfer between layers along with improved surface area and conductivity imparts excellent catalytic activity to covalently cross-linked nano-composites.

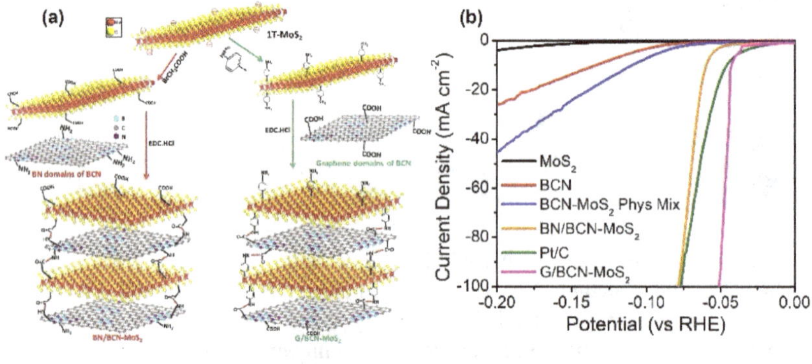


Figure 4. (a) Synthetic approach for the covalent cross-linking of BN/BCN–$MoS_2$ and G/BCN–$MoS_2$ Nanocomposites (EDC = 1-Ethyl-3-(3-dimethylaminopropyl) carbodiimide) (b) Polarization curves showing $H_2$ evolution for $MoS_2$, BCN, BCN–$MoS_2$ (1:2, mixture), BN/BCN–$MoS_2$ (1:2, composite), G/BCN–$MoS_2$ (1:2, composite) samples, including Pt/C (40 wt %) in $N_2$ saturated 0.5 M $H_2SO_4$ with Ag/AgCl (3 M NaCl) reference electrode and Pt coil as counter electrode. Reprinted with permission from Ref. 21. Copyright 2017 American Chemical Society.

Current density of electrocatalyst can be improved by increasing the active surface area which can be achieved by controlling the morphology of the material through nanostructuring. The change in atomic arrange-ment, and phase transformation can lead to fundamental changes in physical and chemical properties and therefore, can enhance the intrinsic activity of the material. A change in phase of semiconducting 2H-$MoS_2$ to the metallic 1T-$MoS_2$ significantly enhance the HER activity.[22]

## 3. Photochemical Hydrogen Evolution

Photocatalysis is akin to natural photosynthesis, where photosystems in plants absorb solar radiation to drive the oxidation of water and reduction of nicotinamide adenine dinucleotide phosphate ($NADP^+$ to NADPH).[23,24] The essence of this process is the utilization of solar energy to drive uphill chemical reactions. Similar to natural photosynthesis, a photocatalyst can absorb solar radiation and drive thermodynamically uphill reactions like water splitting (Figure 5a). As defined by IUPAC, a photocatalyst is a "Catalyst able to produce, upon absorption of light, chemical transform-ations of the reaction partners".[25] Photocatalysts include a wide range of materials such as semiconductors, molecular systems, and dyes; with different structure and morphologies, including nanomaterials, 2D materials, 3D porous structures and complex hierarchical structures.[23,26] We focus on the basics of semiconductor based photocatalytic systems in this chapter.

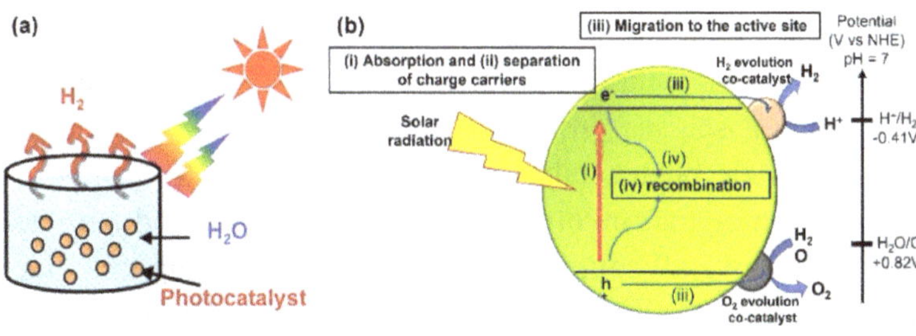


Figure 5. (a) Schematic representation of photochemical hydrogen generation. Reprinted with permission from Ref. 2. Copyright 2008, Royal Society of Chemistry. (b) Thermodynamics and steps involved in photochemical hydrogen generation. Reprinted with permission from Ref. 26. Copyright 2020, John Wiley and Sons. The water redox potentials in these schematics are vs NHE at pH = 7.

### 3.1. *Mechanism*

To drive the water splitting reaction, a semiconductor photocatalyst absorbs photons of energy larger than its band gap and creates electrons in the conduction band and holes in the valence band. These photogenerated electrons and holes are utilized to drive redox reactions (Eqs. (1) and (2)). Thermodynamically, reduction of protons proceeds if the conduction band minima is more negative than the reduction potential of water ($H^+/H_2$; $E^o = 0$ V vs NHE) and oxidation of water proceeds if the valence band maxima is more positive than the oxidation potential of water ($O_2/H_2O$; $E^o = 1.23$ V vs NHE) (Figure 5b).[2] Therefore, a semiconductor's minimum theoretical band gap should be 1.23 eV. The mechanism of photocatalytic hydrogen generation consists of three steps; (i) absorption of light to generate electron and holes, (ii) separation and migration of photo-generated charge carriers, and (iii) redox reactions at the catalyst surface, and are all essential steps that dictate the overall efficiency of the system (Figure 5b). Various factors influence these steps such as crystallinity, surface area, porosity, morphology and cocatalysts. From the kinetic viewpoint, it is crucial to provide an overpotential to overcome kinetic barriers and achieve a high reaction rate. The system's overall efficiency depends on the cumulative effect of efficiencies of all the processes, namely, light absorption, charge separation and migration, transport efficiency and utilization of electrons for the reduction reaction. Maintaining a balance of thermodynamic and kinetic limitations, which depend upon electronic and surface properties, is essential. Therefore, all the effects must be optimized to design an efficient photocatalyst.

Photocatalytic water splitting can proceed in two ways: a one step-process and a two-step Z-scheme process (Figure 6). In the former, the process occurs on a single photocatalyst, while on the latter is inspired by the natural Z-scheme and involves two photocatalysts and an electron mediator.[27] The overall splitting of water is a challenging task since the oxidation step is a 4 $e^-$ process and is generally the kinetically sluggish and rate-determining step. To overcome this problem, sacrificial agents are utilized to expedite the oxidation reaction rate. The sacrificial agents are species such as alcohols, triethanolamine, ascorbic, etc., which can get easily oxidized and remove the sluggish water oxidation step. Sacrificial

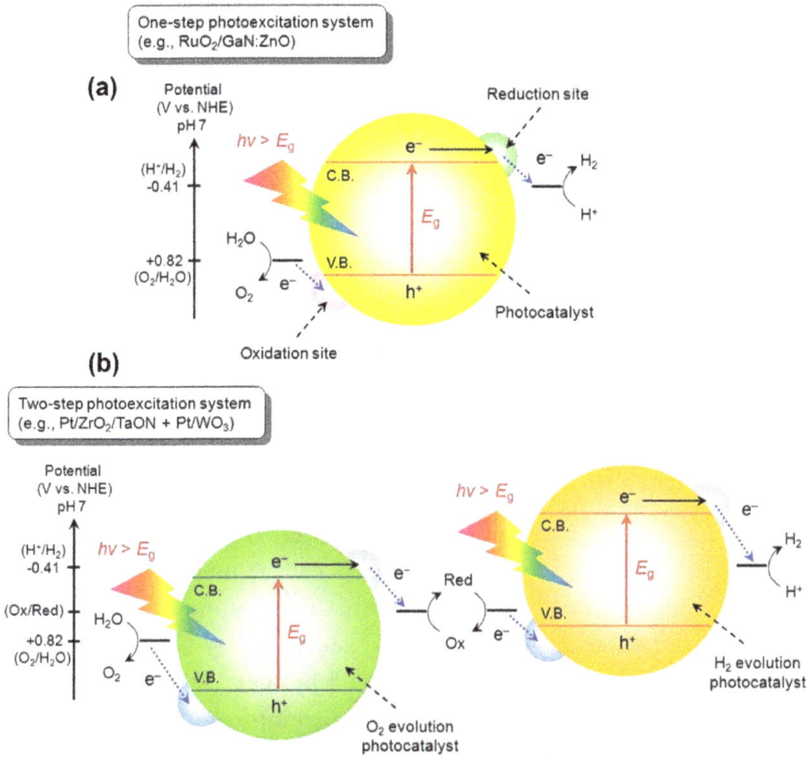


Figure 6. Schematic representation of photochemical hydrogen generation by (a) one-step and (b) two-step photoexcitation systems. The water redox potentials in these schematics are vs NHE at pH = 7. Reprinted with permission from Ref. 27. Copyright 2010, American Chemical Society.

electron donors also reduce photocorrosion by quenching holes.[28] However, care must be taken while using sacrificial agents because these chemicals can themselves degrade to produce hydrogen and contribute to the overall reduction products, which can create inaccuracies in the measurement. It is therefore important to carefully analyse the role of sacrificial agents and their contribution to the overall hydrogen yield in every photocatalytic system.[29,30]

### 3.2. *Materials*

Semiconductor photocatalysts can be screened based on their band gaps and band-edge positions relative to the redox potentials of water. Figure 7 shows the band-edge positions of various semiconductor photocatalysts compared with the redox potentials of water.[2] Another important condition is that the band gap of the semiconductor should be as narrow as possible to harvest more visible light from the solar spectrum. Oxide semiconductors like $TiO_2$, ZnO, and $SrTiO_3$ have been explored for photocatalytic water splitting, but their activity is limited by their large band gaps in the ultraviolet region.[31] However, their band gaps can be tuned by cation or anion substitution. Doping anions such as N and F in $TiO_2$ or ZnO, shifts the absorption edge in the visible region and lead to superior hydrogen evolution activity (Figure 8).[32,33] Apart from oxides, metal sulfides such as CdS (band gap = 2.4 eV) and ZnS (band gap = 3.6 eV) are also well studied.[2,31] The band gap of ZnS can be tuned from UV to the visible region by substituting $Zn^{2+}$ with $Cd^{2+}$, forming solid solutions $Zn_{1-x}Cd_xS$ (Figure 9).[34] A major drawback of using sulfides is photocorrosion, where the sulfides get degraded by their own photogenerated electrons and holes.[2] Two-dimensional transition metal dichalcogenides ($MX_2$, M = transition metal atom and X = chalcogen (S, Se, Te)) have also

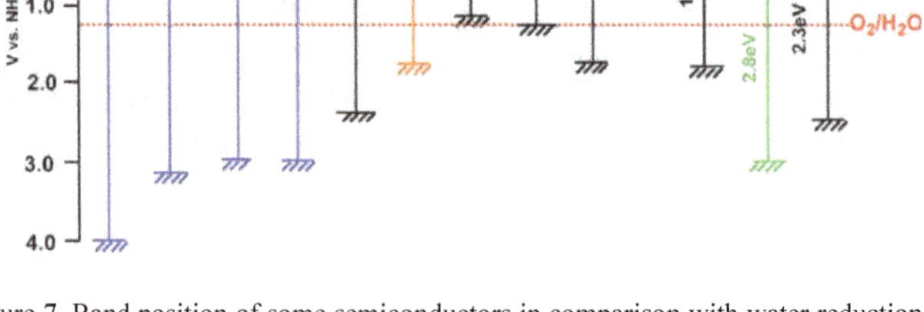


Figure 7. Band position of some semiconductors in comparison with water reduction and oxidation potentials (vs RHE at pH = 7). Reprinted with permission from Ref. 2. Copyright 2008, Royal Society of Chemistry.

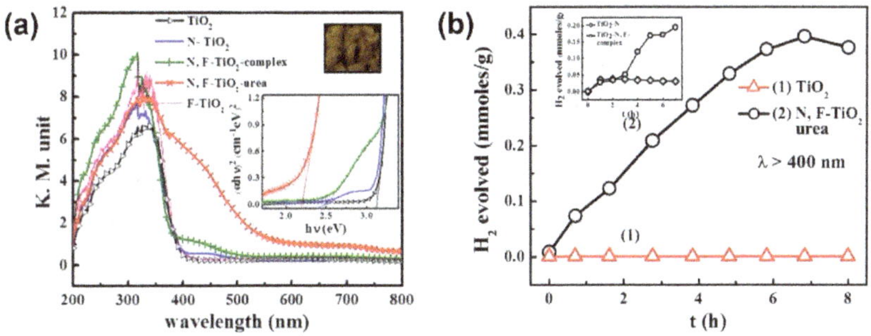


Figure 8. (a) UV-visible spectra and (b) hydrogen evolution for undoped $TiO_2$ and N,F-$TiO_2$ under visible light irradiation. Reprinted with permission from Ref. 33. Copyright 2013, American Chemical Society.

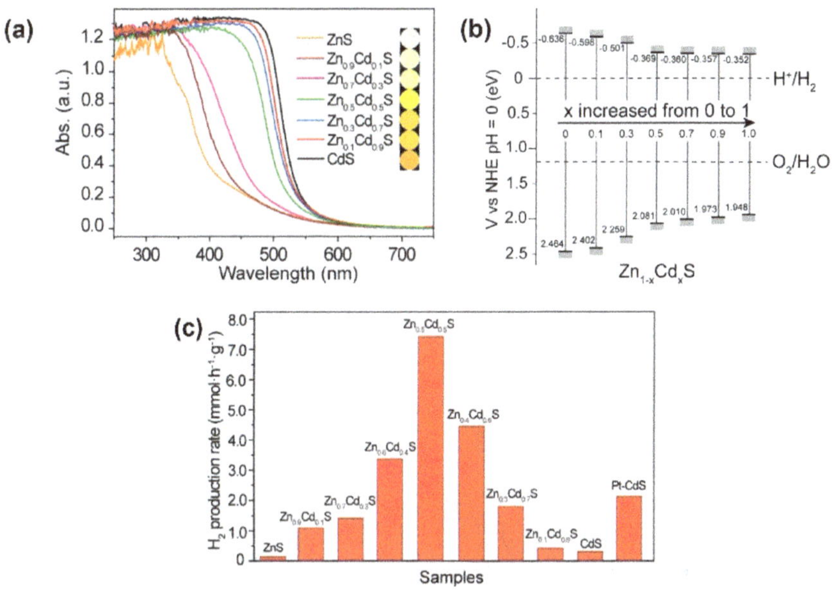


Figure 9. (a) UV-visible diffuse reflectance spectra, (b) conduction and valence band edge potentials and (c) comparison of photocatalytic hydrogen generation for $Zn_{1-x}Cd_xS$ (x = 0, 0.1, 0.1, 0.5, 0.7, 0.9, and 1.0) solid solutions in an aqueous solution of 0.44 M $Na_2S$ and 0.31 M $Na_2SO_3$ under visible light irradiation (> 400 nm). Reprinted with permission from Ref. 34. Copyright 2013, American Chemical Society.

been extensively studied.[35] $MoS_2$, the most studied member, has $\Delta G_H*$ close to Pt and therefore is a good catalyst for HER.[17,36]

Recombination of charge carriers is a major detrimental factor on the catalytic activity. Cocatalysts are deposited on the catalyst surface to increase the number of active sites and decrease the charge carrier recombination.[2,29] The photogenerated electron in the conduction band of the semiconductor are transferred to the fermi level of a metal cocatalyst, if the fermi level lies below the conduction band minima, thereby acting as an electron sink (Figure 10a).[37] Pt is a very efficient cocatalyst, however its use is limited by its high cost and scarcity.[38] Various other oxides, sulfides and phosphides like NiO, NiS, $MoS_2$, $Ni_2P$, $Co_2P$, etc. have been reported as efficient cocatalysts.[38] Charge recombination can also be reduced by the formation of heterostructures, wherein the energy bands of two semiconductors are aligned so that electrons and holes get separated into two different components.[39] Based on the band alignment, these semiconductor heterostructures can be classified into type I, type II and type III (Figure 10b).[40] Type II heterostructure is most favorable for efficient charge carrier separation. Various heterostructures like $MoS_2/TiO_2$,[41]

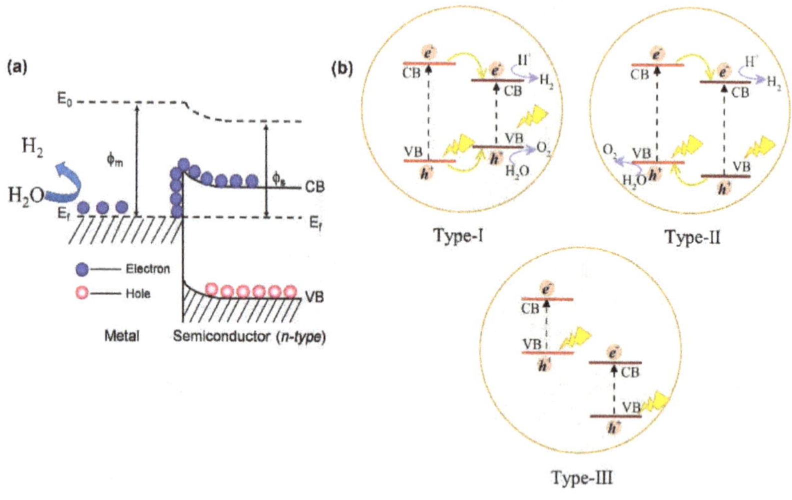


Figure 10. (a) Schematic illustration of mechanism of photochemical $H_2O$ reduction on semiconductor loaded with a metal cocatalyst. Adapted with permission from Royal Society of Chemistry, Ref. 37; permission conveyed through Copyright Clearance Center, Inc. (b–d) Schematic representation of separation of electron–hole pairs in conventional light-responsive semiconductor heterostructures: type I, type II, and type III. Reprinted with permission from Ref. 40, © IOP Publishing.

$ZnIn_2S_4$-Au-$TiO_2$,[42] CdSe/CdS/Pt,[43] ZnO/Pt/CdS,[44] anatase-$TiO_2$/H-rutile-$TiO_2$,[45] etc. have been shown to enhance charge carrier efficiency.

Ternary semiconductors such as $ZnIn_2S_4$, $CuInS_2$ and $CuGaS_2$ have also been studied.[2] Two-dimensional $ZnIn_2S_4$ (ZIS) has a favorable H adsorption at the edge S atom ($\Delta G_H^* = -0.16$ eV) and has been regarded as a promising candidate for photocatalytic water splitting.[46] The photocatalytic activity of ZIS can be enhanced by exposing the active edge sites.[47] However, the recombination of the charge carrier significantly impedes the overall activity, which can be overcome by creating sulfur vacancies ($V_s$) that act as electron traps and concentrate the electron density around $V_s$, thereby reducing the charge recombination and significantly enhancing the photocatalytic activity.[48] The activity can be further enhanced by the deposition of noble metals (like Pt) on the basal plane.[49] A new class of ternary semiconductors which are analogues of CdS with the formula $Cd_4P_2X_3$ (X = Cl, Br and I) exhibit excellent hydrogen evolution activity.[50]

Another strategy for photocatalytic HER is to utilize dye molecules as photosensitizer along with a catalytic material such as $MoS_2$, graphene, etc.[51] The dye molecules undergo a series of transition and transfer an electron to the catalyst which reduces $H^+$ to $H_2$ (Figure 11a). Dye-sensitized HER studies on composites of $MoS_2$ with NRGO exhibit excellent HER activity. Interestingly, the metallic 1T-phase of $MoS_2$ is a better catalyst than 2H-$MoS_2$ due to its increased conductivity.[51] The catalytic activity of 2H-$MoS_2$ can be enhanced by covalently cross-linking with carbon-based supports such as $C_3N_4$ and BCN.[20,21] As discussed in the earlier section, charge transfer occurs from HOMO of the carbon/conducting layer ($C_3N_4$, NRGO or BCN) to the LUMO of the active layer ($MoS_2$) and can occur simultaneously through space and linker bonds. This enhanced charge transfer between layers along with improved surface area and conductivity, imparts excellent photocatalytic activity to covalently cross-linked nanocomposites. The catalytic activity also depends on the linker length, where the activity diminishes with increasing linker length (Figures 11b–d).[20,52]

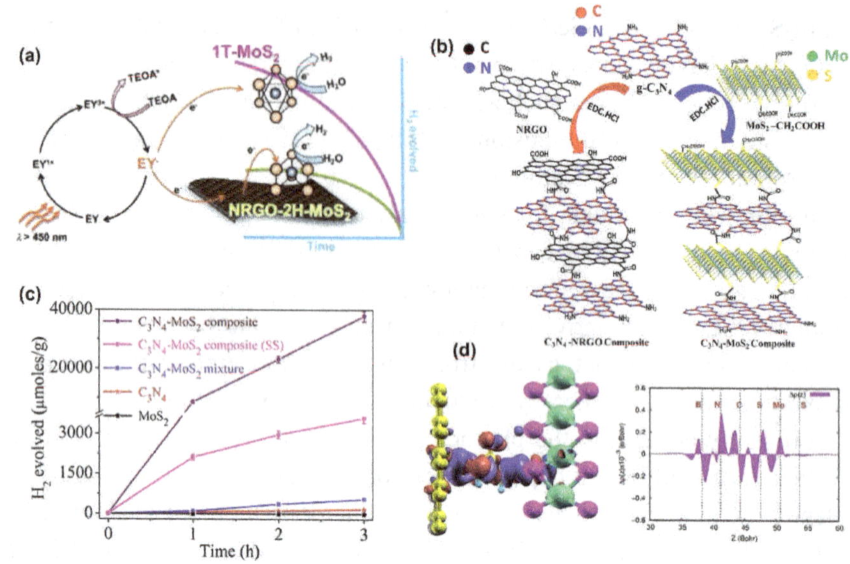

Figure 11. (a) Proposed mechanism for dye-sensitized hydrogen evolution activity of 1T-$MoS_2$ and NRGO-2H-$MoS_2$ (Yellow S, blue-gray Mo). Reproduced with permission from Ref. 51. Copyright 2013 John Wiley and Sons. (b) Synthetic approach for covalent crosslinking of $C_3N_4$–NRGO and $C_3N_4$–$MoS_2$ composites (c) Comparison of the $H_2$ evolution activity of $C_3N_4$, $MoS_2$, physical mixture of $C_3N_4$ and $MoS_2$, $C_3N_4$–$MoS_2$ composite (SS, obtained by solid-state route), and the covalently bonded $C_3N_4$–$MoS_2$ nanocomposite. Reproduced with permission from Ref. 20. Copyright 2017 American Chemical Society. (d) Charge density difference and plane averaged charge density difference for $MoS_2$-BCN composite (C: yellow; Mo: green; S: purple; H: sky blue; red and blue signify charge accumulation and charge depletion, respectively). Reproduced with permission from Ref. 52. Copyright 2013 John Wiley and Sons.

## 4. Conclusion

There is a need for rigorous endeavors to solve the current energy problem by discovering alternate renewable and sustainable energy sources. A viable option is, driving water-splitting reaction to produce hydrogen by electricity produced from renewable sources or by directly utilizing solar energy. As evident from the above discussion, there has been significant advancement in designing better catalyst materials for large-scale utilization of HER. However, there is considerable scope for technological advances in this field towards understanding the mechanism of water-splitting reaction. Designing efficient catalyst materials and electrolyzer design for large-scale utilization is of utmost priority. The current chapter should serve as a basic understanding of electrochemical and photochemical water splitting.

## References


1. B. You and Y. Sun, Innovative Strategies for Electrocatalytic Water Splitting, *Acc. Chem. Res.* **51** (7), 1571-1580, (2018).
2. A. Kudo and Y. Miseki, Heterogeneous Photocatalyst Materials for Water Splitting, *Chem. Soc. Rev.* **38**, 253 (2008).
3. M. M. Ayyub, M. Chhetri, U. Gupta, A. Roy and C. N. R. Rao, Photochemical and Photoelectrochemical Hydrogen Generation by Splitting Seawater, *Chem. Eur. J.* **24**(69), 18455(2018).
4. C. N. R. Rao and S. Dey, Solar Thermochemical Splitting of Water to Generate Hydrogen, *Proc. Natl. Acad. Sci. U S A*, **114**(51), 13385(2017).
5. Wolf. Vielstich, Arnold. Lamm, H. Gasteiger, *Handbook of Fuel Cells: Fundamentals, Technology, Applications*, Wiley, (2003).
6. T. Shinagawa, A. T. Garcia-Esparza and K. Takanabe, Insight on Tafel Slopes from a Microkinetic Analysis of Aqueous Electrocatalysis for Energy Conversion, *Sci. Rep.* **5**, 13801(2015).
7. D. Strmcnik, P. P. Lopes, B. Genorio, V. R. Stamenkovic, N. M. Markovic, Design Principles for Hydrogen Evolution Reaction Catalyst Materials, *Nano Energy* **29**, 29-36(2016).
8. B. E. Conway and G. Jerkiewicz, Relation of Energies and Coverages of Underpotential and Overpotential Deposited H at Pt and Other Metals to the 'Volcano Curve' for Cathodic $H_2$ Evolution Kinetics, *Electrochim. Acta*, **45**(25-26), 4075-4083(2000).
9. D. Strmcnik, P. P. Lopes, B. Genorio, V. R. Stamenkovic and N. M. Markovic, Design Principles for Hydrogen Evolution Reaction Catalyst Materials, *Nano Energy* **29**, 29-36(2016).
10. J. Tafel, Über Die Polarisation Bei Kathodischer Wasserstoffentwicklung, *Zeitschrift für Physikalische Chemie* **50U**, 641-712(1905).
11. Y. Tian, R. Mei, D. zhong Xue, X. Zhang and W. Peng, Enhanced Electrocatalytic Hydrogen Evolution in Graphene via Defect Engineering and Heteroatoms Co-Doping, *Electrochim. Acta* **219**, 781-789(2016).
12. Y. Jia, L. Zhang, A. Du, G. Gao, J. Chen, X. Yan, C. L. Brown, X. Yao, Y. Jia, L. Z. Zhang, X. C. Yan, X. D. Yao, C. L. Brown, A. J. Du, G. P. Gao and J. Chen, Defect Graphene as a Trifunctional Catalyst for Electrochemical Reactions, *Adv. Mater.* **28**(43), 9532-9538(2016).
13. Y. Jiao, Y. Zheng, K. Davey and S. Z. Qiao, Activity Origin and Catalyst Design Principles for Electrocatalytic Hydrogen Evolution on Heteroatom-Doped Graphene, *Nat. Energy* **1**, 16130(2016).

14. H. Jiang, Y. Zhu, Y. Su, Y. Yao, Y. Liu, X. Yang and C. Li, Highly Dual-Doped Multilayer Nanoporous Graphene: Efficient Metal-Free Electrocatalysts for the Hydrogen Evolution Reaction, *J. Mater. Chem. A.* **3**(24), 12642-12645(2015).
15. Y. Ito, W. Cong, T. Fujita, Z. Tang and M. Chen, High Catalytic Activity of Nitrogen and Sulfur Co-Doped Nanoporous Graphene in the Hydrogen Evolution Reaction, *Angew. Chem.* **127**(7), 2159-2164(2015).
16. M. Chhetri, S. Maitra, H. Chakraborty, U. V. Waghmare and C. N. R. Rao, Superior Performance of Borocarbonitrides, $B_xC_yN_z$, as Stable, Low-Cost Metal-Free Electrocatalysts for the Hydrogen Evolution Reaction, *Energy Environ. Sci.* **9**(1), 95-101(2016).
17. T. F. Jaramillo, K. P. Jørgensen, J. Bonde, J. H. Nielsen, S. Horch and I. Chorkendorff, Identification of Active Edge Sites for Electrochemical $H_2$ Evolution from $MoS_2$ Nanocatalysts, *Science* **317**(5834), 100-102(2007).
18. J. Kibsgaard, Z. Chen, B. N. Reinecke and T. F. Jaramillo, Engineering the Surface Structure of $MoS_2$ to Preferentially Expose Active Edge Sites for Electrocatalysis *Nat. Mater,* **11**(11), 963-969(2012).
19. Y. Li, H. Wang, L. Xie, Y. Liang, G. Hong and H. Dai, $MoS_2$ Nanoparticles Grown on Graphene: An Advanced Catalyst for the Hydrogen Evolution Reaction, *J. Am. Chem. Soc.* **133**(19), 7296-7299(2011).
20. K. Pramoda, U. Gupta, M. Chhetri, A. Bandyopadhyay, S. K. Pati and C. N. R. Rao, Nanocomposites of $C_3N_4$ with Layers of $MoS_2$ and Nitrogenated RGO, Obtained by Covalent Cross-Linking: Synthesis, Characterization, and HER Activity, *ACS Appl. Mater. Interfaces.* **9**(12), 10664-10672(2017).
21. K. Pramoda, M. M. Ayyub, N. K. Singh, M. Chhetri, U. Gupta, A. Soni and C. N. R. Rao, Covalently Bonded $MoS_2$-Borocarbonitride Nanocomposites Generated by Using Surface Functionalities on the Nanosheets and Their Remarkable HER Activity, *J. Phys. Chem. C* **122**(25), 13376-13384(2018).
22. D. Voiry, M. Salehi, R. Silva, T. Fujita, M. Chen, T. Asefa, V. B. Shenoy, G. Eda and M. Chhowalla, Conducting $MoS_2$ Nanosheets as Catalysts for Hydrogen Evolution Reaction, *Nano Lett.* **13**(12), 6222-6227(2013).
23. X. Yang and D. Wang, Photocatalysis: From Fundamental Principles to Materials and Applications, *ACS Appl. Energy Mater.* **1**(12), 6657-6693(2018).
24. Y. Tachibana, L. Vayssieres and J. R. Durrant, Artificial Photosynthesis for Solar Water-Splitting, *Nat. Photonics* **6**(8), 511-518(2012).
25. McNaught Alan D, *Compendium of Chemical Terminology*, Oxford: Blackwell Science, 1669(1997).
26. M. M. Ayyub, R. Singh and C. N. R. Rao, Hydrogen Generation by Solar Water Splitting Using 2D Nanomaterials, *Solar RRL* **4**(8), 2000050(2020).
27. K. Maeda and K. Domen, Photocatalytic Water Splitting: Recent Progress and Future Challenges, *J. Phys. Chem. Lett.* **1**(18), 2655-2661(2010).
28. J. Schneider and D. W. Bahnemann, Undesired Role of Sacrificial Reagents in Photocatalysis, *J. Phys. Chem. Lett.* **4**(20), 3479-3483(2013).

29. P. V. Kamat and S. Jin, Semiconductor Photocatalysis: "Tell Us the Complete Story!", *ACS Energy Lett.* **3**(3), 622-623(2018).
30. F. Costantino and P. V. Kamat, Do Sacrificial Donors Donate $H_2$ in Photocatalysis?, *ACS Energy Lett.* **7**(1), 242-246(2022).
31. K. Maeda, Photocatalytic Water Splitting Using Semiconductor Particles: History and Recent Developments, *J. Photochem. Photobiol., C* **12**(4), 237-268(2011).
32. S. R. Lingampalli, K. Manjunath, S. Shenoy, U. V. Waghmare and C. N. R. Rao, $Zn_2NF$ and Related Analogues of ZnO, *J. Am. Chem. Soc.* **138**(26), 8228-8234(2016).
33. N. Kumar, U. Maitra, V. I. Hegde, U. V. Waghmare, A. Sundaresan and C. N. R. Rao, Synthesis, Characterization, Photocatalysis, and Varied Properties of $TiO_2$ Cosubstituted with Nitrogen and Fluorine, *Inorg. Chem.* **52**(18), 10512-10519(2013).
34. Q. Li, H. Meng, P. Zhou, Y. Zheng, J. Wang, J. Yu and J. Gong, $Zn_{1-x}Cd_xS$ Solid Solutions with Controlled Bandgap and Enhanced Visible-Light Photocatalytic $H_2$-Production Activity, *ACS Catal.* **3**(5), 882-889(2013).
35. D. M. Andoshe, J. M. Jeon, S. Y. Kim and H. W. Jang, Two-Dimensional Transition Metal Dichalcogenide Nanomaterials for Solar Water Splitting, *Electron. Mater. Lett.* **11**(3), 323-335(2015).
36. B. Hinnemann, P. G. Moses, J. Bonde, K. P. Jørgensen, J. H. Nielsen, S. Horch, I. Chorkendorff and J. K. Nørskov, Biomimetic Hydrogen Evolution: $MoS_2$ Nanoparticles as Catalyst for Hydrogen Evolution, *J. Am. Chem. Soc.* **127**(15), 5308-5309.
37. H. Wang, L. Zhang, Z. Chen, J. Hu, S. Li, Z. Wang, J. Liu and X. Wang, Semiconductor Heterojunction Photocatalysts: Design, Construction, and Photocatalytic Performances, *Chem. Soc. Rev.* **43**(15), 5234-5244(2014).
38. J. Ran, J. Zhang, J. Yu, M. Jaroniec and S. Z. Qiao, Earth-Abundant Cocatalysts for Semiconductor-Based Photocatalytic Water Splitting, *Chem. Soc. Rev.* **43**(22), 7787-7812(2014).
39. S. J. A. Moniz, S. A. Shevlin, D. J. Martin, Z. X. Guo, J. Tang, Visible-Light Driven Heterojunction Photocatalysts for Water Splitting – a Critical Review, *Energy Environ. Sci.* **8**(3), 731-759(2015).
40. G. Wang, J. Chang, W. Tang, W. Xie and Y. S. Ang, 2D Materials and Heterostructures for Photocatalytic Water-Splitting: A Theoretical Perspective, *J. Phys. D: Appl. Phys.* **55**(29), 293002(2022).
41. P. Wang, Y.-J. Yuan, Q.-Y. Liu, Q. Cheng, Z.-K. Shen, Z.-T. Yu, Z. Zou, P. Wang, Y.-J. Yuan, Q.-Y. Liu, Q. Cheng, Z.-K. Shen, Z.-T. Yu and Z. Zou, Solar-Driven Lignocellulose-to-$H_2$ Conversion in Water Using 2D-2D $MoS_2/TiO_2$ Photocatalysts, *ChemSusChem* **14**(14), 2860-2865(2021).
42. G. Yang, H. Ding, D. Chen, J. Feng, Q. Hao and Y. Zhu, Construction of Urchin-like ZnIn2S4-Au-TiO2 Heterostructure with Enhanced Activity for Photocatalytic Hydrogen Evolution, *Appl. Catal. B* **234**, 260-267(2018).

43. D. V. Talapin, J. H. Nelson, E. V. Shevchenko, S. Aloni, B. Sadtler and A. P. Alivisatos, Seeded Growth of Highly Luminescent CdSe/CdS Nanoheterostructures with Rod and Tetrapod Morphologies, *Nano Lett.* **7**(10), 2951-2959(2007).
44. S. R. Lingampalli, U. K. Gautam and C. N. R. Rao, Highly Efficient Photocatalytic Hydrogen Generation by Solution-Processed ZnO/Pt/CdS, ZnO/Pt/$Cd_{1-x}Zn_xS$ and ZnO/Pt/$CdS_{1-x}$ $Se_x$ Hybrid Nanostructures, *Energy Environ. Sci.* **6**(12), 3589-3594(201).
45. X. Ruan, X. Cui, Y. Cui, X. Fan, Z. Li, T. Xie, K. Ba, G. Jia, H. Zhang, L. Zhang, W. Zhang, X. Zhao, J. Leng, S. Jin, D. J. Singh and W. Zheng, Favorable Energy Band Alignment of $TiO_2$ Anatase/Rutile Heterophase Homojunctions Yields Photocatalytic Hydrogen Evolution with Quantum Efficiency Exceeding 45.6%, *Adv. Energy Mater.* **12**(16), 2200298(2022).
46. Z. Lei, W. You, M. Liu, G. Zhou, T. Takata, M. Hara, K. Domen and C. Li, Photocatalytic Water Reduction under Visible Light on a Novel $ZnIn_2S_4$ Catalyst Synthesized by Hydrothermal Method, *Chem. Commun.* **3**(17), 2142-2143(2003).
47. X. Shi, L. Mao, P. Yang, H. Zheng, M. Fujitsuka, J. Zhang, T. Majima, Ultrathin ZnIn2S4 Nanosheets with Active (110) Facet Exposure and Efficient Charge Separation for Cocatalyst Free Photocatalytic Hydrogen Evolution, *Appl. Catal. B* **265**, 118616(2020).
48. S. Zhang, X. Liu, C. Liu, S. Luo, L. Wang, T. Cai, Y. Zeng, J. Yuan, W. Dong, Y. Pei and Y. Liu, $MoS_2$ Quantum Dot Growth Induced by S Vacancies in a $ZnIn_2S_4$ Monolayer: Atomic-Level Heterostructure for Photocatalytic Hydrogen Production, *ACS Nano* **12**(1), 751-758(2018).
49. X. Shi, C. Dai, X. Wang, J. Hu, J. Zhang, L. Zheng, L. Mao, H. Zheng and M. Zhu, Protruding Pt Single-Sites on Hexagonal $ZnIn_2S_4$ to Accelerate Photocatalytic Hydrogen Evolution, *Nat. Commun.* **13**, 1287(2022).
50. A. Roy, M. Chhetri, S. Prasad, U. V. Waghmare and C. N. R. Rao, Unique Features of the Photocatalytic Reduction of $H_2O$ and $CO_2$ by New Catalysts Based on the Analogues of CdS, $Cd_4P_2X_3$ (X = Cl, Br, I), *ACS Appl. Mater. Interfaces* **10**(3), 2526-2536(2018).
51. U. Maitra, U. Gupta, M. De, R. Datta, A. Govindaraj and C. N. R. Rao, Highly Effective Visible-Light-Induced $H_2$ Generation by Single-Layer 1T-$MoS_2$ and a Nanocomposite of Few-Layer 2H-$MoS_2$ with Heavily Nitrogenated Graphene, Angew. *Chem. Int. Ed.* **52**(49), 13057-13061(2013).
52. R. Singh, U. Gupta, V. S. Kumar, M. M. Ayyub, U. V. Waghmare and C. N. R. Rao, Dependence of the Properties of 2D Nanocomposites Generated by Covalent Crosslinking of Nanosheets on the Interlayer Separation: A Combined Experimental and Theoretical Study, *ChemPhysChem* **20**(13), 1728-1737(2019).

https://doi.org/10.1142/9789811270956_0008

# Chapter 8

## Thermochemical $CO_2$ Conversion to High Energy Dense Fuels

Sathyapal R. Churipard[1,2] and Sebastian C. Peter[1,2,*]

[1]*New Chemistry Unit, Jawaharlal Nehru Centre for Advanced Scientific Research, Bangalore 560064, India*
[2]*School of Advanced Materials, Jawaharlal Nehru Centre for Advanced Scientific Research, Bangalore 560064, India*
[*]*sebastiancp@jncasr.ac.in*

Exploring novel heterogeneous catalysts and processes for the synthesis of high-energy-dense fuels via $CO_2$ hydrogenation provides an alternative and sustainable solution to reduce atmospheric $CO_2$ concentration and meet the world energy demand. Olefins, higher alcohols, and aromatics represent some of the important energy-dense fuels that can be synthesized by $CO_2$ hydrogenation processes. However, synthesizing them in higher yields is highly challenging and requires rational catalyst design and precise control over the active species and processes. The direct synthesis of olefins, higher alcohols and aromatics through $CO_2$ hydrogenation is highly challenging as it demands precise control on C-O and C-C coupling to form long-chain hydrocarbons. Despite these challenges, significant progress has been achieved recently in developing efficient catalysts and understanding the mechanism of these reactions. Several strategies were developed and employed to improve the catalytic activities. In this book chapter, we provide an overview of catalysts and processes employed for the synthesis of olefins, higher alcohols, and aromatics via $CO_2$ hydrogenation. In particular, the catalyst design and catalytic mechanism are explained in detail.

# 1. Introduction

## 1.1. *Energy crisis and global warming issues*

The world has witnessed an unprecedented surge in the usage of fossil fuels to supply the rising energy demand of the world. Today over 80% of world energy demand is supplied by fossil fuels. The coal and hydrocarbons are formed from organic materials through long geological processes (106 to 108 years) that converted the solar energy via photosynthesis from $CO_2$ and water and harnessed them into chemicals. The burning of these fossil fuels to produce energy leads to the concomitant re-emission of $CO_2$ and drastically increases the atmospheric $CO_2$ levels. Every day, 12 million tons of oil, 22 million tons of coal and 10 billion $m^3$ of natural gas are utilized to meet the world energy demand which accounts for the release of approximately 30 billion tons of $CO_2$ into the atmosphere every year. This enormous amount of $CO_2$ released into the atmosphere plays a key role in climate change, global warming and ocean acidification which can pose a serious threat to humankind.[1-10]

The greenhouse gases such as $CO_2$, water vapor, $CH_4$, ozone ($O_3$), nitrous oxide ($N_2O$) and chlorofluorocarbons (CFCs), trap the sunlight on the earth's surface leading to an increase in the earth's temperature. It is evident from several studies that anthropogenic activities since the advent of the industrial revolution are the main cause for the increase in atmospheric $CO_2$ levels and also for the rise in average global temperature.[11] This has triggered intense research to limit atmospheric $CO_2$ levels and convert them into high-value chemicals. The intriguing number of publications on $CO_2$ capture and conversion in recent years testify the compelling need to curtail atmospheric $CO_2$ levels. In this context, carbon capture and storage (CCS) and $CO_2$ capture and utilization (CCU), have become indispensable tools for the mitigation of anthropo-genic $CO_2$ emissions.[6,12-14] In particular, CCU offers promising potential because it not only aims at $CO_2$ capture from the industrial exhaust or the atmosphere but also focuses on using the captured $CO_2$ as a feedstock to produce high-value chemicals for a sustainable supply of energy to the world.[12]

### 1.2. *Existing technologies to reduce $CO_2$*

Various environmental regulations and norms are introduced to reduce $CO_2$ emissions. Several strategies are employed and assessed which include the implementation of greener technologies and green fuels, increasing the efficiency in electric energy production (but it involves high cost and time), and increasing the efficiency of energy utilization. The use of fuel-efficient vehicles and having a responsible attitude towards the usage of energy can also significantly contribute to saving fossil fuel resources. The fuel shift from coal to gas for energy generation is another option to reduce $CO_2$ emissions because the use of coal leads to the production of a comparatively higher amount of $CO_2$ (950g of $CO_2$ per kWh) than natural gas (497g $CO_2$ per kWh) but it is difficult for countries which have low natural gas and abundant coal reserves.[15-18] The use of renewable energy sources such as wind, solar and geothermal energy is another rapidly growing area to address global climate change. Much focus is given to electricity production by improved technologies. Photo-voltaics (PVs) and wind-derived energy are contributing towards the replacement of fossil fuels and have achieved reasonably good success in many countries. The use of biomass for the production of fuels can also help to reduce the dependence on fossil fuels. Biomass has the potential to replace 20% of fossil fuels in the transportation sector 16. Nevertheless, the two major promising routes for mitigating $CO_2$ from the atmosphere are CCS and CCU. In CCS, $CO_2$ is captured and stored as geological sequestration or in the deep ocean which is not desirable. Whereas in CCU, the $CO_2$ is converted into a range of useful products which offers a productive use and recycling of $CO_2$ rather than treating it as waste by storing it.[19,20] The $CO_2$ can be converted into a different class of fuels such as higher alcohols, olefins and aromatics. Table 1 shows the comparison of various fuels synthesized from $CO_2$ with conventional fuels.

The utilization of $CO_2$ for the synthesis of chemicals is an old approach that dates back to the 19$^{th}$ century wherein $CO_2$ is used as feedstocks to produce a range of chemicals such as salicylic acid, urea, carbonates, etc. However, $CO_2$-capture and conversion are the most sought-after recent approaches to curb atmospheric $CO_2$ levels. In particular, $CO_2$ reduction reactions have gained enormous research interest from both academic and

Table 1. Comparison of energy densities of conventional fuels with that of fuels synthesized from $CO_2$.

| Entry | Fuel | Source | Specific energy (kJ/g) | Energy density (KWh/gal) |
|---|---|---|---|---|
| 1 | Methane | $CO_2$ based | 56.0 | 27.0 |
| 2 | Formic acid | $CO_2$ based | - | 6.7 |
| 3 | Propane | $CO_2$ based | 50.4 | 26.8 |
| 4 | Ethanol | $CO_2$ based | 29.7 | 24.7 |
| 5 | Biodiesel | Conventional | 39.6 | 35.0 |
| 6 | Coal | Conventional | 30.2 | 22.9 |
| 7 | Natural gas | Conventional | 53.0 | - |
| 8 | Gasoline | Conventional | 46.5 | 36.6 |
| 9 | Diesel | Conventional | 45.8 | 40.6 |
| 10 | Crude oil | Conventional | 44.0 | - |
| 11 | $H_2$ | Conventional | 141.9 | 10.1 |
| 12 | Wood | Conventional | 14.9 | 11.3 |

industrial research.[21] $CO_2$ conversion to value-added products is possible via various pathways such as electrochemical, photochemical, and thermochemical with the aid of heterogeneous, homogeneous, or enzymatic catalysis. Among these, the thermochemical route has an upper hand from the industrial standpoint owing to its scalability, economic viability, and sustainability of the process.[22,23] In addition, the heterogeneously catalyzed thermochemical process offers advantages such as easy separation of products and catalysts, recycling, scaling-up and stability of the catalysts.[17] Several valuable chemicals can be synthesized by $CO_2$ reduction such as CO, formic acid, methane, olefins, aromatics, alcohols, paraffins, etc. However, to reduce the $CO_2$ emissions substantially from the atmosphere, the most reliable and straightforward solution is to convert it to fuels or commodity chemicals rather than fine chemicals. The demand for fuels and commodity chemicals is significantly higher compared to fine chemicals. For instance, fuel consumption accounts for 30% of total $CO_2$ emissions whereas organic chemicals account for only 4%. Therefore, $CO_2$ hydrogenation to fuels offers a greater impact in reducing the $CO_2$ from the atmosphere rather than converting it to fine chemicals.[1,24] However, despite years of fundamental and applied research on $CO_2$ hydrogenation, there are only a few processes that are applied commercially for the synthesis of important chemicals. Figure 1 shows some of the important energy-dense fuels that can be synthesized from $CO_2$ and their economic values.

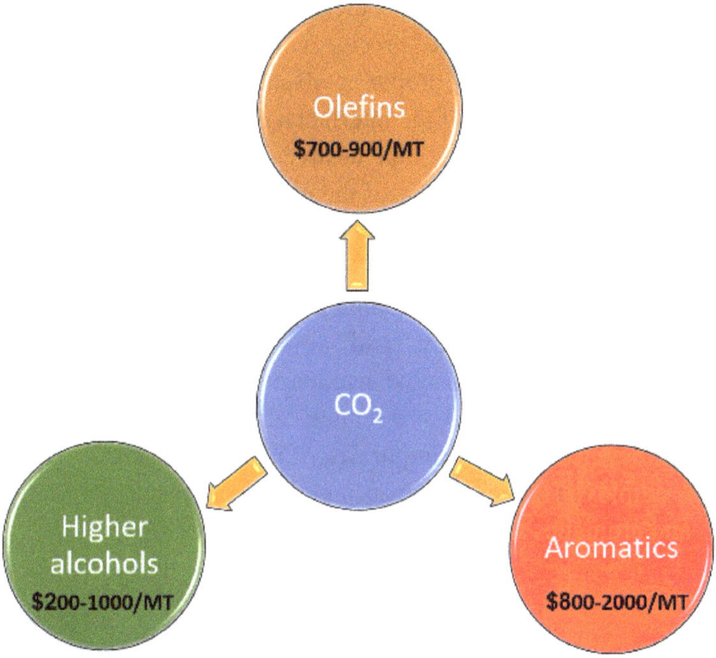


Figure 1. Important energy-dense fuels synthesized via $CO_2$ hydrogenation and their economic values.

## 2. Challenges In Current Processes And Material Design

$CO_2$ molecule has high thermodynamic stability ($\Delta G°$ f = −394.4 kJ mol-1 at 25°C) and strong C = O bond energy (799 kJ mol-1). Consequently, overcoming the chemical inertness of $CO_2$ demands high reaction temperature and optimum reaction conditions to convert $CO_2$ to high-energy-dense fuels. Many of the bifunctional catalysts employed for $CO_2$ hydrogenation to aromatics and olefins are still used as physical mixtures. These catalysts exhibit low single-pass yields towards higher olefins and aromatics. The Anderson−Schulz−Flory (ASF) rule of C-C coupling and hydrocarbon pool mechanism shifts the selectivity towards CO and lower hydrocarbons. Therefore, designing efficient catalysts for selectively producing higher olefins and aromatics in high yield has a huge demand.[1]

The RWGS or methanol synthesis has to be efficiently combined with carbon chain growth in FT synthesis to achieve high selectivity towards heavy unsaturated hydrocarbons. A mere physical mixing of catalysts may not be sufficient to achieve precise control over the selectivity of hydrocarbons but requires a proper catalyst design with well-defined 3D structures. Efficient tuning of the interfacial morphology, particle size and crystalline effect can bring about desired catalytic performance in the reaction. In addition, the stability of the catalyst is a major concern for its commercial applicability. Many of the catalysts employed are prone to deactivation and poisoning and necessitate the development of strategies to reduce these deactivations.[9] Modulating the acidity of the catalysts, creating hierarchical porosity, and developing encapsulated catalyst nanoparticles or shape-controlled nanoparticles on suitable support are some of the efficient strategies to reduce catalyst deactivation and also to improve tolerance against impurities. Different catalytic routes and catalysts employed for $CO_2$ hydrogenation to synthesize higher alcohols, olefins and aromatics are demonstrated in Figure 2.[25-27]

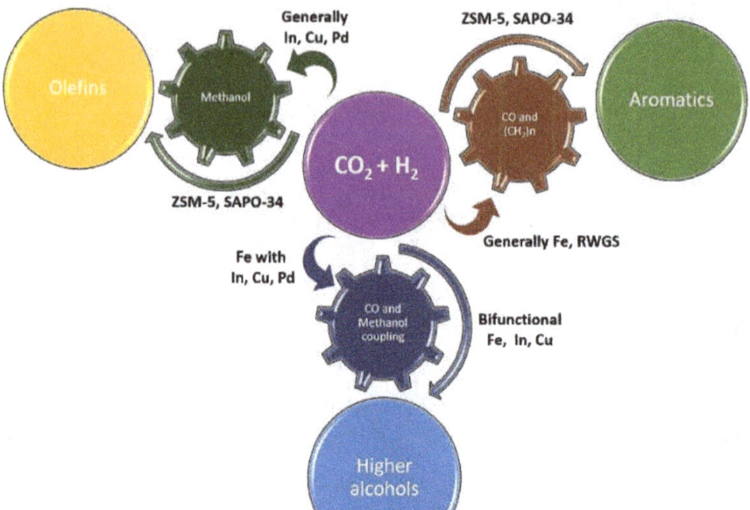


Figure 2. Catalytic routes employed for the synthesis of energy-dense fuels via $CO_2$ hydrogenation.

## 3. Products, Catalysts, and Reaction Mechanisms

### 3.1. *$CO_2$ to olefins*

Conversion of $CO_2$ to olefins is perhaps one of the most attractive routes for converting $CO_2$ to useful chemicals. Especially lower olefins such as ethylene and propylene are the most consumed products in the world due to their application in the production of polymers, plastics, etc. The light olefins can further undergo oligomerization to long-chain hydrocarbons which can be used as fuels. Compared to lower hydrocarbons, higher hydrocarbons have higher energy densities. The higher olefins find immense applications in the synthesis of polyolefins, surfactants, and aliphatic alcohols. They are also used as platform chemical feedstocks.[28,9] The hydrogenation of $CO_2$ to olefins is currently done in 3 routes: (a) MeOH-mediated, (b) Direct hydrogenation of $CO_2$ to olefins over multifunctional catalysts, (c) CO-mediated pathway via RWGS followed by FT synthesis to olefins (FTO). Tuning the selectivity of $CO_2$ hydrogenation to olefins needs a fundamental understanding of the above-mentioned routes. The RWGS route is endothermic and thus it is more favored at high temperatures, whereas the methanol synthesis and $CO_2$ methanation are exothermic and are favored at lower temperatures.[29] This shows that developing a catalyst is challenging for the selective production of olefins. Despite the challenges associated with this reaction, significant progress has been achieved for $CO_2$ hydrogenation to olefins via the FTS route as well as through the methanol route. In the FTS route, CO, Fe and Ru-based catalysts are predominantly used. The Fe and CO-based catalysts are mostly preferred for the synthesis of hydrocarbons from FTS, but they exhibit lower selectivity for light olefins. However, in the methanol route higher selectivity to light olefins can be achieved by the combination of methanol synthesis and methanol to olefins (MTO) over bifunctional catalysts. In particular, silicoaluminophosphate (SAPO) is most widely used catalyst for $CO_2$ hydrogenation to lighter olefins. Especially, SAPO-34 with chabazite (CHA) structure is most preferred owing to its small pores, good hydrothermal stability, and moderate reactivity. The pores of CHA are best suited to harvest light olefin intermediates and the tuning of Si content in the SAPO-34 can also play a

Table 2. Comparison of catalytic performance of various catalysts for $CO_2$ hydrogenation to olefins via a $CO_2$–FTS route.

| Catalysts | Reaction conditions | | | | $CO_2$ conv. (%) | CO sel. (%) | Hydrocarbon distribution (%) | | | | LAOs sel. (%)[e] | $C_{4+}^{=}$ STY[f] | Stability test (h) | Ref |
|---|---|---|---|---|---|---|---|---|---|---|---|---|---|---|
| | GHSV[a] | Temp.[b] (°C) | $H_2/CO_2$[c] | Pressure[d] (bar) | | | $CH_4$ | $C_2^{=}$-$C_3^{=}$ | $C_{4+}^{=}$ | $C_{2+}$ paraffins | | | | |
| $CuFeO_2$ | 1800 | 300 | 3 | 10 | 16.7 | 31.4 | 2.4 | 28.9[g] | 64.9[h] | - | - | - | 100 | 33 |
| $Fe_2Zn$ | 15000 | 330 | 3 | 15 | 43.5 | 9.2 | 11.5 | 16.6 | 60.7 | 11.2 | 54.2 | 340.3 | 100 | 34 |
| Fe/Co-$Y_K$ | 2000 | 300 | 2.65 | 10 | 25.9 | 21.1 | 13.9 | 25.0 | 45.9 | 15.2 | 36.2 | - | 100 | 35 |
| FeMnNa | 12000 | 340 | 3 | 20 | 35.0 | 18.1 | 13.1 | 22.7 | 52.5 | 11.7 | N.G. | 395.9[f] | 100 | 36 |
| Fe/C-$K_2CO_3$-KCl | 2400 | 320 | 3 | 10 | 32.6 | 22.4 | 10.6 | 19.6 | 56.6 | 13.2 | 40.4 | 20.6 | 100 | 37 |
| FeK/SWNTs | 9000 | 340 | 3 | 20 | 52.7 | 9.6 | 13.5 | 14.7 | 47.6 | 24.2 | N.G. | 115.1 | 120 | 38 |
| Fe/C-bio | 2400 | 320 | 3 | 10 | 31.0 | 23.2 | 13.9 | 21.7 | 50.3 | 14.1 | 40 | 17.2 | 100 | 39 |
| Fe-Mn-K | 2400 | 300 | 3 | 10 | 38.2 | 5.6 | 10.4 | 24.2[h] | 61.9[i] | - | - | - | 80 | 40 |
| $FeAlO_{x-5}$ | 2000 | 330 | 1 | 35 | 20.2 | 16.8 | 5.4 | 11.7 | 66.8 | 16.1 | 52.4 | 11.0 | 450 | 41 |
| Na-Zn-Fe | 15000 | 340 | 3 | 25 | 38.0 | 15.0 | 13.0 | 78.0[j] | | 9.0 | | 570.0[f] | 100 | 42 |
| Zn-$CrO_x$/MSAPO | - | 400 | 2.5 | 25 | 17 | - | - | 80[g] | - | - | - | 0.5[j] | - | 43 |
| K-FeCu/ZSM-51 | - | 300 | 2 | 10 | 81.1 | - | - | 30.3 | - | - | - | 0.32[j] | - | 44 |
| $In_2O_3$-$ZrO_2$/SAPO-34 | - | 400 | 3 | 15 | 20.0 | - | - | 80-90 | - | - | - | 0.22-0.25 | - | 45 |

[a]Gas hourly space velocity; [b]Reaction temperature; [c]$H_2/CO_2$ ratio; [d]Reaction pressure; [e]Selectivity of linear a-olefins among hydrocarbons; [f]STY: space-time yield of $C_{4+}$ olefins; [g]STY of total olefins; [h]$C_2$-$C_4$ olefin selectivity; [i]$C_{5+}$ olefin selectivity; [j]Total olefin selectivity; [j] STY in L $h^{-1}g^{-1}$.

major role in olefin selectivity.[30,31] Some of the best catalysts reported in the literature and their catalytic performance for $CO_2$ hydrogenation to olefins via FTS-route and MeOH-route are listed in Table 2.

The $CO_2$ hydrogenation to olefins proceeds majorly through $CO_2$-FTS because the Fe-based catalyst is active for both RWGS and FTS. The catalyst exists in different phases and the activation and reaction induce several structural transformations in the structure of the catalyst. It leads to the formation of different iron phases such as iron carbides, metallic iron, and iron oxides. The $Fe_3O_4$ phase is active for RGWS and converts $CO_2$ to CO.[26] The adsorbed CO* can dissociate into surface carbon and surface oxygen species during the FTS reaction. Later, the diffusion of surface carbon species into the Fe lattice forms $Fe_7C_3$, $\chi$-$Fe_5C_2$, 0-$Fe_3C$, $\varepsilon$-$Fe_3C$ and $\varepsilon$-$Fe_{2.2}C$ phases in different reaction conditions. Among these Fe carbides formed, $\chi$-$Fe_5C_2$ plays a major role in FTS reaction and is responsible for chain growth and olefin selectivity.

Recently, Zhang *et al.* studied the $CO_2$ hydrogenation to $C_{2+}$ olefins on Na and Zn promoted iron catalyst.[32] The mechanistic studies revealed that ZnO is the active species that catalyzes the RWGS reaction. The adsorption of $CO_2$ and $H_2$ on ZnO leads to the formation of activated $CO_2$* and H* species which subsequently gets hydrogenated to form CO* and OH* species. The OH* species interacts with another H* to form water. Whereas, the CO* formed near the proximity of $\chi$-$Fe_5C_2$ will undergo hydrogenation (FTO reaction) to form $CH_x$ ($1 \leq x \leq 3$). The $CH_x$ thus formed will be converted to $C_nH_m$ intermediates through C-C coupling which upon hydrogenation finally leads to the formation of $C_2$-$C_{12}$ olefins. On the other hand, the CO* formed away from ZnO species initially desorbs to the gaseous phase and diffuses to $\chi$-$Fe_5C_2$ and undergoes a series of transformations mentioned above to form $C_2$-$C_{12}$ olefins (Figure 3). The presence of $Na^+$ on the surface of $\chi$-$Fe_5C_2$ facilitates the desorption of $C_2$-$C_{12}$ olefins, regulates the hydrogenation of $C_nH_m$ intermediates and improves the selectivity towards $C_2$-$C_{12}$ olefins. The proximity between ZnO (RWGS catalyst) and $\chi$-$Fe_5C_2$ (FTO catalyst) also plays a decisive role in controlling the selectivity of products. Increasing the proximity between these catalysts significantly enhanced the olefin selectivity from 70% to 78% along with suppressing the $CH_4$ selectivity.

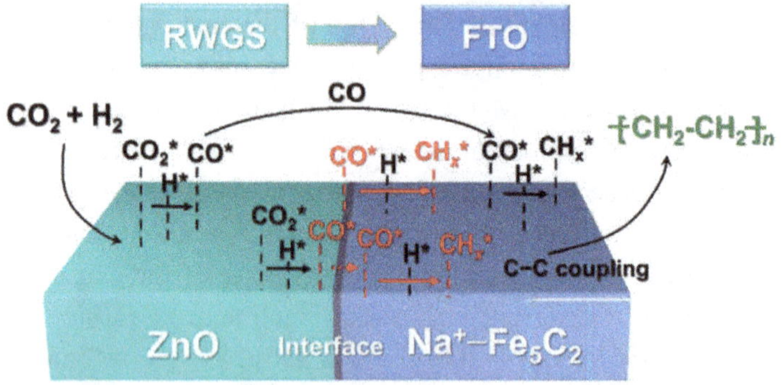


Figure 3. Schematic illustrating the reaction mechanism for hydrogenation of $CO_2$ to olefins over the Na-Zn promoted Fe Catalyst. Reproduced with permission from Ref. 32 Copyright 2021 Elsevier.

### 3.2. *$CO_2$ to aromatics*

Aromatics are one of the most significant platform molecules for the chemical industry often produced by energy-intensive processes like naphtha reforming. However, the depleting crude oil reserves demand an alternative route to produce aromatics. The synthesis of aromatics via catalytic $CO_2$ hydrogenation provides a non-fossil fuel route and is a viable way to address the dual challenges of the impending energy and climate crises.[9,46,47] In the past decades, much focus was given to $CO_2$ hydrogenation to methanol and dimethyl ether (DME). However, recently conversion of $CO_2$ to aromatics has received tremendous interest. Nevertheless, the sluggish kinetics and catalyst deactivation by carbon deposition remains a major bottleneck in the process. Overcoming the chemical inertness and thermodynamic stability of $CO_2$, its conversion to aromatics demands an efficient catalyst with superior stability.[47,48] The direct synthesis of aromatics from $CO_2$ hydrogenation was first reported in 1991 by Kuei *et. al.* over a combination of fused iron Fischer-Tropsch synthesis (FTs) catalysts and HZSM-5 in a single-stage reactor.[49] They could achieve a maximum aromatic selectivity of 21.4% (exclusive of CO) with 38% $CO_2$ conversion.

Currently, $CO_2$ hydrogenation to aromatics is majorly done via two routes: (a) $CO_2$-MeOH and (b) $CO_2$-FT route. In the methanol-mediated route, the synthesis of aromatics involves the use of bifunctional catalysts which include methanol synthesis catalysts (such as $ZnAlO_x$, Zn-$ZrO_2$, $ZnCrO_x$, $Cr_2O_3$ and $In_2O_3$), which converts $CO_2$ and $H_2$ to methanol through formate or CO-mediated pathway and active catalyst for methanol to aromatics (MTA) such as zeolites. ZSM-5 zeolite is the most preferred catalyst because of its well-dispersed active sites, tunable acid strength/amount, well-defined porous structure, and its ability to facilitate the carbon chain length. Methanol formed is converted to aromatics inside the pores of zeolites via the hydrocarbon pool mechanism ($nCH_3OH \rightarrow C_nH_{2n-6} + nH_2O + 3H_2$(MTA).[50,51] The $CO_2$ hydrogenation to methanol process is thermodynamically favorable at high pressure and low temperature (473–573 K) often on noble metal-based catalysts, on the other hand, MTA is favoured at low pressure and high temperatures (673–823).[52] Therefore, single-pass $CO_2$ conversion to aromatics will be more feasible than operating it separately as the single-pass can eliminate the tedious reactor design for the two-step process. Lately, metal oxide-based catalysts, especially $In_2O_3$ and $Ga_2O_3$ have shown superior activity and higher selectivity than the noble metals-based catalysts at high temperatures. This offers a possibility to synthesize MeOH in reaction conditions that favor aromatics formation.[48,53,54,55] Although the $CO_2$–MeOH route gives a $CO_2$ conversion of 9–34% with 56–80% selectivity to hydrocarbons, the major challenge in MeOH-mediated route is the formation of high CO (> 35% selectivity) due to RWGS ($CO_2 + H_2 \rightarrow CO + H_2O$) which is more favorable in the reaction conditions employed. The co-feeding of CO was considered as an efficient strategy to suppress the undesirable CO formation, however, it still suffers from low reaction rates and low space-time yield (STY) which makes this route inefficient compared to the $CO_2$-FTS route.[56]

To overcome the high CO formation through RWGS, Zhou *et al.* performed $CO_2$ hydrogenation over bifunctional catalysts comprising different metal oxides such as $In_2O_3$–$ZrO_2$, Cu–Zn–Al, $ZnGa_2O_4$, ZnO–$ZrO_2$ and $ZnAl_2O_4$ and zeolite (HZSM-5) prepared via coprecipitation methods.[57] Among these, the ZnO–$ZrO_2$/ZSM-5 catalyst demonstrated the highest STY for aromatics with low CO selectivity. This catalyst was further explored with different synthesis methods such as sol-gel and hard template routes. It was observed that the ZnO–$ZrO_2$ aerogel catalyst synthesized by the sol-gel method exhibited high surface area and enhanced oxygen vacancies which resulted in higher aromatic STY during $CO_2$ hydrogenation compared to

$ZnO–ZrO_2$ catalysts prepared by other methods. The highest aromatic selectivity of 76% and STY of 29.9 mg $h^{-1}g^{-1}$ were achieved at 16% $CO_2$ conversion using $ZnO–ZrO_2$/ZSM-5 catalyst. The summary of catalysts reported for $CO_2$ hydrogenation to aromatics is listed in Table 3.

Another promising route for the synthesis of aromatics from $CO_2$ hydrogenation is modified Fischer-Tropsch synthesis (FTs). It involves FT synthesis reactions and reverse water gas shift (RWGS) and is extensively used for the direct synthesis of gasoline range hydrocarbons from $CO_2$ over multifunctional metal/zeolite catalysts.[58] Iron-based catalysts are predominantly used for modified-FT synthesis owing to the dual capability of Fe-based catalysts to facilitate both RWGS and FTs reactions.[59] In addition, the presence of promoters such as K or Na can further enhance the activity of Fe-based catalysts for alkene synthesis.[60,61] For instance, the influence of Na doping on $Fe_3O_4$ towards $CO_2$ hydrogenation to light olefins was explained by Wei *et al.* wherein the presence of Na enhances the surface basicity and improves the carbonization of iron-based catalysts which leads to the enhanced olefin production.[62]

Similarly, the effect of different promoters was studied by Gascon *et al.* Among the different promoters used, K was the most effective in enhancing the olefin selectivity. K increases the $CO_2$ and CO uptake and meanwhile minimizes the $H_2$ affinity. It also enables the catalyst to retain the Fe active phases such as iron oxide and iron carbides which results in excellent catalytic activity.[63] The nature and textural properties of zeolites used can also play a significant role in altering the aromatic selectivity in the $CO_2$-FT route for aromatics.[26] Although the H-ZSM-5 topology is best suited for the synthesis of aromatics from alkenes or MeOH, the presence of undesirable acidic properties and narrow channels often leads to over-hydrocracking and favors undesired products such as short-chain hydrocarbons. This limitation can be addressed by creating the hierarchical porous structure via etching in an alkali medium which improves the mass transfer efficiency as well as modulates the acidic property and improves the aromatic yield during the reaction.[64,65]

The catalytic mechanism for $CO_2$ hydrogenation to aromatics via the $CO_2$–FTS route over iron-based catalyst involves several complex mechanisms, which include phase transformations of the catalyst in operando conditions. The iron-based catalysts employed for $CO_2$ hydrogenation can exist in

Table 3. Summary of catalytic performance reported for $CO_2$ hydrogenation to aromatics via $CO_2$-MeOH and $CO_2$-FTS routes.

| Catalyst | Reaction Condition | | | | $CO_2$ | CO | Hydrocarbon distribution (%) | | | Aromatic | Light | Aromatic | Stability | Ref. |
|---|---|---|---|---|---|---|---|---|---|---|---|---|---|---|
| | GHSV[a] | Temp. (°C) | $H_2/CO_2$ | Pressure (Bar) | con. (%) | sel. (%) | $CH_4$ | $C_2$-$C_4$ | $C_{5+}$ aliphatic | sel.[b] (%) | Aromatic[c] (%) | STY[d] | test (h) | |
| **Aromatics synthesis via $CO_2$–MeOH route** | | | | | | | | | | | | | | |
| $Cr_2O_3$/ZnZSM-5@Si | 1200 | 350 | 3[e] | 30 | 27.6 | 12.8 | 1.6 | 18.9 | 9.2 | 61.3 | 63.9 | 31.7 | - | 68 |
| $Cr_2O_3$/HZSM-5(40) | 1200 | 350 | 3 | 30 | 33.6 | 41.2 | 3.0 | 18.8 | 7.7 | 41.5 | - | 26.1 | - | 68 |
| ZnCrOx/HZSM-5 | 3000 | 320 | 3 | 50 | 19.9 | 70.2 | 30.3 | | 13.2 | 16.8 | - | 15.7 | 100 | 69 |
| ZnZrOx/HZSM-5 | 1020 | 315 | 3 | 30 | 17.5 | 23.9 | 1.2 | 29.8 | 7.7 | 45.3 | 20.4 | 12.6 | 120 | 70 |
| ZnAlOx/Si-HZSM-5 | 6000 | 320 | 3 | 30 | 5.0 | 47.0 | 1.0 | 29.5 | 2.5 | 35.5 | 29.9 | 16.6 | - | 71 |
| $ZnCr_2O_4$/ZSM-5 | 1200 | 350 | 3[e] | 40 | 23.1 | 27.8 | 0.5 | 12.3 | 1.8 | 61.6 | 44.2 | 26.7 | 100 | 72 |
| $ZnCr_2O_4$/ZSM-5 | 1200 | 350 | 3 | 40 | 19.8 | 65.4 | 2.8 | 15.3 | 2.0 | 27.6 | 50.0 | 10.2 | - | 72 |
| ae-Zn–$ZrO_2$/HZSM-5 | 2400 | 340 | 3 | 40 | 15.9 | 34.0 | 0.3 | 19.4 | 4.3 | 50.1 | 31.4 | 29.9 | 52 | 37 |
| $Cr_2O_3$/HZSM-5(40) | 1200 | 350 | 3[e] | 30 | 34.5 | 11.4 | 1.5 | 16.4 | 6.2 | 67.2 | 17.4 | 43.5 | 100 | 68 |
| ZnO–$ZrO_2$/HZSM-5 | 2700 | 340 | 3 | 30 | 19.1 | 42.5 | 0.6 | 29.4 | | 42.0 | 12.0 | 16..1 | 100 | 73 |
| ZnZrO/HZSM-5 | 1200 | 320 | 3 | 40 | 14.1 | 44.0 | 0.4 | 19.6 | 7.0 | 40.1 | 11.0 | 10.6 | 100 | 74 |
| **Aromatics synthesis via $CO_2$–FTS route** | | | | | | | | | | | | | | |
| $Fe_3O_4$ | 4800 | 340 | 3 | 10 | 27.0 | 35.9 | 43.2 | 5.7 | - | - | - | - | - | 75 |
| Na/Fe | 4800 | 320 | 3 | 10 | 29.9 | 24.7 | 13.2 | 22.7 | - | - | - | - | - | 75 |
| $ZnFeO_x$–4.25Na/S-HZSM5 | 4000 | 320 | 3 | 30 | 36.2 | 11.0 | 8.2 | 16.5 | 15.3 | 53.4 | 60.0 | 120.8 | 100 | 75 |
| $ZnFeO_x$–4.25Na/S-HZSM5-Si | 4000 | 320 | 3 | 30 | 36.9 | 11.1 | 9.7 | 27.7 | 21.6 | 36.4 | 70.7 | 83.9 | - | 75 |
| 6.25Cu$Fe_2O_3$/HZSM-5-c | 1000 | 320 | 3 | 30 | 57.3 | 3.5 | 10.5 | 8.6 | 23.4 | 54.6 | 31.3 | 48.9 | 120 | 76 |
| FeK/a-$Al_2O_3$&P/ZSM-5 | 3000 | 400 | 1 | 30 | 36.4 | 10.2 | 10.8 | 39.6 | 10.1 | 35.5 | - | 121.1 | - | 77 |
| $FeK_{1.5}$/HSG&HZSM-5(50) | 26000 | 340 | 3 | 20 | 35.0 | 39.0 | 3.5 | 4.4 | 24.0 | 41.5 | 67.0 | 589.8 | 120 | 78 |
| Fused Fe/HZSM-5 | 600 | 350 | 2 | 20 | 38.1 | 14.5 | 10.1 | 48.5 | 20.0 | 18.3 | 92.3 | 8.7 | 100 | 79 |
| 1Na–Fe/HZSM-5(30) | 4000 | 340 | 3 | 30 | 45.3 | 11.3 | 9.8 | 24.6 | 9.8 | 49.8 | 40.2 | 141.0 | 48 | 80 |
| Na–$Fe_3O_4$/HZSM-5(160) | 4000 | 320 | 1 | 30 | 22.0 | 20.1 | 4.0 | 16.6 | 30.5 | 39.1 | 42.1 | 107.6 | -[f] | 81 |
| $Fe_2O_3$@$KO_2$/ZSM-5 | 5000 | 375 | 3 | 30 | 48.9 | 12.8 | 13.9 | 45.9 | 15.3 | 21.7 | 66.0 | 82.9 | 150 | 82 |
| Na–$Fe_3O_4$/HZSM-5(25)- Si-1-12% | 4000 | 320 | 2 | 30 | 28.5 | 15.0 | 6.2 | 32.7 | 18.1 | 36.6 | 75.0 | 86.9 | 50 | 83 |
| NaFe/HZSM-5 | 4800 | 340 | 3 | 10 | 30.9 | 26.4 | 26.5 | 29.2 | 3.0 | 30.4 | 62.3 | 70.5 | 40 | 84 |
| NaFe@C/HZSM-5-0.2 M | 9000 | 320 | 3 | 30 | 33.3 | 13.3 | 4.2 | 10.4 | 34.6 | 43.5 | - | 203.7 | 60 | 85 |
| 6.25Cu$Fe_2O_3$/HZSM-5 | 1000 | 320 | 3 | 30 | 55.4 | 4.4 | 12.5 | 9.4 | 15.4 | 59.2 | 54.2 | 51.2 | - | 76 |

[a]Unit in mL $gcat^{-1}$ $h^{-1}$; [b]Aromatic selectivity in total products including CO; [c]Light aromatics content in total aromatics; [d]STY: space time yield of aromatics (mg$CH_2$ $gcat^{-1}$ $h^{-1}$); [e]The data is obtained when the feed gas contains 3 6 vol% CO except $CO_2$ and $H_2$.

different oxidation states such as $\alpha$-$Fe_2O_3$, $\gamma$-$Fe_2O_3$, and $Fe_3O_4$. The mixture of different phases of iron such as iron oxides, iron carbides and metallic iron are identified to co-exist after activation and during the catalytic reaction.[9,66,67] The mechanism of this reaction can be probed by a combination of characterization tools such as XPS, XRD, EXAFS and Mossbauer spectroscopy. The presence of $Fe_5C_2$ and $Fe_3O_4$ was confirmed on the spent catalyst which is an active species for FTS and RWGS.[26] A pictorial representation of the reaction scheme for the synthesis of aromatics via the $CO_2$-FTS route on NaFe/HZSM-5 is shown in Figure 4. The first step of the reaction is the reduction of $CO_2$ to CO on the $Fe_3O_4$ phase of the catalyst via RWGS which is subsequently hydrogenated to $C_2$-$C_5$ olefins on $\chi$-$Fe_5C_2$ sites through FTS. These olefins produced on Fe-based catalysts diffuse into zeolite pores and undergo oligomerization on Bronsted acidic sites (BAS) to produce $C_{6+}$ olefins which further gets converted to naphthenes and undergo a series of hydrogen transfer and cyclization reactions to produce light aromatics. The aromatics formed can also undergo isomerization and transalkylation, which lead to the formation of a wide range of aromatic products. A fraction of aromatics can also produce PAHs and form coke which ultimately leads to catalysts'

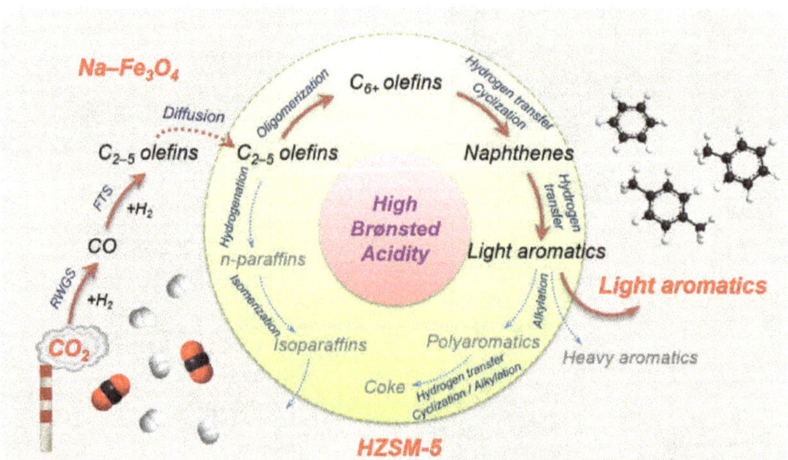


Figure 4. Reaction scheme for the synthesis of aromatics via $CO_2$ hydrogenation over NaFe/HZSM-5 composite catalysts. Reproduced with permission from Ref. 86. Copyright 2021 Elsevier.

deactivation. The nature of BAS on the zeolite dictates the multistep olefin oligomerization process.

Higher alcohols play a significant role in the chemical industry and are extensively used in the chemical, pharmaceutical and energy sectors. Owing to its high-octane number and enhancement of combustion efficiency of the engines, higher alcohols find application as fuel additives.[14,87,88,89] The $CO_2$ hydrogenation to $C_1$ molecules has achieved tremendous progress. However, the direct hydrogenation of $CO_2$ to higher alcohols is still challenging because of the high energy barrier for C-C coupling. The selective formation of higher alcohols from $CO_2$ hydro-genation is more challenging owing to the presence of many consecutive and parallel reactions and the high energy barrier associated with CO insertion in comparison with $CH_x$ hydrogenation. The alcohol synthesized by this process follows the ASF distribution rule. The first report on $CO_2$ hydrogenation to ethanol was reported in 1999 by Inui *et al.* using a multi-functional FT type composite catalyst having Fe and Cu-based catalysts. The improved catalytic activity is ascribed to the optimization of reduction property upon composite formation, but the understanding of the reaction mechanism remains vague.[90] While designing a catalyst for higher alcohol synthesis the catalyst must have a high kinetic barrier for methane formation which will reduce the methane selectivity.

The thermodynamic studies revealed that low temperature favors $CO_2$ conversion and alcohol formation whereas with an increase in temperature, the $CO_2$ conversion and alcohol selectivity decreases because the formation of alcohol is an exothermic reaction. RWGS is an endothermic reaction that will lead to increased CO formation at a higher temperature. High pressure also enhances alcohol synthesis because the formation of alcohol leads to volume contraction while the RWGS is isovolumetric.[91,92,93,94] The higher alcohol synthesis by $CO_2$ hydrogenation is predominantly studied on 4 classes of catalysts such as Cu-, Mo-, Rh-, and Co-based catalysts. In addition, catalytic systems such as Ru, Pt, Pd, and Au have also been explored for HA synthesis. Various catalyst systems were reported for $CO_2$ hydrogenation to HA synthesis and their catalytic performance is summarized in Table 4.

The catalytic mechanism of HA synthesis involves a multistep process. In the first step, the $CO_2$ hydrogenation will lead to the formation of $C_1$

Table 4. Summary of catalytic activity of various catalysts for $CO_2$ to higher alcohols (HA) synthesis.

| Catalysts | $T^a$ (°C) | $P^b$ (Bar) | $R^c$ | $GHSV^d$ ($cm^3\ gcat^{-1}\ h^{-1}$) | $X^e_{CO2}$ (%) | $S^f_{CO}$ (%) | $S^g_{HC}$ (%) | $S^h_{MeOH}$ (%) | $S^i_{HA}$ (%) | $Y^j_{HA}$ (%) | Ref. |
|---|---|---|---|---|---|---|---|---|---|---|---|
| $RhLi/SiO_2$ | 240 | 50 | 3 | 6000 | 7.0 | 15.7 | 63.5 | 5.2 | 15.5 | 1.09 | 95 |
| $RhMnLi/SiO_2$ | 240 | 50 | 3 | 6000 | 16.1 | 2.4 | 88.6 | 1.3 | 7.7 | 1.24 | 95 |
| $Rh/CeO_2$-$SiO_2$ | 240 | 50 | 3 | 6000 | 8.3 | 33.5 | 51.4 | 8.8 | 6.1 | 0.51 | 96 |
| $RhFeli/TiO_2$ | 250 | 30 | 3 | $6000^k$ | 15.7 | 12.5 | 53.9 | 2.2 | 31.3 | 4.91 | 97 |
| $RhFe/TiO_2$ | 270 | 20 | 1 | 8000 | 9.2 | 28.4 | 63.4 | 1.3 | 6.4 | 0.59 | 98 |
| $RhCe/SiO_2$ | 260 | 50 | 3 | - | 14.0 | 22.3 | 64.6 | 7.3 | 5.4 | 0.76 | 99 |
| $RhFe/SiO_2$ | 260 | 50 | 3 | 6000 | 26.7 | 19.7 | 34.7 | 29.4 | 16.0 | 4.27 | 100 |
| $CuZnFeZrO_2K$ | 320 | 30 | 3.0 | 3600 | 24.2 | 29.6 | 50.3 | 17.9 | 2.2 | 0.53 | 101 |
| CuFeCoK | 350 | 60 | 3.0 | 5000 | 19.98 | 71.3 | $22.2^l$ | 2.4 | 4.0 | 0.80 | 102 |
| CuZnFeZrK | 320 | 30 | 3.0 | 3600 | 25.5 | 53.1 | 23.6 | 14.3 | 8.9 | 2.27 | 101 |
| CuZnK | 300 | 60 | 3.0 | 5000 | 25.1 | 80.4 | 8.7 | 6.9 | 4.0 | 1.0 | 103 |
| CuZnK | 350 | 60 | 3.0 | 5000 | 27.2 | 84.3 | $8.2^l$ | 6.2 | 1.33 | 0.36 | 102 |
| CuZnFeK | 300 | 60 | 3.0 | 5000 | 42.3 | 6.9 | 56.4 | 4.7 | 31.9 | 13.51 | 103 |
| $Fe/Mo_2C$ | 200 | 40 | 3 | - | 38.0 | 6.8 | 18.6 | 58 | 16 | n.a | 104 |
| $Co/Mo_2C$ | 200 | 40 | 3 | - | 35.0 | 9.5 | 17.1 | 46 | 25 | n.a | 104 |
| CoMoS | 340 | 104 | 3 | $0.43^k$ | 32.0 | 57.5 | 16.5 | 20.0 | 6.0 | 1.92 | 105 |
| $IrMo/SiO_2$ | 200 | 4.9 | 2 | 2000 | 11.9 | 15.9 | 67.8 | 9.6 | 6.1 | 0.73 | 106 |
| KMoCoS | 320 | 5.0 | 3 | 3200 | 8.1 | 69.8 | 12.2 | 13.2 | 4.8 | 0.39 | 107 |
| $Mo_2C$ | 200 | 4 | 3 | - | 20.0 | 4.9 | 25.8 | 53 | 16 | n.a | 104 |
| $\alpha$-$CoMoO_4/K$ | 250 | 3 | 1 | 1200 | 7.2 | 27.8 | 31.6 | 35.7 | 4.8 | 0.35 | 108 |

[a]Reaction temperature; [b]Reaction pressure; [c]$H_2/CO_2$ ratio; [d]Gas hourly space velocity; [e]$CO_2$ conversion; [f]Selectivity to CO; [g]Selectivity to hydrocarbons; [h] Selectivity to methanol; [i]Selectivity to higher alcohols; [j]Yield of higher alcohols; [k]Space velocity in $h^{-1}$.

intermediates such as CO, COO, $CO_3$, HCOH and $CH_x$ either in the gas phase or on the surface of the catalyst. These $C_1$ intermediates will couple to form $C_2$ species. Subsequent hydrogenation of $C_2$ species leads to the formation of $C_2$ hydrocarbons and oxygenates. Finally, higher alcohols are formed through various routes as shown in Eqs. (1) to (4). A schematic illustration of the proximity effect and reaction pathway for hydrogenation of $CO_2$ to HA over the CZA/K-CMZF multifunctional catalyst is shown in Figure 5. The synergetic effect and proximity of active sites also play a decisive role in determining the catalytic performance in the $CO_2$ hydrogenation to HA.

1. $CO/CO_n^* + CH_X^* \longrightarrow CO_nCH_x \longrightarrow CH_3CH_2OH$ (n=1,2 or 3) Eq. 1

2. $CH_X^* + CH_X^* \longrightarrow CH_XCH_X^* \xrightarrow{CO/CO_2} COCH_XCH_X^* \xrightarrow{H_2} CH_3CH_2CH_2OH$

$CH_XCH_X^* \downarrow CH_2{=}CH_2 \xrightarrow{H_2O} CH_3CH_2OH$ Eq. 2

3. $HCOH^* + HCOH^* \xrightarrow{-H_2O} CHCOH^* \longrightarrow CH_3CH_2OH$ Eq. 3

4. $HCOO^* + CH_X^* \longrightarrow CH_3COO^* \longrightarrow CH_3CH_2OH$ Eq. 4

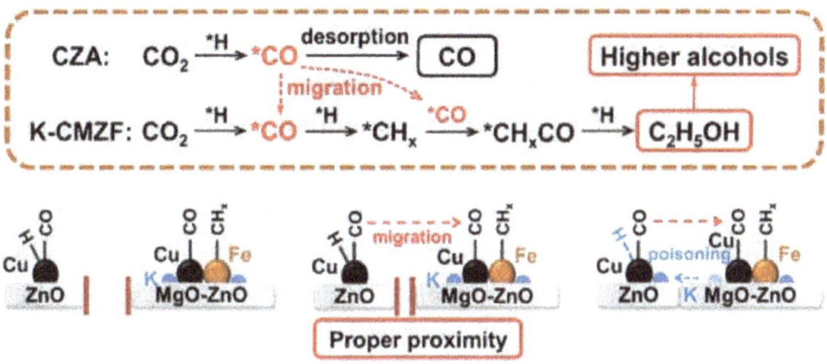


Figure 5. Schematic illustration of proximity effect and reaction pathway for hydrogenation of $CO_2$ to HA over the CZA/K-CMZF multifunctional catalyst. Reproduced with permission from Ref. 94. Copyright 2021. American Chemical Society.

## 4. Summary and Outlook

The rise in $CO_2$ levels in the atmosphere has become a global issue and minimizing them is one of the biggest challenges in front of the scientific community today. $CO_2$ hydrogenation to fuels and high-value chemicals offers a promising potential to address this issue. However, there are several challenges in $CO_2$ hydrogenation to high energy-dense fuels such as olefins, alcohols, and aromatics. Despite these challenges, tremendous progress is achieved in recent years in terms of catalyst design, understanding the mechanism as well as in scaling up the process for sustainable fuel synthesis from $CO_2$. In the present book chapter, we have reviewed some of the recent developments in the thermocatalytic hydrogenation of $CO_2$ to high energy-dense fuels. The tandem catalytic routes such as FTS and MeOH to produce olefins, higher alcohols and aromatics are discussed in detail with a special emphasis on catalyst design, reaction mechanism, and catalytic performance of some of the best catalysts. The product selectivity in $CO_2$ hydrogenation to fuels involves a complex interplay of many parameters which makes it highly challenging to obtain the desired products. For instance, the RWGS reaction is endothermic and is favored at high temperatures, whereas the methanol synthesis and $CO_2$ methanation are exothermic and are favoured at low temperatures. Therefore, proper tuning of selectivity of desired products in $CO_2$ hydrogenation is dictated by the combined effect of thermodynamics, catalyst design and thermodynamics of the reaction.

Although there is an exceptional advancement in the field of $CO_2$ hydrogenation to fuels, there is still scope for improvement in this field in terms of catalyst stability and commercialization of these processes. The use of hierarchical zeolites is a good area to be explored while targeting the superior stability of the catalyst and getting high selectivity towards higher hydrocarbons/olefins. The presence of small pores in microporous zeolites often limits the product distribution to lower olefins and hydrocarbons by mass transport limitation and leads to the over-cracking of higher hydrocarbons to small-chain products. The use of hierarchical zeolites can improve the mass transfer and can enhance the production of higher hydrocarbons. Exploring hydride-based catalysts for $CO_2$ hydrogenation is another expanding field that needs to be focused on in the future. In addition, proper techno-economic analysis has to be done to assess the $CO_2$ hydrogenation to fuels on a commercial scale. The use of renewable hydrogen and renewable energy would make these processes more economically feasible.

## References

1. M. I. Alam, R. Cheula, G. Moroni, L. Nardi, M. Maestri, Mechanistic and Multiscale Aspects of Thermo-catalytic $CO_2$ Conversion to $C_1$ Products, *Catal. Sci. Technol.* **11**, 6601-6629, (2021).
2. A. Saravanan, P. Senthil Kumar, Dai-Viet N. Vo, S. Jeevanantham, V. Bhuvaneswari, V. A. Narayanan, P. Yaashikaa, S. Swetha and B. Reshma, A Comprehensive Review on Different Approaches for $CO_2$ Utilization and Conversion Pathways, *Chem. Eng. Sci.* **236**, 116515, (2021).
3. X. Gao, T. Atchimarungsri, Q. Ma, T-S, Zhao, N. Tsubaki, Realizing Efficient Carbon Dioxide Hydrogenation to Liquid Hydrocarbons by Tandem Catalysis Design. *Energy Chem.* **2**, 100038, (2020).
4. S. Roy, A. Cherevotan, S. C. Peter, Thermochemical $CO_2$ Hydrogenation to Single Carbon Products: Scientific and Technological Challenges, *ACS Energy Lett.* **3**, 1938-1966, (2018).
5. A. Cherevotan, J. Raj, S. C. Peter, An Overview of Porous Silica Immobilized Amines for Direct Air $CO_2$ Capture, *J. Mater. Chem.*, A, **9**, 27271-27303, (2021).
6. S. C. Peter, B. Ray, S. R. Churipard, An Overview of the Materials and Methodologies for $CO_2$ Capture under Humid Conditions. *J. Mater. Chem. A* **9**, 26498-26527, (2021).
7. A. Cherevotan, J. Raj, L. Dheer, S. Roy, S. Sarkar, R. Das, C. P. Vinod, S. Xu, P. Wells and U. V. Waghmare, and S. C. Peter, Operando Generated Ordered Heterogeneous Catalyst for the Selective Conversion of $CO_2$ to Methanol, *ACS Energy Lett.* **6**, 509-516, (2021).
8. K. S. Kanakikodi, S. R. Churipard, R. Bai, S. P. Maradur, Upgrading of Lignocellulosic Biomass-derived Furfural: An Efficient Approach for the Synthesis of Bio-fuel Intermediates over $\gamma$-alumina Supported Sodium Aluminate. *Mol. Catal.* **510**, 111716, (2021).
9. J. Wei, R. Yao, Y. Han, Q. Ge, J. Sun, Towards the Development of the Emerging Process of $CO_2$ Heterogenous Hydrogenation into High-value Unsaturated Heavy Hydrocarbons, *Chem. Soc. Rev.* **50**, 10764-10805, (2021).
10. C. Aranthady, T. Jangid, K. Gupta, A. K. Mishra, S. Kaushik, V. Siruguri, G. M. Rao, G. V. Shanbhag, N. G. Sundaram, Selective $SO_2$ Detection at Low Concentration by Ca Substituted $LaFeO_3$ Chemiresistive Gas Sensor: A Comparative Study of $LaFeO_3$ Pellet vs Thin Film, *Sensors and Actuators B: Chemical* **329**, 129211, (2021).

11. D. Bagchi, S. Sarkar, A. K. Singh, C. P. Vinod, and S. C. Peter, Potential- and Time-Dependent Dynamic Nature of an Oxide-Derived PdIn Nanocatalyst during Electrochemical $CO_2$ Reduction, *ACS Nano* **16**, 4, 6185–6196, (2022).
12. S. Das, J. Pérez-Ramírez, J. Gong, N. Dewangan, K. Hidajat, B. C. Gates, S. Kawi, Core–shell Structured Catalysts for Thermocatalytic, Photocatalytic, and Electrocatalytic Conversion of $CO_2$, *Chem. Soc. Rev.* **49**, 2937-3004 (2020).
13. S. C. Peter, Reduction of $CO_2$ to Chemicals and Fuels: A Solution to Global Warming and Energy Crisis. *ACS Energy Lett.* **3**, 1557-1561, (2018).
14. S.R. Churipard, K. S. Kanakikodi, J. R. Choudhuri, S. P. Maradur, Polyoxotungstate ($[PW_{11}O_{39}]^{7-}$) Immobilized on Mesoporous Polymer for Selective Liquid-phase Oxidation of Alcohols using $H_2O_2$, *RSC Adv.* **10**, 35988-35997, (2020).
15. M. D. Porosoff, B. H. Yan, J. G. Chen, Catalytic Reduction of $CO_2$ by $H_2$ for Synthesis of CO, Methanol and Hydrocarbons: Challenges and Opportunities. *Energ Environ Sci.* **9**, 62-73, (2016).
16. M. Aresta, A. Dibenedetto, A. Angelini, Catalysis for the Valorization of Exhaust Carbon: From $CO_2$ to Chemicals, Materials, and Fuels. Technological Use of $CO_2$, *Chem. Rev.* **114**(3), 1709–1742, (2014).
17. S. R. Churipard, P. Manjunathan, P. Chandra, G. V. Shanbhag, R. Ravishankar, P. V. Rao, G. S. Ganesh, A. Halgeri, S. P. Maradur, Remarkable Catalytic Activity of a Sulfonated Mesoporous Polymer (MP-$SO_3$-H) for the Synthesis of Solketal at Room Temperature, *New J. Chem.* **41**, 5745-5751, (2017).
18. K. S. Kanakikodi, S. R. Churipard, A. Halgeri, S. P. Maradur, Solid Acid Catalyzed Carboxymethylation of Bio-derived Alcohols: An Efficient Process for the Synthesis of Alkyl Methyl Carbonates, *Sci. Reports* **10**, 1-12, (2020).
19. D. Bagchi, J. Raj, A. K. Singh, A. Cherevotan, S. Roy, K. Sai Manoj, C. P. Vinod, S. C. Peter, Structure-Tailored Surface Oxide on Cu–Ga Intermetallics Enhances $CO_2$ Reduction Selectivity to Methanol at Ultralow Potential, *Adv. Mater.* **34**, 2109426, (2022).
20. R. Das, K. Das, B. Ray, C. P. Vinod, S. C. Peter, Green Transformation of $CO_2$ to Ethanol using Water and Sunlight by the Combined Effect of Naturally Abundant Red Phosphorus and $Bi_2MoO_6$, *Energy Environ. Sci.* **15**, 1967-1976, (2022).
21. Y. L. Zheng, H. C. Liu, Y. W. Zhang, Engineering Heterostructured Nanocatalysts for $CO_2$ Transformation Reactions: Advances and Perspectives. *ChemSusChem.* **13**, 6090-6123, (2020).
22. S. Roy, A. Cherevotan and S. C. Peter, Thermochemical $CO_2$ Hydrogenation to Single Carbon Products: Scientific and Technological Challenges, *ACS Energy Lett.* **3**, 1938-1966, (2018).
23. S. Saeidi, N. A. S. Amin, M. R. Rahimpour, Hydrogenation of $CO_2$ to Value-added Products — A Review and Potential Future Developments. *J. $CO_2$ Util.* **5**, 66-81, 2014.

24. M. D. Porosoff, B. Yan, J. G. Chen, Catalytic Reduction of $CO_2$ by $H_2$ for Synthesis of CO, Methanol and Hydrocarbons: Challenges and Opportunities, *Energy Environ. Sci.* **9**, 62-73. (2016).
25. M. D. Garba, M. Usman, S. Khan, F. Shehzad, A. Galadima, M. F. Ehsan, A. S. Ghanem, M. Humayun, $CO_2$ Towards Fuels: A Review of Catalytic Conversion of Carbon Dioxide to Hydrocarbons, *J. Environ. Chem. Eng.* **9**, 104756, (2021).
26. X. Cui, P. Gao, S. Li, C. Yang, Z. Liu, H. Wang, L. Zhong, Y. Sun, Selective Production of Aromatics Directly from Carbon Dioxide Hydrogenation, *ACS Catal.* **9**, 5, 3866–3876, (2019).
27. G. Song, M. Li, P. Yan, M. A. Nawaz, D. Liu, High Conversion to Aromatics via $CO_2$-FT over a CO-reduced Cu-$Fe_2O_3$ Catalyst Integrated with HZSM-5, *ACS Catal.* **10**, 11268-11279, (2020).
28. Z. Ma, M.D. Porosoff, Development of Tandem Catalysts for $CO_2$ Hydrogenation to Olefins, *ACS Catal.* **9**, 2639-2656, (2019).
29. G. Prieto, Carbon Dioxide Hydrogenation into Higher Hydrocarbons and Oxygenates: Thermodynamic and Kinetic Bounds and Progress with Heterogeneous and Homogeneous Catalysis, *ChemSusChem.* **10**, 1056-1070, (2017).
30. D. Goud, R. Gupta, R. Maligal-Ganesh, S.C. Peter, Review of Catalyst Design and Mechanistic Studies for the Production of Olefins from Anthropogenic $CO_2$, *ACS Catal.* **10**, 14258-14282, (2020).
31. F. Bleken, M. Bjørgen, L. Palumbo, S. Bordiga, S. Svelle, K.P. Lillerud, U. Olsbye, The Effect of Acid Strength on the Conversion of Methanol to Olefins over Acidic Microporous Catalysts with the CHA Topology, *Top. Catal.* **52**, 218-228, (2009).
32. Z. Zhang, H. Yin, G. Yu, S. He, J. Kang, Z. Liu, K. Cheng, Q. Zhang, Y Wang, Selective Hydrogenation of $CO_2$ and CO into Olefins over Sodium-and zinc-promoted Iron Carbide Catalysts, *J. Catal.* **395**, 350-361, (2021).
33. Y. H. Choi, Y. J. Jang, H. Park, W.Y. Kim, Y. H. Lee, S. H. Choi, J. S. Lee, Carbon Dioxide Fischer-Tropsch Synthesis: A New Path to Carbon-Neutral Fuels. *Appl. Catal. B: Environ.* **202**, 605-610, (2017).
34. C. Zhang, M. Xu, Z. Yang, M. Zhu, J. Gao, Y. -F. Han, Uncovering the Electronic Effects of Zinc on the Structure of $Fe_5C_2$-ZnO Catalysts for $CO_2$ Hydrogenation to Linear α-Olefins, *Appl. Catal. B: Environ.* **295**, 120287, (2021).
35. L. Guo, Y. Cui, H. Li, Y. Fang, R. Prasert, J. Wu, G. Yang, Y. Yoneyama and N. Tsubaki, Selective Formation of Linear-alpha Olefins (LAOs) by $CO_2$ Hydrogenation over Bimetallic Fe/Co-Y Catalyst, *Catal. Commun.* **130**, 105759, (2019).
36. Y. Xu, P. Zhai, Y. Deng, J. Xie, X. Liu, S. Wang and D. Ma, Highly Selective Olefin Production from $CO_2$ Hydrogenation on Iron Catalysts: A Subtle Synergy between Manganese and Sodium Additives, *Angewandte Chemie* **132**, 21920-21928, (2020).
37. C. Zhou, J. Shi, W. Zhou, K. Cheng, Q. Zhang, J. Kang and Y. Wang, Highly Active ZnO-$ZrO_2$ Aerogels Integrated with H-ZSM-5 for Aromatics Synthesis from Carbon Dioxide, *ACS Catal.* **10**, 302-310, (2019).

38. S. Wang, T. Wu, J. Lin, Y. Ji, S. Yan, Y. Pei, S. Xie, B. Zong and M. Qiao, Iron–Potassium on Single-walled Carbon Nanotubes as Efficient Catalyst for $CO_2$ Hydrogenation to Heavy Olefins, *ACS Catal.* **10**, 6389-6401, (2020).
39. L. Guo, J. Sun, X. Ji, J. Wei, Z. Wen, R. Yao, H. Xu and Q. Ge, Directly Converting Carbon Dioxide to Linear α-olefins on Bio-promoted Catalysts, *Commun. Chemistry* **1**, 1-8, (2018).
40. B. Yao, T. Xiao, O. A. Makgae, X. Jie, S. Gonzalez-Cortes, S. Guan, A. I. Kirkland, J. R. Dilworth, H. A. Al-Megren and S. M. Alshihri, Transforming Carbon Dioxide into Jet Fuel using an Organic Combustion-synthesized Fe-Mn-K Catalyst, *Nature Commun.* **11**, 1-12, (2020).
41. M. K. Khan, P. Butolia, H. Jo, M. Irshad, D. Han, K.-W. Nam and J. Kim, Selective Conversion of Carbon Dioxide into Liquid Hydrocarbons and Long-chain α-olefins over Fe-amorphous $AlO_x$ Bifunctional Catalysts, *ACS Catal.* **10**, 10325-10338, (2020).
42. Z. Zhang, H. Yin, G. Yu, S. He, J. Kang, Z. Liu, K. Cheng, Q. Zhang and Y. Wang, Selective Hydrogenation of $CO_2$ and CO into Olefins over Sodium-and zinc-promoted Iron Carbide Catalysts, *J. Catal.* **395**, 350-361, (2021).
43. F. Jiao, J. Li, X. Pan, J. Xiao, H. Li, H. Ma, M. Wei, Y. Pan, Z. Zhou and M. J. S. Li, Selective Conversion of Syngas to Light Olefins, *Science* **351**, 1065-1068, (2016).
44. S.-H. Kang, J. W. Bae, P. Sai Prasad and K.-W. Jun Fischer–Tropsch Synthesis using Zeolite-supported Iron Catalysts for the Production of Light Hydrocarbons, *Catal. Lett.* **125**, 264-270, (2008).
45. J. Gao, C. Jia, B. J. Liu, Direct and Selective Hydrogenation of $CO_2$ to Ethylene and Propene by Bifunctional Catalysts, *Catal. Sci. Technol.* **7**, 5602-5607, (2017).
46. X. Yang, R. Wang, J. Yang, W. Qian, Y. Zhang, X. Li, Y. Huang, T. Zhang and D. Chen, Exploring the Reaction Paths in the Consecutive Fe-based FT Catalyst–zeolite Process for Syngas Conversion, *ACS Catal.* **10**, 3797-3806, (2020).
47. Y. Wang, S. Kazumi, W. Gao, X. Gao, H. Li, X. Guo, Y. Yoneyama, G. Yang and N. Tsubaki, Direct Conversion of $CO_2$ to Aromatics with High Yield via a Modified Fischer-Tropsch Synthesis Pathway, *Appl. Catal. B: Environ.* **269**, 118792, (2020).
48. Z. Li, Y. Qu, J. Wang, H. Liu, M. Li, S. Miao and C. Li, Highly Selective Conversion of Carbon Dioxide to Aromatics over Tandem Catalysts, *Joule* **3**, 570-583, (2019).
49. C. K. Kuei and M. D. J. Lee C. K. Kuei, Hydrogenation of Carbon Dioxide by Hybrid Catalysts, Direct Synthesis of Aromatics from Carbon Dioxide and Hydrogen, *The Canadian J. Chem. Eng.* **69**, 347-354, (1991).
50. J. S. Martínez-Espín, K. De Wispelaere, T. V. Janssens, S. Svelle, K. P. Lillerud, P. Beato, V. Van Speybroeck and U. Olsbye Hydrogen Transfer Versus Methylation: On the Genesis of Aromatics Formation in the Methanol-to-hydrocarbons Reaction over H-ZSM-5, *ACS Catal.* **7**, 5773-5780, (2017).

51. S. Müller, Y. Liu, F. M. Kirchberger, M. Tonigold, M. Sanchez-Sanchez and J. A. Lercher, Hydrogen Transfer Pathways during Zeolite Catalyzed Methanol Conversion to Hydrocarbons, *J. Am. Chem. Soc.* **138**, 15994-16003, (2016).
52. Y. Hartadi, D. Widmann and R. J. Behm, $CO_2$ Hydrogenation to Methanol on Supported Au Catalysts under Moderate Reaction Conditions: Support and Particle Size Effects, *ChemSusChem* **8**, 456-465, (2015).
53. X. Zhang, A. Zhang, X. Jiang, J. Zhu, J. Liu, J. Li, G. Zhang, C. Song and X. J. Guo, Utilization of $CO_2$ for Aromatics Production over ZnO/$ZrO_2$-ZSM-5 Tandem Catalyst, *J. $CO_2$ Util.* **29**, 140-145, (2019).
54. Y. Ni, Z. Chen, Y. Fu, Y. Liu, W. Zhu and Z. Liu, Selective Conversion of $CO_2$ and $H_2$ into Aromatics, *Nat. Cummun.* **9**, 1-7, (2018).
55. H. Bahruji, M. Bowker, G. Hutchings, N. Dimitratos, P. Wells, E. Gibson, W. Jones, C. Brookes, D. Morgan and G. J. Lalev, Pd/ZnO Catalysts for Direct $CO_2$ Hydrogenation to Methanol, *J. Catal.* **343**, 133-146, (2016).
56. Y. Wang, L. Tan, M. Tan, P. Zhang, Y. Fang, Y. Yoneyama, G. Yang and N. Tsubaki, Rationally Designing Bifunctional Catalysts as an Efficient Strategy to Boost $CO_2$ Hydrogenation Producing Value-added Aromatics, *ACS Catal.* **9**, 895-901, (2018).
57. C. Zhou, J. Shi, W. Zhou, K. Cheng, Q. Zhang, J. Kang and Y. Wang, Highly Active ZnO-$ZrO_2$ Aerogels Integrated with H-ZSM-5 for Aromatics Synthesis from Carbon Dioxide, *ACS Catal.* **10**, 302-310, (2019).
58. J. Wei, R. Yao, Q. Ge, Z. Wen, X. Ji, C. Fang, J. Zhang, H. Xu and J. Sun, Catalytic Hydrogenation of $CO_2$ to Isoparaffins over Fe-based Multifunctional Catalysts, *ACS Catal.* **8**, 9958-9967, (2018).
59. M. K. Gnanamani, G. Jacobs, H. H. Hamdeh, W. D. Shafer, F. Liu, S. D. Hopps, G. A. Thomas and B. H. Davis, Hydrogenation of Carbon Dioxide over Co–Fe Bimetallic Catalysts, *ACS Catal.* **6**, 913-927, (2016).
60. L. Guo, J. Sun, X. Ji, J. Wei, Z. Wen, R. Yao, H. Xu and Q. Ge, Directly Converting Carbon Dioxide to Linear α-olefins on Bio-promoted Catalysts, *Commun. Chem.* **1**, 1-8, (2018).
61. W. D. Shafer, G. Jacobs, U. M. Graham, H. H. Hamdeh and B. H. Davis, Increased $CO_2$ Hydrogenation to Liquid Products using Promoted Iron Catalysts, *J. Catal.* **369**, 239-248, (2019).
62. J. Wei, J. Sun, Z. Wen, C. Fang, Q. Ge, H. Xu, New Insights into the Effect of Sodium on $Fe_3O_4$-based Nanocatalysts for $CO_2$ Hydrogenation to Light Olefins, *Cat. Sci. Technol.* **6**, 4786-4793, (2016).
63. A. Ramirez, L. Gevers, A. Bavykina, S. Ould-Chikh and J. Gascon, Metal Organic Framework-derived Iron Catalysts for the Direct Hydrogenation of $CO_2$ to Short Chain Olefins, *ACS Catal.* **8**, 9174-9182, (2018).
64. Y. Wang, D. Wang, M. Tan, B. Jiang, J. Zheng, N. Tsubaki, M. Wu, Monodispersed Hollow $SO_3H$-functionalized Carbon/silica as Efficient Solid Acid Catalyst for Esterification of Oleic Acid, *ACS Appl. Mater. Interface* **7**, 26767-26775, (2015).

65. X. Wang, J. Feng, Y. Bai, Q. Zhang and Y. Yin., Synthesis, Properties, and Applications of Hollow Micro-/nanostructures, *Chem Rev.* **116**, 10983-11060, (2016).
66. E. de Smit and B. M. Weckhuysen, The Renaissance of Iron-based Fischer–Tropsch Synthesis: On the Multifaceted Catalyst Deactivation Behaviour, *Chem. Soc. Rev.* **37**, 2758-2781, (2008).
67. S. Abelló and D. Montané, Exploring Iron-based Multifunctional Catalysts for Fischer–Tropsch Synthesis: A Review, *ChemSusChem* **4**, 1538-1556, (2011).
68. Y. Wang, L. Tan, M. Tan, P. Zhang, Y. Fang, Y. Yoneyama, G. Yang and N. Tsubaki, Rationally Designing Bifunctional Catalysts as an Efficient Strategy to Boost $CO_2$ Hydrogenation Producing Value-added Aromatics. *ACS Catal.* **9**, 895-901, (2018).
69. J. Zhang, M. Zhang, S. Chen, X. Wang, Z. Zhou, Y. Wu, T. Zhang, G. Yang, Y. Han and Y. Tan. Hydrogenation of $CO_2$ into Aromatics over a ZnCrOx–zeolite Composite Catalyst, *Chem. Commun.* **55**, 973-976, (2019).
70. T. Wang, C. Yang, P. Gao, S. Zhou, S. Li, H. Wang and Y. Sun, ZnZrOx Integrated with Chain-like Nanocrystal HZSM-5 as Efficient Catalysts for Aromatics Synthesis from $CO_2$ Hydrogenation, *Appl. Catal. B: Environ.* **286**, 119929, (2021).
71. Y. Ni, Z. Chen, Y. Fu, Y. Liu, W. Zhu and Z. Liu, Selective Conversion of $CO_2$ and $H_2$ into Aromatics, *Nature Commun.* **9**, 1-7, (2018).
72. W. Gao, L. Guo, Y. Cui, G. Yang, Y. He, C. Zeng, A. Taguchi, T. Abe, Q. Ma and Y. Yoneyama. Selective Conversion of $CO_2$ into Para-Xylene over a $ZnCr_2O_4$-ZSM-5 Catalyst, *ChemSusChem* **13**, 6541-6545, (2020).
73. Zhang, A. Zhang, X. Jiang, J. Zhu, J. Liu, J. Li, G. Zhang, C. Song and X. Guo. Utilization of $CO_2$ for Aromatics Production over ZnO/$ZrO_2$-ZSM-5 Tandem Catalyst. *J. $CO_2$ Util.* **29**, 140-145, (2019).
74. Z. Li, Y. Qu, J. Wang, H. Liu, M. Li, S. Miao and C. Li, Highly Selective Conversion of Carbon Dioxide to Aromatics over Tandem Catalysts, *Joule* 3, 570-583, (2019).
75. X. Cui, P. Gao, S. Li, C. Yang, Z. Liu, H. Wang, L. Zhong and Y. Sun, Selective Production of Aromatics Directly from Carbon Dioxide Hydrogenation, *ACS Catal.* **9**, 3866-3876, (2019).
76. G. Song, M. Li, P. Yan, M. A. Nawaz and D. Liu, High Conversion to Aromatics via $CO_2$-FT over a CO-reduced Cu-$Fe_2O_3$ Catalyst Integrated with HZSM-5, *ACS Catal.* **10**, 11268-11279, (2020).
77. C. Dai, X. Zhao, B. Hu, J. Zhang, Q. Hao, H. Chen, X. Guo and X. Ma, Hydrogenation of $CO_2$ to Aromatics over Fe–K/Alkaline $Al_2O_3$ and P/ZSM-5 Tandem Catalysts, *Ind. Eng. Chem. Res.* **59**, 19194-19202, (2020).
78. L. Ai, J. Su, M. Wang and J. Jiang, Bamboo-structured Nitrogen-doped Carbon Nanotube Coencapsulating Cobalt and Molybdenum Carbide Nanoparticles: An Efficient Bifunctional Electrocatalyst for Overall Water Splitting, *ACS Sus. Chem. Eng.* **6**, 9912-9920, (2018).

79. C. K. Kuei and M. D. Lee, Hydrogenation of Carbon Dioxide by Hybrid Catalysts, Direct Synthesis of Aromatics from Carbon Dioxide and Hydrogen. *The Canadian J. Chem. Eng.* **69**, 347-354, (1991).
80. C. Wen, J. Jiang, C. Chiliu, Z. Tian, X. Xu, J. Wu, C. Wang and L. Ma, Single-step Selective Conversion of Carbon Dioxide to Aromatics over Na-$Fe_3O_4$/hierarchical HZSM-5 Zeolite Catalyst, *Energy & Fuels* **34**, 11282-11289, (2020).
81. J. Wei, Q. Ge, R. Yao, Z. Wen, C. Fang, L. Guo, H. Xu and J. Sun, Directly Converting $CO_2$ into a Gasoline Fuel, *Nat. Commun.* **8**, 1-9, (2017).
82. A. Ramirez, A. Dutta Chowdhury, A. Dokania, P. Cnudde, M. Caglayan, I. Yarulina, E. Abou-Hamad, L. Gevers, S. Ould-Chikh and K. De Wispelaere, Effect of Zeolite Topology and Reactor Configuration on the Direct Conversion of $CO_2$ to Light Olefins and Aromatics, *ACS Catal.* **9**, 6320-6334, (2019).
83. J. Wei, R. Yao, Q. Ge, D. Xu, C. Fang, J. Zhang, H. Xu and J. Sun, Precisely Regulating Brønsted Acid Sites to Promote the Synthesis of Light Aromatics via $CO_2$ Hydrogenation. *Appl. Catal. B: Environ.* **283**, 119648, (2021).
84. Y. Xu, C. Shi, B. Liu, T. Wang, J. Zheng, W. Li, D. Liu and X. Liu, Selective Production of Aromatics from $CO_2$, *Catal. Sci. & Tech.* **9**, 593-610, (2019).
85. Y. Wang, S. Kazumi, W. Gao, X. Gao, H. Li, X. Guo, Y. Yoneyama, G. Yang and N. Tsubaki, Direct Conversion of $CO_2$ to Aromatics with High Yield via a Modified Fischer-Tropsch Synthesis Pathway. *Appl. Catal. B: Environ.* **269**, 118792, (2020).
86. J. Wei, R. Yao, Q. Ge, D. Xu, C. Fang, J. Zhang, H. Xu and J. B. E. Sun, Precisely Regulating Brønsted Acid Sites to Promote the Synthesis of Light Aromatics via $CO_2$ Hydrogenation, *Appl. Catal. B: Environ.* **283**, 119648, (2021).
87. Ye, X., Yang, C., Pan, X., Ma, J., Zhang, Y., Ren, Y., Liu, X., Li, L. and Huang, Y., Highly Selective Hydrogenation of $CO_2$ to Ethanol via Designed Bifunctional Ir1–$In_2O_3$ Single-atom Catalyst, *J. Am. Chem. Soc.* **142**, 19001-19005, (2020).
88. Goldemberg, J., Ethanol for a Sustainable Energy Future, *Science* **315**, 808-810, (2007).
89. K. Stangeland, H. Li and Z. Yu, Thermodynamic Analysis of Chemical and Phase Equilibria in $CO_2$ Hydrogenation to Methanol, Dimethyl ether, and Higher Alcohols, *Ind. Eng. Chem. Res.* **57**, 4081-4094, (2018).
90. H. T. Luk, C. Mondelli, D. C. Ferré, J. A. Stewart and J. Pérez-Ramírez, Status and Prospects in Higher Alcohols Synthesis from Syngas, *Chem Soc Rev.* **46**, 1358-1426, (2017).
91. T. Inui, T. Yamamoto, M. Inoue, H. Hara, T. Takeguchi and J.-B. J. A. C. A. G. Kim, Highly Effective Synthesis of Ethanol by $CO_2$-hydrogenation on Well Balanced Multi-functional FT-type Composite Catalysts, *Appl. Catal. A: General* **186**, 395-406, (1999).
92. F. Zeng, C. Mebrahtu, X. Xi, L. Liao, J. Ren, J. Xie, H. J. Heeres and R. Palkovits, Catalysts Design for Higher Alcohols Synthesis by $CO_2$ Hydrogenation: Trends and Future Perspectives, *Appl. Cat. B: Environ.* **291**, 120073, (2021).

93. K. Stangeland, H. Li, Z. J. I. Yu and E. C. Research, Thermodynamic Analysis of Chemical and Phase Equilibria in $CO_2$ Hydrogenation to Methanol, Dimethyl Ether, and Higher Alcohols, *Ind. Eng. Chem. Res.* **57**, 4081-4094, (2018).
94. D. Xu, H. Yang, X. Hong, G. Liu and S. Edman Tsang, Tandem Catalysis of Direct $CO_2$ Hydrogenation to Higher Alcohols, *ACS Catal.* **11**, 8978-8984, (2021).
95. H. Kusama, K. Okabe, K. Sayama and H. Arakawa, $CO_2$ Hydrogenation to Ethanol Over Promoted Rh/$SiO_2$ Catalysts. *Catal. Today* **28**, 261-266, (1996).
96. H. Kusama, K. Okabe, K. Sayama and H. Arakawa, Rhodium Catalysts Promoted by Cerium Oxide in the $CO_2$ Hydrogenation to Ethanol. *J. of The Japan Petroleum Institute* **42**, 178-179, (1999).
97. C. Yang, R. Mu, G. Wang, J. Song, H. Tian, Z.-J. Zhao and J. Gong, Hydroxyl-Mediated Ethanol Selectivity of $CO_2$ Hydrogenation, *Chem. Science* **10**, 3161-3167, (2019).
98. M. R. Gogate and R. J. Davis, Comparative Study of CO and $CO_2$ Hydrogenation over Supported Rh–Fe Catalysts, *Catal. Commun.* **11**, 901-906, (2010).
99. H. Kusama, K. Okabe, H. Arakawa, Effect of Ce Additive on $CO_2$ Hydrogenation over Rh/$SiO_2$ Catalysts, *J. of The Japan Petroleum Institute* **44**, 384-391, (2001).
100. H. Kusama, K. Okabe, K. Sayama and H. Arakawa, Ethanol Synthesis by Catalytic Hydrogenation of CO2 over Rh/FeSiO2 Catalysts, *Energy* **22**, 343-348, (1997).
101. W. Guo, W. G. Gao, H. Wang and J. J. Tian, Higher Alcohols Synthesis from $CO_2$ Hydrogenation over $K_2O$-Modified $CuZnFeZrO_2$ Catalysts, *Adv. Mater. Res.* **827**, 20-24, (2014).
102. Guo, H., Li, S., Peng, F., Zhang, H., Xiong, L., Huang, C., Wang, C. and Chen, X. Roles Investigation of Promoters in K/Cu–Zn Catalyst and Higher Alcohols Synthesis from $CO_2$ Hydrogenation over a Novel Two-stage Bed Catalyst Combination System, *Catal. Lett.* **145**, 620-630, (2015).
103. S. Li, H. Guo, C. Luo, H. Zhang, L. Xiong, X. Chen and L. Ma, Effect of Iron Promoter on Structure and Performance of K/Cu–Zn Catalyst for Higher Alcohols Synthesis from $CO_2$ Hydrogenation, *Catal. Letters* **143**, 345-355, (2013).
104. Y. Chen, S. Choi and L. T. Thompson, Low Temperature $CO_2$ Hydrogenation to Alcohols and Hydrocarbons over Mo2C Supported Metal Catalysts. *J. Catal.* **343**, 147-156, (2016).
105. D. L. Nieskens, D. Ferrari, Y. Liu and R. Kolonko Jr, The Conversion of Carbon Dioxide and Hydrogen into Methanol and Higher Alcohols. *Catal. Commun.* **14**, 111-113, (2011).
106. M. Kishida, K. Yamada, H. Nagata and K. Wakabayashi, $CO_2$ Hydrogenation for $C_{2+}$ Alcohols Synthesis over Silica-supported Ir-Mo Catalysts, *Chem. Lett.* **23**, 555-556, (1994).
107. S. Liu, H. Zhou, L. Zhang, Z. Ma and Y. Wang, Activated Carbon-Supported Mo-Co-K Sulfide Catalysts for Synthesizing Higher Alcohols from $CO_2$. *Chem. Eng. & Technol.* **42**, 962-970, (2019).

108. A. Calafat, F. Vivas and J. Brito. Effects of Phase Composition and of Potassium Promotion on Cobalt Molybdate Catalysts for the Synthesis of Alcohols from $CO_2$ and $H_2$, *Appl. Catal. A: Gen.* **172**, 217-224, (1998).

https://doi.org/10.1142/9789811270956_0009

# Chapter 9

# Porous Materials for Sustainable Energy Applications: Promises and Challenges

Abhishek Garg,[‡] Subhajit Laha,[‡] Muthusamy Eswaramoorthy[*] and Tapas Kumar Maji[†]

*Chemistry and Physics of Materials Unit, New Chemistry Unit and School of Advanced Materials, Jawaharlal Nehru Centre for Advanced Scientific Research, Jakkur, Bangalore 560064, Karnataka, India*
*[*]eswar@jncasr.ac.in (M.E), [†]tmaji@jncasr.ac.in (T.K.M)*

Energy is omnipresent. To ensure sustainable energy storage and conversion, it is crucial to develop carriers that store/harvest energy through mass and electron transfer. The ease of mass/electron transfer with unique structural and chemical functionalities of porous materials has attracted energy research community. In this chapter, we review the recent advancement in energy storage and conversion by engaging porous materials and highlighting the correlation between structure, functionalities, and composition. By reporting the promises and challenges of the different classes of porous materials for versatile utilities, we also underline on the demand and methodologies for future development of materials and technologies for sustainable energy applications.

## 1. Introduction

The dramatic depletion of fossil fuel as a corollary to the blossoming economy is expected to bring the world in the edge of energy and environment crisis unless otherwise steps are taken to mitigate the impending disaster. In this direction, huge efforts are being made not only to tap the renewable energy sources but also to store and convert the renewal energy on demand. Developing novel materials which could act

[‡]Authors contributed equally to this chapter.

as energy reservoirs and facilitate energy harvesting by efficient mass and electron diffusion are essential to realize these goals.

In this direction, porous materials by their unique features offer solutions to energy-saving and conversion modules, thereby addressing some critical global concerns on the sustainability of fuel banks and environmental crises. In recent years, porous materials which can facilitate prompt mass and electron migration have been reported for photo- and electrochemical energy applications. The key elements for high performance and wide applicability of porous materials have been determined by porosity, structure, composition, and morphology which is essentially crafted by organic, inorganic, or hybrid building units. This chapter presents the recent upsurge of porous materials with solitary design strategies and dynamic pore functionalities for energy-efficient separation, storage, and conversion. Our discussion will be restricted to a few pivotal classes of porous materials such as zeolites, mesoporous oxides, porous carbon, porous coordination polymer, pillared clays, etc., and their derivatives and composites for sustainable energy applications (Figure 1).

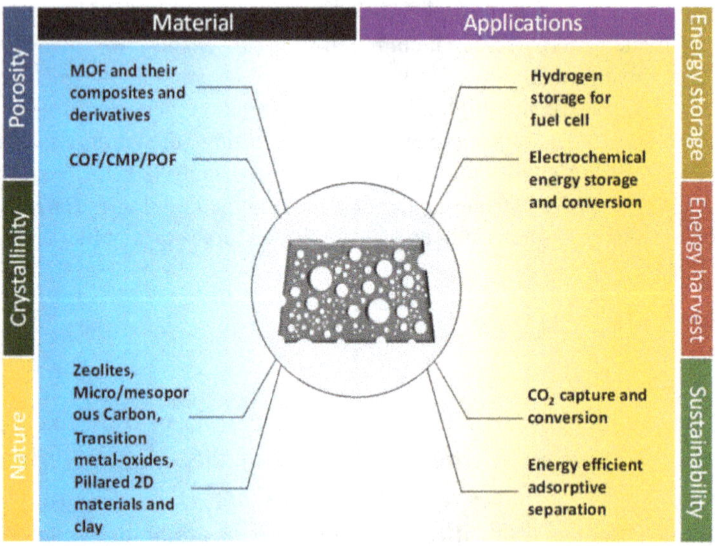


Figure 1. Schematic representing different porous materials and their respective applications.

## 2. Hydrogen Storage

In search for alternative energy storage, hydrogen could be considered as most sustainable, clean, and affordable as it is capable to produce carbon-neutral energy cycle by producing water as the only by-product. The simple molecule, $H_2$ possesses higher gravimetric energy density compared to conventionally used gasoline (comparatively higher volumetric energy against $H_2$).[1,2] Nevertheless, it is extremely critical and risky to store and transport $H_2$ which limits its usage in large scale.
Furthermore, the compression and liquification of hydrogen gas to obtain high volumetric energy density requires much expansive and potentially unsafe extreme operating conditions (compression requires 700 bar pressure whereas liquification has been performed under 20 K). To aid the development of materials for hydrogen storage, the US Department of Energy (DoE) set the limit of technical storage with ultimate target of 6.5 wt.% and 50 g $L^{-1}$.[3] Over recent years, several porous materials and their related composites, in particular metal-organic frameworks (MOFs), covalent-organic frameworks (COFs), porous organic polymers (POPs), zeolites, and carbon-based materials have shown promising hydrogen storage capacities owing to their optimum porosity, and tuneable high surface areas. In 2003, Yaghi's group[4] pioneered the exploration of MOFs for hydrogen storage. Later on, there has been significant evolution of gravimetric/volumetric deliverable $H_2$-capacities in MOFs, with design strategies derived through both experimental and theoretical standpoints. For example, strong bonding interactions with exposed low-valent metal nodes to enhance binding energies (high isosteric heat of adsorption), such as Cu(I), V(II), Mn (II), Cr (III), Ho (III), leads to improved $H_2$-storage capacities under ambient or near ambient conditions.[5-7] Note that the deliverable hydrogen capacity can be expanded by using a combined temperature and pressure swing operation condition, based on tank design engineering, proposed by DoE. Apart from that, MOFs with high-valent metal clusters such as $Al_3$, $Fe_3$ and $Zr_6$, building units with carboxylate linkers impart good structural stability in the framework implemented for commercial gas tanks. The durable Zr-MOFs (NU-1101, -1102, -1103), with high BET surface area (3,720 to 6,245 $m^2$ $g^{-1}$) collected from $Zr_6$ clusters and tetra carboxylate linkers showcase high volumetric $H_2$ uptake

43.2 g L$^{-1}$ (12%) under the pressure and temperature swing operating conditions (77 K/100 bar →160 K/5 bar; Figures 2 and 3).[8,9]

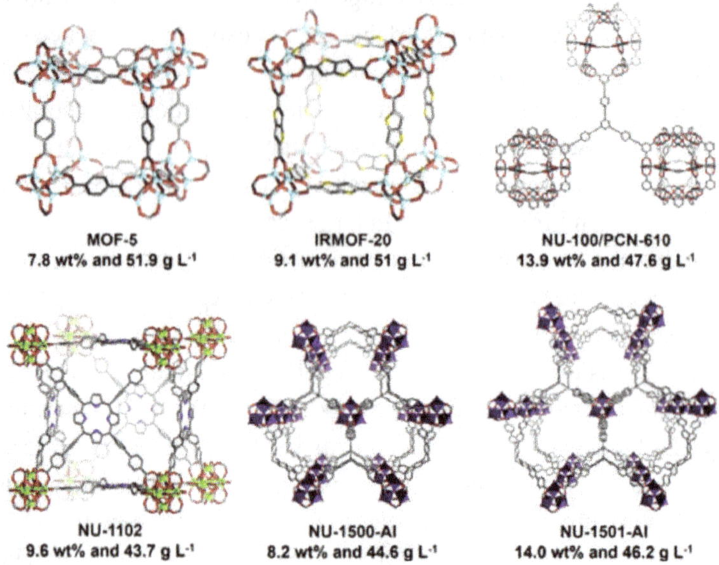


Figure 2. Crystal structure of representative MOF with their deliverable H2 capacities. Reproduced with permission from Ref. 3.

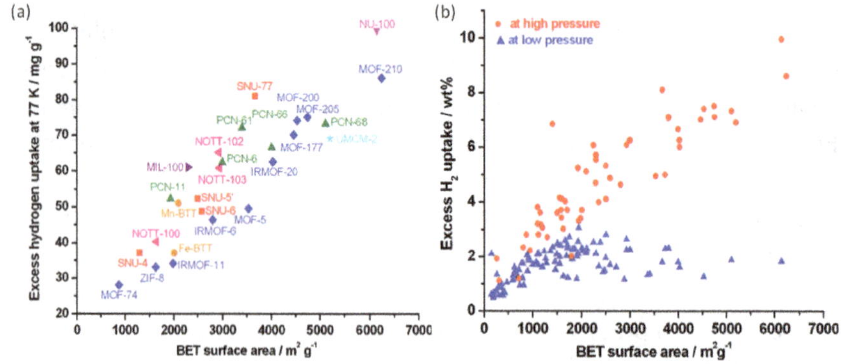


Figure 3. (a) High pressure H$_2$ uptake capacities at 77 K vs BET surface areas for highly porous MOFs. (b) A plot showing the relationship between H$_2$ uptake capacities at 77 K and BET surface areas of various MOFs comparing low and high pressures is 1 atm and high pressures are in the range of 1090 bar. Reproduced with permission from Ref. 5.

Like MOFs, COFs are also employed as physical sorbents for hydrogen storge. COFs, for being crystalline and highly porous, exhibit relatively higher gravimetric uptake against MOFs as they are consisting of lighter elements with relative lower density (absence of metal). However, such materials limit in promising $H_2$-storage as a consequence of lack of micro-porosity and active binding sites, result in lower binding interactions upon adsorption. Such drawbacks have been primarily addressed by introducing optimum metal-doping to improve $H_2$-storage under ambient temperature. For example, Li-doped COF-105 and COF-108 improve their storage capacity relatively double to their pristine analogues, attaining 6.84 and 6.73 wt.% of gravimetric uptake.[10] Similar materials, constructed through rigid blocks and conjugated π-bonds (microporous polymer (CMPs), hyper-cross-linked polymers (HCPs), porous aromatic frame-works), exhibit moderate storage capacities (~0.9–7 wt.%) under cryogenic temperature.[10,11] Although these materials result in high surface areas. but alike COFs, they do exhibit weak physiosorbed $H_2$ capacity.

## 3. Electrochemical Energy Storage and Conversion

### 3.1. *Electrolysis of water*

Electrocatalytic water splitting refers to the dissociation of water to produce $H_2$ at the cathode and $O_2$ at the anode. It is considered as a green technology because water is the reactant and electron from renewable sources of energy acts as reductant. A typical water electrolyzer consists of a catalyst layer coated over the current collector dipped in an electrolyte. Various strategies have been developed to design catalysts depending upon the operating conditions of the electrolyzer. There have been three main technologies developed till now: (1) Proton exchange membrane (PEM) electrolyzer; (2) Alkaline electrolyzer; (3) Solid oxide electrolyzer. Solid oxide electrolyzer suffers from high energy consumption due to its operation at high temperature. In the PEM electrolyzer, the electrolysis is performed under acidic conditions leading to the advantages of high proton conductivity and greater hydrogen production rate.[12] However, the instability of other non-noble metal-based catalyst under operating pH condition limits the OER catalysts to various noble metal and noble metal oxide catalyst which in turn suffer from high cost.[13] The alkaline electrolyzer, on the other hand, has an advantage of broad selection of non-noble metal and metal oxide catalysts. However, it suffers from sluggish hydrogen production (~2 to 3 orders of magnitude) due to additional water dissociation step.[14] This motivates researchers to develop efficient catalysts with the features of low cost, high durability and wide pH range operationality.

In order to develop an effective electrocatalyst, the following four factors should be considered:

1. Reduce the activation energy barrier to improve the activity of the surface catalytic sites.[15]
2. Enhance the electronic conductivity of materials thereby quick charge transfer can occur from the catalyst surface to the electrode thus, lowering polarization potential.[16]
3. Improve the surface area of the materials to achieve high current densities at low overpotentials.[17]
4. Improve the mass diffusion to enhance the flow of reactants to the active surface and provide easy release of products.[18]

The former two aspects are material specific related to its intrinsic activity, while the latter two can be easily tuned depending upon the employed synthetic conditions. This provides a greater control over the last two properties which will be discussed in the subsequent sub-sections.

### 3.2. *Hydrogen production*

A hydrogen evolution reaction (HER) can proceed through two mechanisms: Volmer-Heyrovsky and Volmer-Tafel. Volmer step [Eqs. (1) and (4)] involves the capture of a proton by an electron on the catalytic surface site to produce adsorbed hydrogen as an intermediate. Depending upon the adsorbed hydrogen coverage over the catalyst, it can either follow Heyrovsky or Tafel pathway. The low coverage leads to the interaction of adsorbed hydrogen and proton from the electrolyte to produce hydrogen via Heyrovsky pathway [Eqs. (2) and (5)], whereas the high coverage leads to the combination of two adsorbed hydrogen atoms to evolve $H_2$ gas via Tafel pathway [Eqs. (3) and (6)].[19]

**Under acidic conditions:**

| | | |
|---|---|---|
| Volmer step: | $H^+ + e^- \rightarrow H_{ad}$ | (1) |
| Heyrovsky step: | H+ + e- + Had → H2 | (2) |
| Tafel step: | 2Had → H2 | (3) |

**Under alkaline/neutral conditions:**

| | | |
|---|---|---|
| Volmer step: | H2O + e- → OH- + Had | (4) |
| Heyrovsky step: | H2O + e- + Had → OH- + H2 | (5) |
| Tafel step: | 2Had → H2 | (6) |

According to Sabatier principle, there should be an optimum free energy of hydrogen adsorption. Pt lies at the top of volcano plot with $\Delta G_{H*}$~0 making it an ideal catalyst for HER. However, the high-cost limits its industrial application. This can be overcome by utilizing a porous carbon matrix such as carbon black which helps to increase Pt dispersion and thus enhance its atomic utilization efficiency.[20-22] With a cheaper alternative to Pt, Ru possesses a similar bond strength with hydrogen (~65 kcal $mol^{-1}$). Javeed *et al.*[23] utilized nitrogenated holey two-dimensional carbon structure (SSA 280 $m^2$ $g^{-1}$) to disperse Ru nanoparticles (Figure 4). The synthesized catalyst, Ru@$C_2N$, possesses enhanced SSA of 400 $m^2$ $g^{-1}$ with a pore volume of 0.338 $cm^3$ $g^{-1}$ and an average pore diameter of 3.37 nm. As a result, it showed enhanced HER performance in both acidic ($\eta_{10}$ = 22 mV) and alkaline ($\eta_{10}$ = 17 mV) conditions. In this way, the atomic utilization efficiency of these noble metals has been enhanced using porous materials. However, the poor lifetime, limited reserves, and the cost of these composite catalysts still limit their commercial viability.

Thus, various non-noble transition metal-based catalysts (such as Fe, Co, Ni, etc.) and porous carbon-based materials (such as heteroatom doped graphene and their composites) have attracted significant attention for HER process. Zheng *et al.*[24] for the first time utilized graphitic carbon nitride (g-$C_3N_4$) and N-doped graphene, which showed an overpotential of 240 mV at 10 mA $cm^{-2}$. Later, Lin *et al.*[25] synthesized N-doped

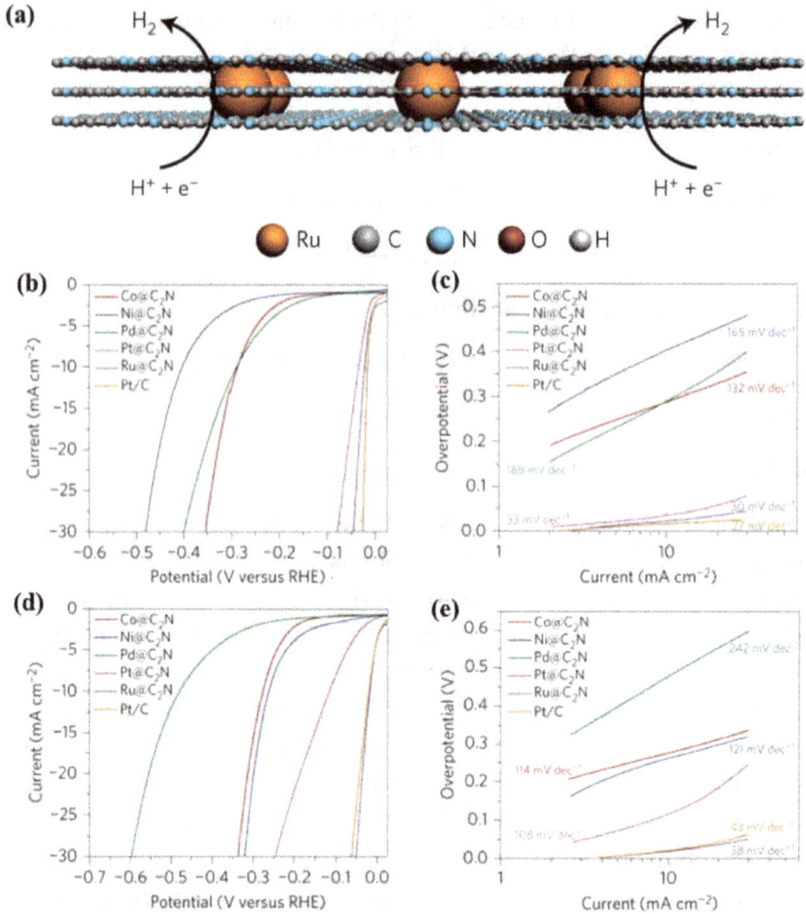


Figure 4. (a) Schematic representing the structure of the Ru@$C_2N$ electrocatalyst. (b, c) LSV curve and Tafel plot of Co@$C_2N$, Ni@$C_2N$, Pd@$C_2N$, Pt@$C_2N$ and Pt/C in 0.5 M $H_2SO_4$. (d, e) LSV curve and Tafel plot of Co@$C_2N$, Ni@$C_2N$, Pd@$C_2N$, Pt@$C_2N$ and Pt/C in 1 M KOH. Reproduced with permission from Ref. 23.

mesoporous graphene with an overpotential of 239 mV at 10 mA $cm^{-2}$. Although graphene and its composites can effectively catalyze these reactions, they suffer from aggregation due to strong π–π interactions hindering their mass transport properties. Dheeraj *et al.*[26] utilized a template-based strategy to effectively synthesize a N-doped tubular porous carbon (Figure 5). They chose halloysite clay as the soft template which

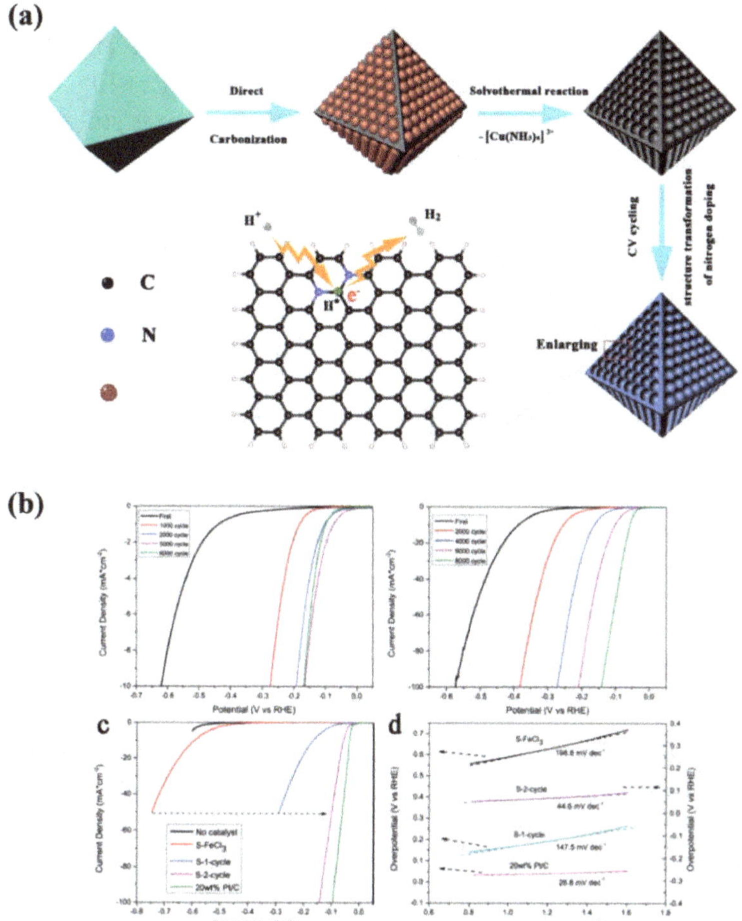


Figure 6. (a) Schematic representing of the synthesis of Cu/C composites with Cu on the surface and edges. (b) Electrocatalytic performance of the corresponding catalysts. Reproduced with permission from Ref. 27.

was coated by polydopamine through self-polymerization of dopamine hydrochloride under basic conditions (pH 8.5). This was followed by pyrolysis at 800°C and subsequently washed with HF and HCl to remove template and metal species. The resulting carbon possesses a specific surface area of 570 $m^2$ $g^{-1}$. The abundant micropores and mesopores (> 15 nm) with a total pore volume of 1.879 $cm^3$ $g^{-1}$ provides a balance between confinement effect and the mass transport properties to the material. As a result, it showed an HER overpotential of 276 mV at 10 mA $cm^{-2}$ with an exchange current density of $1.5 \times 10^{-2}$ mA $cm^{-2}$. However, the performance of these metal free catalysts is still far beyond the commercial reach due to highly unfavorable positive Gibbs free energy ($\Delta G_{H*}$), leading to weaker adsorption of the intermediate state (H*). Both experimental and density functional theory (DFT) proved that single nitrogen doping in a six membered ring of the nitrogen-graphene lattice is insufficient in tuning the $\Delta G_{H*}$. In this regard, Lin *et al.*[27] synthesized graphene analogous material with a dual graphitic-N doping in a six membered carbon ring (Figure 6). The dual graphitic nitrogen doped material shows a very low overpotential of 57 mV at 10 mA $cm^{-2}$ which is comparable to most of the metal-based catalyst and even better in some cases. However, the stability of this catalyst was not discussed. This puts up a question about the stability of these catalysts. In this way, new strategies can be designed to tune the $\Delta G_{H*}$ value of carbon-based catalysts close to zero via porosity and/or surface engineering.

### 3.3. *Oxygen evolution reaction (OER)*

The reduction of water is often hampered by the oxygen evolution reaction (OER), a slow half-reaction in water electrolysis. It is relatively easy to produce $H_2$ on the cathode as compared to the water oxidation on the anode. This is due to the involvement of various intermediate species (Figure 7). Hence, it is critical to develop effective and stable electro-catalysts for water oxidation to reduce electrolysis energy costs associated with the electrolysis. There have been a variety of electrocatalysts explored for catalyzing OER to a greater degree. Even though noble metal-based catalysts, such as $IrO_2$, $RuO_2$, and others, possess high activity, they experience poor stability due to dissolution of the catalyst in the electrolyte, as well as high prices. Various non-noble metals (Fe, Co, Ni, Mn, etc.) based materials thus gained continuing interest. It is possible to alter their chemical composition to tune the performance of their surface-active sites. Sisir *et al.*[28] synthesized bimetallic NiCoP nanoporous catalyst and showed the effect of introduction of Ni on the electronic structure of Co and P atoms. The substitution of Ni in NiCoP leads to the lowering of the p-band center of P atom from −1.97 eV to −1.55 eV, thus favoring the adsorption of the species with P atom. On the other hand, catalysts such as layered double hydroxides (LDHs) although possesses highly efficient active sites suffers from poor electrical conductivity ($10^{-13}$–$10^{-17}$ S $cm^{-1}$). This leads to hindered charge transport across the electrode/catalyst and catalyst/reactant interfaces.

In acidic electrolytes, the oxygen intermediates are formed as follows:[29]

$$* + H_2O \rightarrow *OH + H^+ + e^- \tag{7}$$

$$*OH \rightarrow *O + H^+ + e^- \tag{8}$$

Afterwards, the two possible pathways are:

$$2*O \rightarrow * + * + O_2 \tag{9}$$

or

$$*O + H_2O \rightarrow *OOH + H^+ + e^- \tag{10}$$

$$*OOH \rightarrow * + O_2 + H^+ + e^- \tag{11}$$

Similarly for alkaline electrolytes, *O is generated first

$$*OH + OH^- \rightarrow *O + H_2O + e^- \tag{12}$$

Subsequently, the pathways are:

$$2*O \rightarrow * + * + O_2 \tag{13}$$

or

$$*O + OH^- \rightarrow *OOH + e^- \tag{14}$$

$$*OOH + OH^- \rightarrow * + O_2 + H_2O + e^- \tag{15}$$

Soumita *et al.*[30] showed the improved performance of NiFe-layered double hydroxides for OER by the introduction of 1-T rich $MoS_2$ (Figure 7). This leads to improved OER performance such as lower overpotentials ($\eta_{10}$ = 190 mV) and Tafel slope (31 mV $dec^{-1}$). Increasing the exposure to these highly active sites and mass diffusion of reactants and products species, to and away from the surface, can be facilitated by creating high porosity on the catalyst substrate. A general method is to develop porosity in these materials by thermal decomposition, electrochemical deposition, template-assisted, etc. Zhang *et al.*[31] demonstrated that metal-free N and P co-doped porous carbon (NPMC) prepared by the thermal decomposition

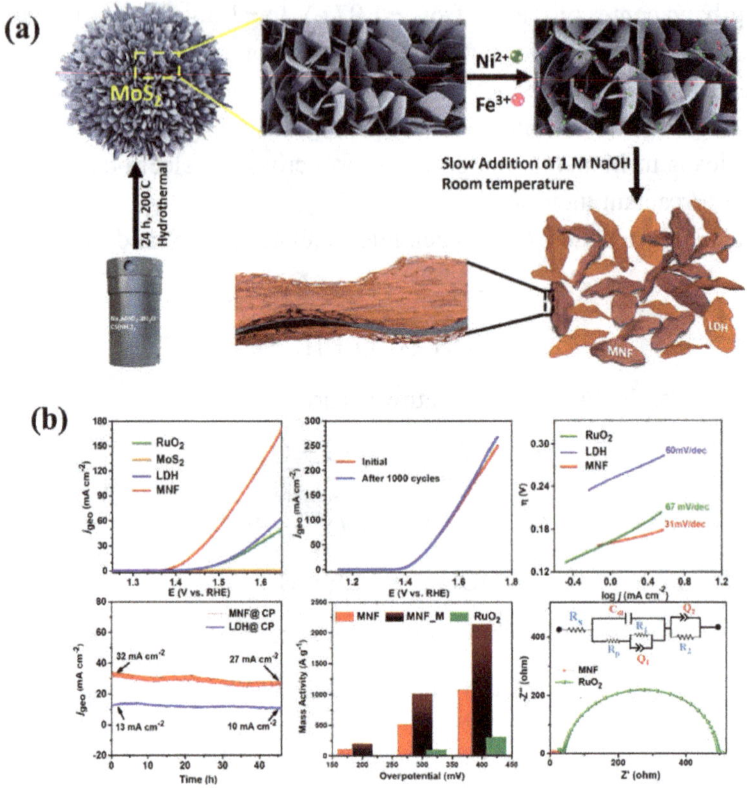


Figure 7. (a) Schematic representing the synthesis of NiFe LDH/$MoS_2$ composite. (b) Electrocatalytic performance of NiFe LDH/$MoS_2$ composite. Adapted with permission from Ref. 30.

of PANI aerogel display a mesoporous structure with pore size of less than 10 nanometers. The NPMC synthesized at 1000°C possesses the specific surface area of 1568 $m^2$ $g^{-1}$ with a pore volume of 1.1 $cm^3$ $g^{-1}$. This promotes the diffusion of $O_2$ species to the surface leading to the rapid increase in current density beyond 1.3 V vs RHE for the OER in 0.1 M KOH. By utilizing a template-aided technique, Hu *et al.*[32] synthesized stereoscopic holes in graphitic sheets with N, S co-doped content. The mixture of melamine, nickel sulphate complex and KCl was pyrolyzed under Ar, followed by etching using aqua regia solution resulting in the mesoporous structure with a diameter between 3 and 30 nm and a pore volume of 1.4 $cm^3$ $g^{-1}$. The synthesized catalyst exhibited a current density of 10 mA $cm^{-2}$ at 1.56 V vs RHE and a low Tafel slope of 71 mV $dec^{-1}$ in 0.1 M KOH. In spite of their promising technoeconomic potential, carbon-based porous materials are not stable over the long run. Especially in acidic media, the anodic conditions of the OER can potentially cause the oxidation of carbon-based catalysts, resulting in oxygen species accumulating on the surface.[33] It may result in a decrease in electrical conductivity, as well as the electrode's decomposition in some cases.[34] This limits the application of porous carbon based materials for oxygen evolution reaction. In an early study, Tuysuz *et al.*[35] utilized template-assisted strategy to synthesize mesoporous $Co_3O_4$ with varying specific surface areas (SSA). By utilizing KIT-6 as the silica template, a mesoporous $Co_3O_4$ with a specific surface area of 156, 113 and 17 $m^2$ $g^{-1}$ and a crystallite size of 6, 8, and 10 nm, respectively, was obtained. With increasing porosity, the OER current density increases and the water oxidation potential decreases. Lu *et al.*[36] further enhanced the OER activity of $mCo_3O_4$ by the introduction of electronegative Au nanoparticles due to the change in the oxidation state of cobalt to highly active $Co^{4+}$. However, increasing the Au loading from 2.5% to 10% lead to decreased activity which was attributed to the decrease in the specific surface area of the composite from 109 $m^2$ $g^{-1}$ to 75 $m^2$ $g^{-1}$.

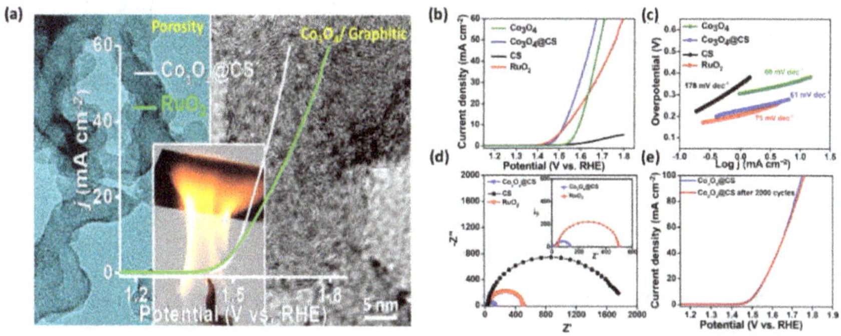


Figure 8. (a) Schematic representing the synthesis of NiFe LDH/$MoS_2$ composite. (b) Electrocatalytic performance of NiFe LDH/$MoS_2$ composite. Adapted with permission from Ref. 37.

In this direction, Dheeraj *et al.*[37] utilized a simple combustion-based strategy to synthesize well dispersed $Co_3O_4$ nanoparticles over highly graphitized porous carbon (Figure 8). The porous structure in the meso-macro length scale (BET SSA of 165 $m^2$ $g^{-1}$) ensures easy access of the electrolyte with uniform dispersion of the nanoparticles (~ 2 nm size) leading to enhanced OER performance (require potential of 70 mV lower than $RuO_2$ to achieve a current density of 30 mA $cm^{-2}$) even at low metal loadings (~1 wt. % Co). This proves the efficacy of the porosity in increasing the OER activity. Besides transition metal oxides, transition metal phosphides, sulphides, nitrides, etc. based porous catalysts have also been studied extensively. However, these catalysts undergo reconstruction during the process leading to significant change in performance which needs further understanding.

### 3.4. *Metal air batteries*

In order to buffer the unpredictable energy production and supply from renewable sources in the future energy network, electrochemical energy storage devices are essential components. With the advent of lithium-ion batteries, we have witnessed a profound improvement in every aspect of our lives. There have been continuous efforts to increase their energy and power density. This is further increased with the need to electrify transportation. As a potential candidate for energy storage, metal air batteries show an energy density three to thirty times higher than lithium-ion batteries. Table 1 shows the theoretical energy density and specific energy of various metal-air systems.[38] The commercialization of alkali metals-based batteries is limited due to their abundance and safety concerns. With their high energy density and specific energy, aqueous magnesium and aluminum batteries are susceptible to corrosion problems, limiting their

Table 1. Theoretical energy density and specific energy (including oxygen) calculated using specific mass and volume of the discharged products.

| $M/O_2$ battery | Energy density (Wh $L^{-1}$) | Specific energy (Wh $Kg^{-1}$) |
|---|---|---|
| $Si/O_2$ | 9930 | 3750 |
| $Li/O_2$ | 7990 | 3460 |
| $Al/O_2$ | 6790 | 2790 |
| $Mg/O_2$ | 6670 | 2850 |
| $Zn/O_2$ | 6100 | 1090 |
| $Na/O_2$ | 4430 | 1580 |

cyclability. Out of all the metal air batteries, Zn-air battery has received significant attention because of low cost, abundance, and safer operations under aqueous condition.[39,40] Hence, the scope of this chapter will be limited to Zn-air batteries.

### 3.4.1. *Working principle*

The fabrication of Zn-air battery typically involves four main components: a bifunctional cathode catalyst coated on a gas diffusion layer, a Zn anode, a separator and an alkaline electrolyte. During discharging, Zn anode gets oxidized to $Zn^{2+}$ which combines with hydroxides in the electrolyte to form $Zn(OH)_4^{2-}$ [Eq. (16)], which, upon supersaturation, forms ZnO [Eq. (17)]. The released electrons move from the outer circuit to the air cathode producing hydroxides on reduction of $O_2$ gas [Eq. (18)].
Zinc electrode reaction:

$$Zn + 4OH^- \leftrightarrow Zn(OH)_4^{2-} + 2e^- \qquad E^0 = 1.199\ V\ vs\ RHE \qquad (16)$$

Zn(OH)42- ↔ ZnO + 2OH- + H2O (17)

Air electrode reaction:

O2 + 2H2O + 4e- ↔ 4OH- E0 = 0.401 V vs RHE (18)

The reverse process occurs upon charging where ZnO gets reduced to Zn over metal anode and the hydroxide gets oxidized to evolve oxygen gas (Figure 9).

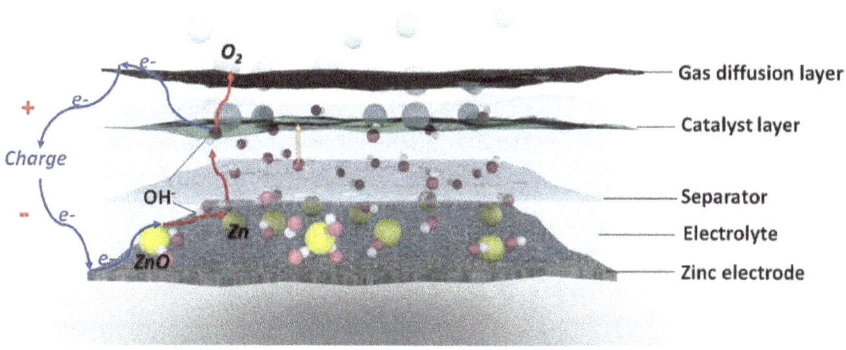


Figure 9. Schematic representation of charging status in rechargeable Zinc air batteries. Adapted with permission from Ref. 41.

### 3.4.2. *Performance descriptors*

Daniel *et al.*[41] evaluated the performance characteristics of alkaline Zn anodes on the basis of seven performance descriptors (Table 2).

Table 2. Performance metrics for a Zinc-air battery.

| Category | Descriptor |
|---|---|
| Cell properties | $m_{AM}/m_{anode}$ |
| | $Q_{AM}/V_E$ |
| Retention | $N_c$ |
| | $Q_c$ |
| Utilization | $X_{AM}$ |
| | $Q_{dis}$ |
| Combined | $N_C Q_{dis}$ |

AM = active material mass; madditives = mass of the additives; $V_E$ = volume of added electrolyte; $N_c$ = number of cycles; $Q_c$ = coulombic efficiency; $Q_{dis}$ = discharge capacity of anode; $Q_0$ = actual capacity of anode; $W_{grav}$ = specific energy.

These are calculated as,

$$m_{anode} = m_{AM} + m_{additives}$$

$$Q_c = (Q_{dis}/Q_{charge}) * 100$$

$$X_{AM} = (Q_{dis}/Q_0) * 100$$

$$W_{grav} = Q_{dis} * V$$

### 3.4.3. *Electrode material development*

The development of electro rechargeable Zn-air batteries is hindered due to the lack of efficient, robust air catalysts as well as the lack of stable, cyclable Zn anodes. Li *et al.*[42] developed CoO/CNT hybrid oxygen reduction reaction catalyst and Ni-Fe layered double hydroxide oxygen evolution catalyst. The CoO/CNT catalyst showed an onset potential of 20 mV more negative than Pt/C, however at high current density of 100 mA cm$^{-2}$ it surpassed Pt/C with a potential of 20 mV more positive. The efficacy of this catalyst was further proven by fabricating a primary zinc air battery. The battery results in an open circuit voltage of 1.4 V and a high current density of 200 mA $cm^{-2}$ at a potential of 1.0 V. Upon galvanostatically discharging at 5 mA $cm^{-2}$ and 50 mA $cm^{-2}$, it showed a durability of 22 h and 12 h, respectively (Figure 10).

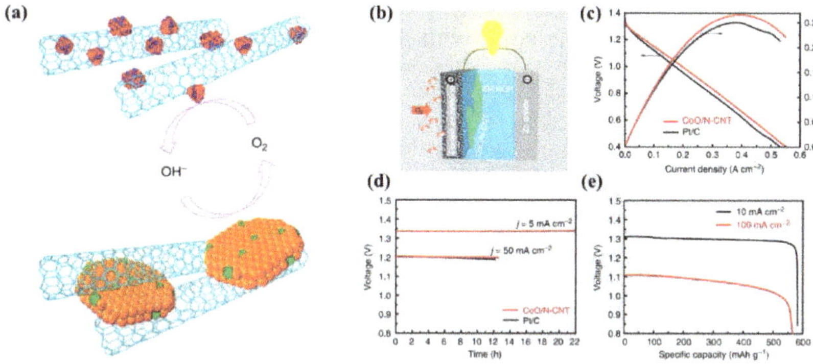


Figure 10. (a) Schematic illustration of the structure of the CoO/CNT hybrid; (b, c, d, e) Rechargeable Zn-air battery performance including the charge discharge profiles and the galvanostatic long term cyclability. Reproduced with permission from Ref. 42.

The performance was regained by using a fresh Zn metal and changing the electrolyte. However, fabricating a rechargeable battery with a combination of CoO/CNT and Ni-Fe LDH catalyst showed rapid decline in performance especially when run at high current densities (20 mA $cm^{-2}$) for longer run time (20 h per cycle). There was an increase in ORR overpotential by 200–250 mV for the second cycle. The authors suggested that the oxidation of the ORR catalyst occurred during positive OER potentials which leads to the detachment of CoO particles from the CNT. This promotes aggregation of the nanoparticles resulting in increased ORR potential. This was tackled by using a tri-electrode configuration whereby the OER and ORR catalysts were loaded on different electrodes. During discharge, the Zn anode was connected to the ORR electrode, and during charging, it was connected to the OER electrode. This avoids the exposure of ORR catalyst to positive potentials, resulting in only 20 mV increase in overpotential over 100 h operation at 20 mA $cm^{-2}$. However, this configuration comes at the expense of high weight and thus, lower energy density (Wh/kg). Therefore, it is appealing to develop an efficient and corrosion resistant bifunctional electrocatalysts.

Metal organic frameworks, MOFs, with their high surface area and active metal centre, serve as an excellent bifunctional catalyst.[43,44] Therefore, they can serve as an ideal candidate for metal air batteries. Sambhaji *et al.*[45] synthesized hexaiminobenzene based metal–organic framework (Mn/Fe-HIB-MOF) via the reaction of hexaaminobenzene (obtained through amination of chloro-anilic acid) with the metal ions $Fe^{2+}$ and $Mn^{2+}$ (Figure 11).

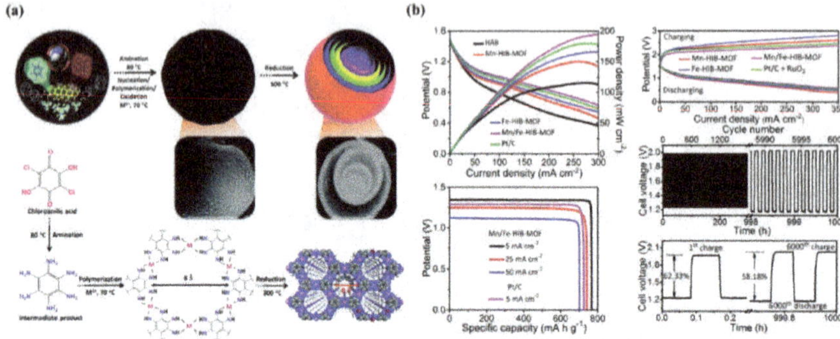


Figure 11. (a) Schematic illustration and reaction mechanism of the synthesis strategy of 3D M-HIB-MOFs; (b) Rechargeable Zn-air battery performance including the charge discharge profiles and the galvanostatic long term cyclability. Reproduced with permission from Ref. 45.

The as-synthesized material displayed an excellent current density of 300 mA $cm^{-2}$ at a discharge voltage of 0.65 V and an open circuit voltage of 1.48 V. The battery displayed an energy density of 1027, 930, and 788 W $h/kg_{Zn}$ (close to theoretical 1086 $Wh/kg_{Zn}$) at a current density of 5, 25 and 50 mA $cm^{-2}$, respectively. The high surface area (2298 $m^2$ $g^{-1}$), large pore volume (1.63 $cm^3$ $g^{-1}$) with a pore size of 0.8 nm promoted better electrolyte and $O_2$ diffusion in the pores, while high electrical conductivity (359 S $cm^{-1}$) caused by good crystallinity stimulate better charge transfer. Similarly, various other reports have also shown the feasibility of MOFs for air batteries.[46-48] However, in spite of the better performance by few MOFs, most of them suffer from poor electrical conductivity, low stability and harsh synthetic conditions.[49,50] Thus, pristine conductive MOFs with better stability for Zn-air batteries are still at exploration stage.

Porous carbon-based materials with the advantages of low cost, tuneable surface functionalities, high surface area, etc. are proven effective for

Zn-air batteries. A pioneering work in 2014 by Tian *et al.*[51,52] showed the OER and ORR bifunctionality of CNT core/N-doped carbon shell and graphene/CNT composites. Many reports then emerged showing the efficacy of carbon-based composites for bifunctional catalysis.[53-55] To facilitate oxygen redox catalysis, activated carbon, carbon nanotubes, graphene, and heteroatom doped carbons have been used as state-of-the-art catalyst support materials. They provide high conductivity and promote the dispersion of transition metal-based compounds. The doped carbon provides better ORR kinetics and the transition metal-based compounds are accountable for efficient OER. Li *et al.*[56] utilized solution-based growth mechanism to anchor $Mn_3O_4$ nanoparticles over ionic liquid modified reduced graphene oxide (r-GO-IL). r-GO-IL provides better stability to nanoparticles and promotes facile electron transport. However, carbon suffers from electrochemical corrosion (oxidation) at potential > 0.207 V vs RHE, resulting in decreased electrical conductivity and leads to detachment of the supporting catalyst promoting aggregation. As a result, the durability and performance of air electrodes degrades. A simple strategy to tackle this is to encapsulate the supporting catalyst in the porous carbon layers. Due to the weak coordinate bonds in MOFs, they can be decomposed at high temperatures providing hollow nanostructures with various morphologies. Various monometallic, bimetallic as well as single atom catalysts are synthesized through this route. In this direction, Tarandeep *et al.*[57] synthesized core shell $Co_3O_4$@Co/NCNT via the carbonization of a cobalt-based MOF $[\{Co(bpe)_2N(CN)_2\}.N(CN)_2.5H_2O]_n$. A zinc-air battery associated with this catalyst achieved a power density of 135 mW/cm$^2$ with trifunctional activity (ORR, OER, HER; Figure 12).

Recently, He *et al.*[58] fabricated a zinc air battery with a remarkable cyclic stability exceeding 620 h and 430 h at 5 and 10 mA cm$^{-2}$, respectively. They had encapsulated CoNi alloy nanoparticles in N and S co-doped graphitic carbon. The resulting catalyst has nanoporous structure in μm-length scale preventing CoNi agglomeration and thus higher stability. This provides an opportunity to develop effective strategies so as to mitigate the stability issues.

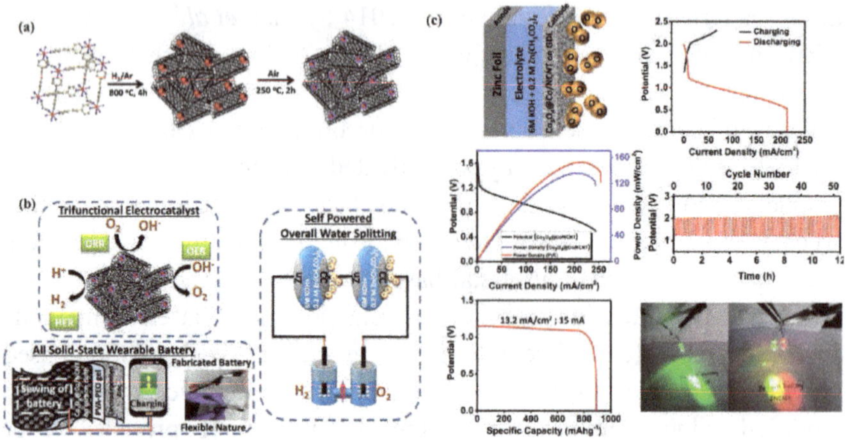


Figure 12. (a) Schematic illustration of the synthesis strategy of $Co_3O_4$@Co/NCNT Nanocomposite; (b) Rechargeable Zn-air battery performance showing the charge discharge profiles and the galvanostatic long term cyclability. Reproduced with permission from Ref. 57.

## 4. $CO_2$ Capture and Conversion

Global energy demand is being met largely by burning fossil fuel, and anthropogenic $CO_2$ is the major key by-product of such combustion. Recent statistics reveals that the atmospheric $CO_2$ concentration has enhanced to more than 400 ppm and expected to reach 600 ppm in the next 50 years.[59] In the past decades, several porous substances, including MOFs, have been employed for $CO_2$ capture and conversion applications. The engagement of MOFs has undergone three development stages: (a) High $CO_2$ uptake capacity with (b) tuneable selectivity towards separation from other competitive gases such as $CO_2$ from $CO_2/H_2$ (pre-combustion C-capture), $CO_2/CH_4$ (natural gas purification), or $CO_2/N_2$ gas mixtures (post-combustion C-capture), and (c) attempts to explore possible $CO_2$ transformation reaction by optimizing its catalytic performance and conversion efficiency through photo and/or electro-catalytic pathways.[60-62] One of the triggering factors for MOF-based $CO_2$ adsorption is recognizing or accessing the active site of the MOF known as open metal site (OMS) or unsaturated metal centre (UMC). The concept originated in the early 20's and in 2005, HKUST-1 demonstrated an outstanding $CO_2$ storage capacity of 10 mmol $g^{-1}$ under ambient temperature and above 40 bar pressure.[63,64] For instance, temperature-swing $CO_2$ adsorption experiments were performed in MOF-177 and MOF-74(Mg).[65] The latter one was marked by an outstanding selectivity (Selectivity factor 148.1) at a typical flue gas temperature. MOF-74(Mg) showed typically high $CO_2$ uptake capacity even more than previously most studied zeolite NaX (SF 87.4) and further targeted for pre-/post- and natural gas separation.[66,67] On the other hand, the competitive adsorption of water molecules over $CO_2$ restricts its application in the real domain. Therefore, hydrophobic and super-hydrophobic frameworks are incorporated for practical $CO_2$ adsorption in presence of water.[68] Recently, several alkylamine functionalities have been covalently tethered to the linker of IRMOF-74-III predominantly for selective $CO_2$ capture under humid conditions.[69,70] Another very recent technology is adsorbing $CO_2$ from the air and generating a concentrated stream for industrial or in further sequestration.

Few laboratories recently integrated such technology from lab to industrial scale production to control $CO_2$ level in a confined space like indoor, spacecraft and submarines.[71,72] In the current decade, we have demonstrated the growth and stabilization of MOF nanoparticles on layered materials namely nanoclay,[73] aminoclay,[74-76] graphene oxide,[77] and functional graphene oxide.[78] These composites mitigate the drawback of individual components and provide novel characteristics with high hydrothermal stability and major improvement in $CO_2$ uptake.

To date, MOF accomplishes in photocatalytic and electrocatalytic reduction, heterogeneous hydrogenation of $CO_2$ and conversion to organic products. The successful $CO_2$ conversion demands (a) high thermodynamic and kinetic stability, (b) high negative potential required for multi-electron redox potential (Eqs. (19)–(23)), (c) good number of basic sites to activate acidic $CO_2$ and (d) sufficient metal-active sites for activation. In assistance to photo-and-electrocatalytic influence the state-of-art pathways are outlined below;[79]

$$CO_2 + 2e^- + 2H^+ \text{ ® } COOH \qquad (E = -0.61\ V) \qquad (19)$$

$$CO2 + 2e- + 2H+ \text{ ® } CO + H2O \qquad (E = -0.53\ V) \qquad (20)$$

$$CO2 + 4e- + 4H+ \text{ ® } HCHO + H2O \qquad (E = -0.48\ V) \qquad (21)$$

CO2 + 6e- + 6H+ ® CH3OH + H2O (E = -0.38 V) (22)

CO2 + 8e- + 8H+ ® CH4 + 2H2O (E = -0.24 V) (23)

MOFs acted as a photocatalytic reactor in a single system through the combination of a photosensitizer, photocatalyst and $CO_2$ sorbent for effective reduction. Ru, Re, Ir-based porphyrin-based materials are well studied for $CO_2$ conversion.[80] Similarly, Ru-based MOF complexes became a trendy category of photosensitizer exhibiting exclusive reduction with high-rate of catalytic conversion. One of the representative examples is found in a Zr-based MOF composite UiO-67-$Ru^{II}$. The post-synthetic exchange method was adapted for incorporating photoreactive $Ru^{II}$-CO ([$Ru^{II}$(bpy)(terpy)(CO)]$(PF_6)_2$) complex into the pore of UiO-67.[81] Re is an elite pick as a light-harvesting agent for MOF-based photocatalytic $CO_2$ reduction.

Currently, Karmakar *et al.* have demonstrated the confinement of a photosensitizer, [$Ru(bpy)_2Cl_2$], and a molecular catalyst, [$Re(CO)_5Cl$], within a Zr-MOF pore (MOF-808) through post-synthetic modified method to achieve solar-driven $CO_2$ reduction.

Such covalent grafting offers improved charge separation with a maximum production rate of 440 μmol $g^{-1}$ $h^{-1}$ (Selectivity > 99%, QE = 0.11) $CO_2$-to-CO conversion (Figure 13).[82] Hybrids of ternary nanocomposites, by combining $TiO_2$, $Cu_2O$ and $Cu_3(BTC)_2$, causes around 3-times improvement in photogenerated holes that improved highly efficient conversion of $CO_2$ to $CH_4$.[83] Apart from MOF, there are only a handful examples available as it is extremely challenging for non-metallic organic system to cross high kinetic barrier for multielectron transfer. Barman *et al.* have recently designed a redox-active conjugated microporous polymer (CMP) by composing donor-acceptor duo (TPA-PQ; TPA, tris(4-ethynylphenyl) amine, an electron doner with a redox-active acceptor phenanthraquinone or PQ) that leading towards visible light-driven $CO_2$ to $CH_4$ reduction with excellent yield (32.2 mmol $g^{-1}$ in 16 h) and high selectivity (> 97%; Figure 14).[84]

Porous carbon, due to its low cost and comparatively easier synthetic procedure, carbon-based materials have gained significant attention. However, porous carbon as a whole is inert without any surface modification. Thus, various doped carbon has been explored. Song *et al.*[85]

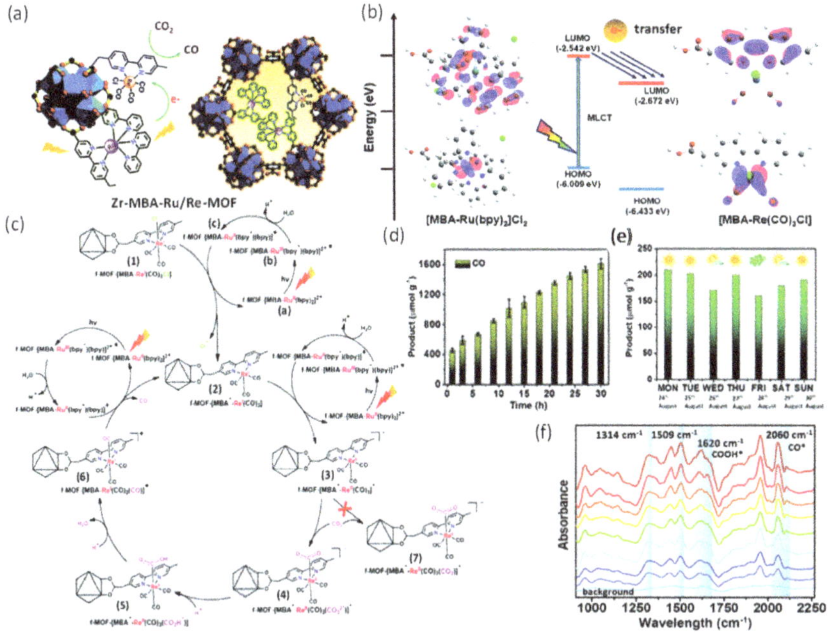


Figure 13. (a) Schematic for the construction of Zr-MBA-Ru/Re-MOF and immobilization of the molecular photosensitizer and catalyst toward $CO_2$ reduction. (b) The corresponding HOMO–LUMO positioning of the Ru-photosensitizer [MBA-Ru(bpy)$_2$]Cl$_2$ and Re-catalyst [MBA-Re(CO)$_3$Cl], representing (c) the possible charge transfer mechanism of $CO_2$-to-CO photo-conversion. (d) CO production as a function of irradiation time and (e) sunlight driven $CO_2$ reduction (one week) with Zr-MBA-Ru/Re-MOF in aqueous medium for 6 h as a catalyst in water. (f) Time-dependent *in situ* diffuse reflectance FT-IR spectra for $CO_2$ reduction in a mixture of $CO_2$ and $H_2O$ vapor. Reproduced with permission from Ref. 82.

reported an ethanol FE of 77% at the potential of −0.56 V (vs. RHE) using N-doped mesoporous carbon. The high activity is attributed to the presence of both pyridinic and pyrrolic N sites within the porous matrix facilitating C-C bond formation (Figure 15).

## 5. Energy-Efficient Separations

Industrial feedstock separation, especially hydrocarbons, majorly found in crude oil or petroleum gas is considered as complex and significant in recent days. The existing chemical separation technologies, operating

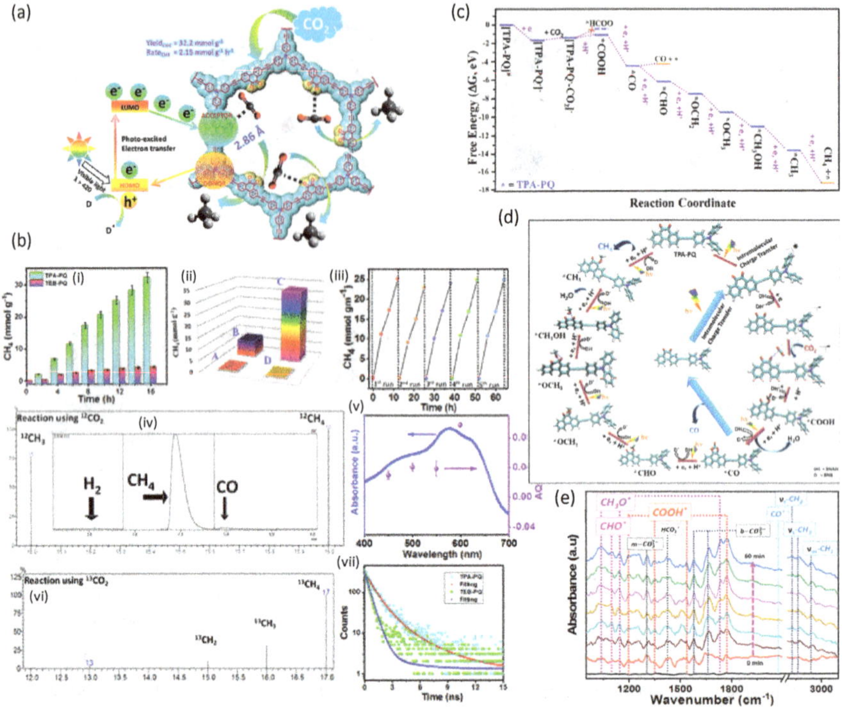

Figure 14. (a) Schematic of visible-light-driven $CO_2$ reduction involving a donor–acceptor CMP with fast, selective, and efficient $CH_4$ production. (b) (i) Time-dependent $CO_2$ reduction product formation in MeCN–$H_2O$ (2:1) medium in the presence of BNAH and TEA involving TPA-PQ and TEB-PQ as photocatalysts. (ii) Product formation under different conditions after 16 h of irradiation. (A) TPA-PQ in $H_2O$ medium; (B) TPA-PQ in $H_2O$ medium in the presence of BNAH and TEA; (C) TPA-PQ in MeCN–$H_2O$ (2:1) medium in the presence of BNAH and TEA; (D) TPA-PQ using TEA as sacrificial electron donor in MeCN medium. (iii) Recyclability plot of $CH_4$ production during 12 h of photocatalysis for each run. (iv) Mass spectra of $^{12}CH_4$ (top) and $^{13}CH_4$ (bottom) during photochemical reduction of $CO_2$ involving $^{12}CO_2$ and $^{13}CO_2$ as reactant, respectively. Inset shows a typical chromatogram during the catalysis. (v) Apparent quantum efficiency (AQE) at different monochromatic wavelengths during photocatalysis under similar reaction conditions. (vi) Time-resolved photoluminescence spectra (TRPL) of TPA-PQ ($\lambda_{em}$ at 530 nm) and TEB-PQ ($\lambda_{em}$ at 450 nm) using a 510 and 405 nm laser excitation source for TPA-PQ and TEB-PQ, respectively. Reproduced with permission from Ref. 84.

through cryogenic distillation or solvent extraction process at higher temperature-pressure demand, tugs 15% of total manufacturing energy cost.[86] On the contrary, adsorptive separation potentially fulfils the

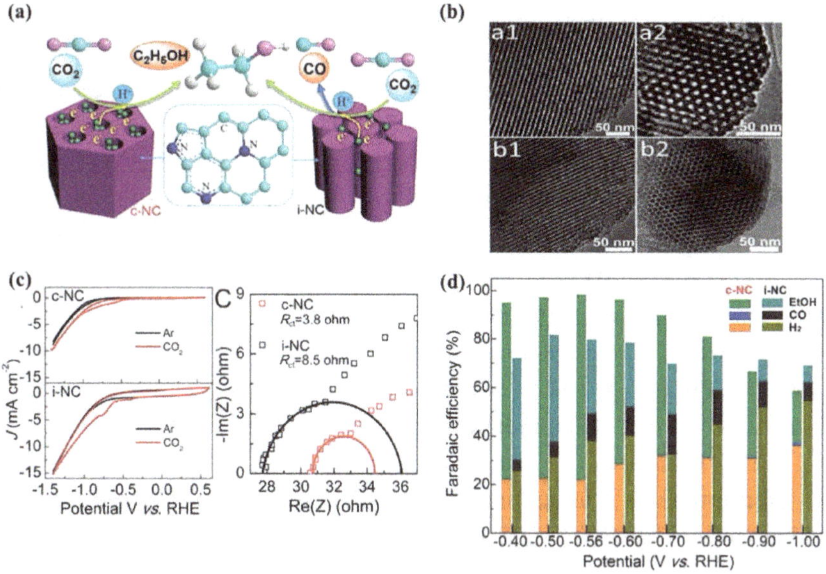


Figure 15. (a) Schematic illustration of the electroreduction of $CO_2$ using c-NC and i-NC; (b) TEM images; (c, d) Electrocatalytic $CO_2$ reduction performance of c-NC and i-NC catalysts. Reproduced with permission from Ref. 85.

energy-efficient separation economy, and saves or saving 80% of utilized energy compared to the conventional techniques.[86] In the early days of porous materials, they were dedicated to only gas storage for potential usage in fuel cells, gas tanks, and several stationary power facilities. MOF as a clean energy carrier provides a wide vacancy for methane and carbon dioxide storage towards efficient adsorbed natural gas (ANG) and carbon capture, and sequestration/storage (CSS) technologies. Along with customary flue ($N_2/CO_2$) and natural gas ($CO_2/CH_4$) separation, olefin, paraffin, alkyne, alkene or aromatic isomer-purification-pulls renewed interest due to exclusive industrial demand.[87] For such exclusive separations, MOF showed notable performance because of ambient state separation technology and exclusive molecular recognition properties driven by several kinetic and thermodynamic factors. In the early days, MIL and MOF-74 series attracted substantial attention for hydrocarbon storage and separation due to the presence of open metal sites.[88,89] MOF-74(Mg) exhibited higher binding affinity to both $C_2H_4$ and $C_3H_6$ compared to $C_2H_6$ and $C_3H_8$, respectively, introduced MOF as a potential material for olefin/paraffin separation.[90]

One of the remarkable examples demonstrated by Long *et al.* developed $Fe_2$(dobdc) MOF with unsaturated Fe(II) sites which showed a high ethylene/ethane separation factor, based on stronger interaction between unsaturated metal centre and π-electrons of $C_2H_4$.[91] In recent times, Chen *et al.* exposed that further oxygen modification of $Fe_2$(dobdc) i.e., $Fe_2O_2$(dobdc), showed a high $C_2H_6$ uptake (74.3 mL/g) with reverse selectivity concerning pristine one. The reported selectivity factor was nearly 4 with a high isosteric heat of adsorption (66.8 kJ/mol).[92] Further, Chen and his co-workers developed an energy-efficient and efficient synergistic sorbent separation strategy (SSST) by integrating three MOFs such as TIFSIX-2-Cu-i, Zn-atz-ipa and SIFSIX-3-Ni, with their own different favourable target molecules, obtain ultra-pure ethylene from the mixture of $C_2H_2/CO_2/C_2H_4/C_2H_6$.[93] In a recent work, Laha *et al.* unfold the

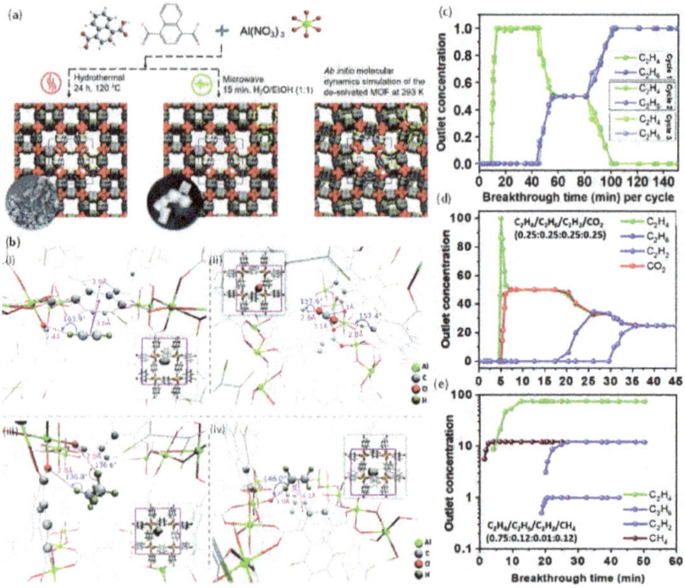

Figure 16. (a) The synthesis protocols of Al-MOF with morphological and structural architecture. (b) Locations of the highest binding affinity sites determined by DFT optimization inside the large pore for (i) $C_2H_2$, (ii) $CO_2$, (iii) $C_2H_6$ and $C_2H_4$, respectively. Experimental breakthrough separation for (c) binary and (d, e) quaternary gas mixtures. Reproduced with permission from Ref. 94.

quaternary gas ($C_2H_4/C_2H_6/C_2H_2/CO_2$ and $C_2H_4/C_2H_6/C_2H_2/CH_4$) separation with Al-MOF judiciously synthesized through microwave heating and mixed-solvent treatment. The tailored MOF results in improved porosity with high hydrocarbon uptake capacity showing simultaneous trapping of $C_2H_6$ and $C_2H_2$ towards efficient purification of $C_2H_4$ (Figure 16).[94] Additionally, two-dimensional layered materials showed promising performance as membranes due to easy permeation through molecular channels. In this context, graphene and graphene oxide-based membranes have opened a new dimension, primarily focusing on molecular separation, water purification and desalination.

In this context, Achari *et al.* demonstrated the purpose of 2D materials, especially graphene oxide and $MoS_2$ and their molecular channels for selective permeation of $H_2/CO_2$ separation. The selective preference of membranes towards $H_2/CO_2$ combination has been investigated as a consequence of alteration in phase and structural transformation with time and temperature (Figure 17).[95] Porous polymers, especially MOF and COF, have been integrated with such 2D materials aiming for mixed-matrix membranes including their thin-film and stand-alone analogues, and are extensively studied for $H_2/CO_2$, $N_2/CO_2$, $CO_2/CH_4$ separation.[96,97]

Gas-phase adsorption and separation is preferably studied by most of the MOF community, while isolation and purification of liquid mixtures such as chiral compounds, bio-based water-dispersible liquids and industrially important aromatic hydrocarbons are broadly studied in the recent decade. The most important structural isomer separation of this section is extracting each component from BTEX mixture. Benzene, toluene, ethylbenzene, and xylenes (ortho, meta and para) are important constituents in petroleum product e.g., gasoline or petrol.[98] They are highly important industrial substances, used in manufacturing adhesives, paint, rubber manufacturing units. In this context, Laha *et al.* demonstrated a 'dynamic chemical clip' in a supramolecular framework for selective sorting of liquid isomers governing via thermodynamic and kinetic preferences (Figure 18).[99] The interplay of several supramolecular interactions project exclusive selectivity of meta-xylene, among other $C_8$ isomers, eventually unveiled by crystallographic and theoretical analyses of each isomer. This finding emphasizes the impact of flexible dynamic space for remarkable sorting and separation over rigid porous structures. Additionally, first-row benchmark MOFs such

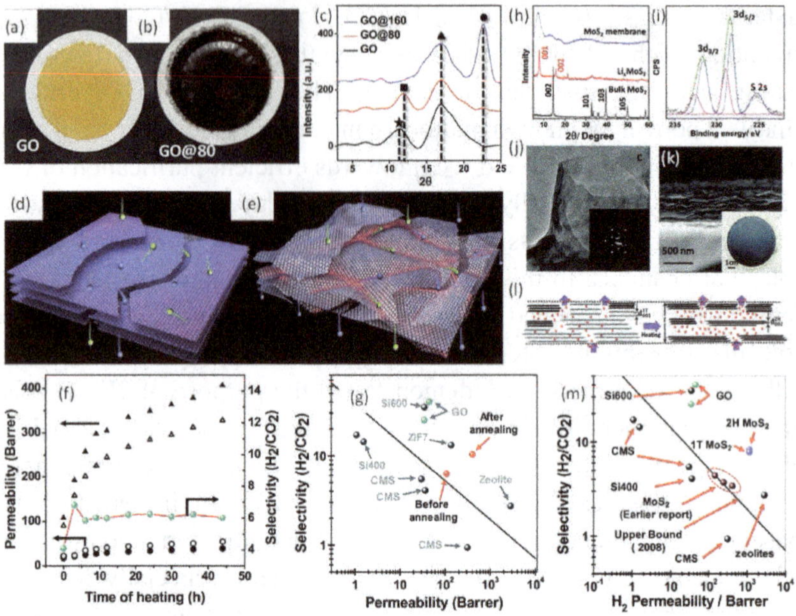


Figure 17. (a, b) Optical images of the graphene oxide membrane before and after annealing at 80°C for 42 h. (c) XRD profiles for the as prepared GO membranes along with membranes heated at 80°C and 160°C. (d, e) A schematic representation of gas permeation mechanism through the as-synthesized and domain formed GO membranes. (f, g) The permeability and selectivity of $H_2$ and $CO_2$ as individual gases and as a 1:1 gas mixture through a GO membrane. (h) XRD pattern of bulk $MoS_2$ (black) and lithiated $MoS_2$ (Red). Blue graph is the XRD pattern recorded for the as-prepared membrane. (i) Fitted XPS spectra of the $Mo^{4+}$ 3d region of the as-prepared 1T membrane (black dots). Contributions from the 2H and 1T phases are indicated by red and blue lines, respectively. Green lines represent the resultant curve. (j) TEM image of dispersion of 1T $MoS_2$, showing single layered structure. Inset: Electron diffraction pattern recorded from single layer 1T $MoS_2$ dispersion. (k) Cross-sectional FESEM image of the $MoS_2$ membrane showing layered structure. Inset: Optical image of the as-prepared $MoS_2$ membrane on an AAO disc. (l) Schematic of gas permeation pathway across $MoS_2$ membranes before and after heating at 160°C. (m) Comparison of ideal H2/CO2 gas separation performance of the as-prepared MoS2 membrane (1T) and $MoS_2$ membrane heated at 160°C (2H) with other membranes reported in the literature. Reproduced with permission from Ref. 95.

as ZIFs, HKSUT-1, UiO-66, MIL and its derivatives are further utilized for such aromatic hydrocarbon separation with adequate selectivity.[100-102]

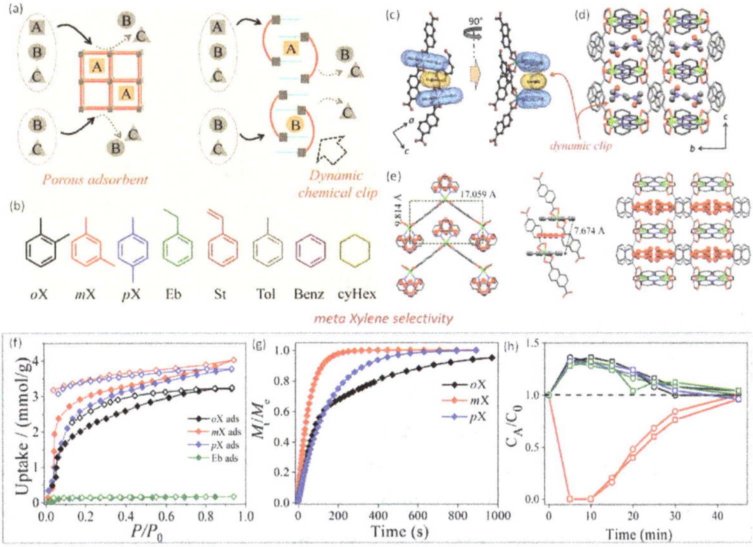


Figure 18. (a) An illustration of the "dynamic chemical clip" for separation of individual chemical isomers (A, B and C), in comparison to the conventional porous adsorbents. (b) Chemical isomers sorted by the dynamic chemical clipping process. (c) View of the sandwiched nanospace created by o-phen from the neighboring 1D polymer filled by (d) guest DMF (shown in ball and stick model) which has been further replaced by m-xylene (mX) during crystallization. (f) The adsorption isotherm and (g) kinetic measurements represents meta selectivity over oX and pX. (h) Breakthrough experiment for C8 isomers with retention times of C8 isomers: oX (black), mX (red), pX (blue), Eb (green); empty square and circle symbols represent two independent experiments. Reproduced with permission from Ref. 99.

## 6. Outlook

The utilization of porous materials for wide range of energy utilities has strikingly progressed towards innovative energy technologies. Such materials actively participate in photo- and electrochemical energy harvesting, in terms of molecular diffusion, electron and charge transportation, selective sorting, boosting conductivity, and durability. Despite the efforts, a few factors are required to be taken care of in understanding gap between such material and existing commercial technology to meet the current energy global requirements.

1. High $H_2$ storage capacity of such advanced material could perform better but requires the fruitful collaboration of chemistry, material science, engineering and commercial liabilities to run hydrogen-powered fuel-cell automobiles.
2. For electrolytic applications, the material with high conductivity is of paramount importance, while a large library of such materials, especially MOFs, falls under insulator category.
3. The material with ordered pores gives rise to fast Faradaic redox processes effectively provides faster kinetics and prolonged life cycle essential for battery-type electrodes.
4. Durability against high temperature, moisture and chemicals are essential to make sure that the materials could be used under multi-variant conditions by avoiding structural deformation.
5. In recent times, such materials have been found attractive as electro-catalyst. To facilitate improved exposure of active sites (improved surface-to-volume ratio), ultrafine nanoparticles or even single atoms are found remarkably efficient.
6. The adsorptive-based separation under ambient temperature and pressure could mitigate the lion's share of energy expenditure of chemical industries. The modulation of size, shape, porosity, chemical functionality with high recyclability, fast adsorption-desorption and without appreciable loss of target molecules can make such material translatable for commercial purpose.
7. Porous materials with defined pore, functionality or structural interpenetration ensures better capture and selectivity of $CO_2$ with successful conversion to wide range of value-added products than existing technology. However, the difficulty arises with the supply of millions of tonnes production of porous materials against relatively inexpensive commercial amines.
8. As discussed earlier, the $CO_2$ emission to the atmosphere is largely responsible for climate change. The exclusively designed porous materials that could adsorb atmospheric $CO_2$ directly under ambient condition in presence of nitrogen, oxygen and water molecules must be encouraged.

## References


1. Durbin, D. and Malardier-Jugroot, C. Review of Hydrogen Storage Techniques for on Board Vehicle Applications. *International Journal of Hydrogen Energy* **38**, 14595-14617 (2013).
2. Allendorf, M. D. *et al.* An Assessment of Strategies for the Development of Solid-state Adsorbents for Vehicular Hydrogen Storage. *Energy & Environmental Science* **11**, 2784-2812 (2018).
3. Chen, Z., Kirlikovali, K. O., Idrees, K. B., Wasson, M. C. and Farha, O. K. Porous Materials for Hydrogen Storage. *Chem* **8**, 693-716, doi:https://doi.org/10.1016/j.chempr.2022.01.012 (2022).
4. Rosi, N. L. *et al.* Hydrogen Storage in Microporous Metal-Organic Frameworks. *Science* **300**, 1127-1129, doi:10.1126/science.1083440 (2003).
5. Suh, M. P., Park, H. J., Prasad, T. K. and Lim, D.-W. Hydrogen Storage in Metal–Organic Frameworks. *Chemical Reviews* **112**, 782-835, doi:10.1021/cr200274s (2012).
6. Chen, Z. *et al.* Balancing Volumetric and Gravimetric Uptake in Highly Porous Materials for Clean Energy. *Science* **368**, 297-303, doi:10.1126/science.aaz8881 (2020).
7. Jaramillo, D. E. *et al.* Ambient-Temperature Hydrogen Storage via Vanadium(II)-Dihydrogen Complexation in a Metal–Organic Framework. *Journal of the American Chemical Society* **143**, 6248-6256, doi:10.1021/jacs.1c01883 (2021).
8. García-Holley, P. *et al.* Benchmark Study of Hydrogen Storage in Metal–Organic Frameworks under Temperature and Pressure Swing Conditions. *ACS Energy Letters* **3**, 748-754, doi:10.1021/acsenergylett.8b00154 (2018).
9. Gómez-Gualdrón, D. A. *et al.* Understanding Volumetric and Gravimetric Hydrogen Adsorption Trade-off in Metal–Organic Frameworks. *ACS Applied Materials & Interfaces* **9**, 33419-33428, doi:10.1021/acsami.7b01190 (2017).
10. Ben, T. *et al.* Targeted Synthesis of a Porous Aromatic Framework with High Stability and Exceptionally High Surface Area. *Angewandte Chemie International Edition* **48**, 9457-9460, doi:https://doi.org/10.1002/anie.200904637 (2009).
11. Germain, J., Fréchet, J. M. J. and Svec, F. Nanoporous Polymers for Hydrogen Storage. *Small* **5**, 1098-1111, doi:https://doi.org/10.1002/smll.200801762 (2009).
12. Hu, C., Zhang, L. and Gong, J. Recent Progress Made in the Mechanism Comprehension and Design of Electrocatalysts for Alkaline Water Splitting. *Energy & Environmental Science* **12**, 2620-2645 (2019).
13. Sardar, K. *et al.* Water-Splitting Electrocatalysis in Acid Conditions Using Ruthenate-Iridate Pyrochlores. *Angewandte Chemie* **126**, 11140-11144 (2014).
14. Strmcnik, D. *et al.* Improving the Hydrogen Oxidation Reaction Rate by Promotion of Hydroxyl Adsorption. *Nature Chemistry* **5**, 300-306 (2013).

15. Zhang, W., Qi, J., Liu, K. and Cao, R. A Nickel-based Integrated Electrode from an Autologous Growth Strategy for Highly Efficient Water Oxidation. *Advanced Energy Materials* **6**, 1502489 (2016).
16. Chen, P. *et al.* Metallic Co4N Porous Nanowire Arrays Activated by Surface Oxidation as Electrocatalysts for the Oxygen Evolution Reaction. *Angewandte Chemie* **127**, 14923-14927 (2015).
17. Xu, L. *et al.* Plasma-engraved Co3O4 Nanosheets with Oxygen Vacancies and High Surface Area for the Oxygen Evolution Reaction. *Angewandte Chemie* **128**, 5363-5367 (2016).
18. Wan, S. *et al.* Hierarchical Co (OH) F Superstructure Built by Low-dimensional Substructures for Electrocatalytic Water Oxidation. *Advanced Materials* **29**, 1700286 (2017).
19. Markovića, N. M., Sarraf, S. T., Gasteiger, H. A. and Ross, P. N. Hydrogen Electrochemistry on Platinum Low-index Single-crystal Surfaces in Alkaline Solution. *Journal of the Chemical Society, Faraday Transactions* **92**, 3719-3725 (1996).
20. Paul, R., Zhu, L., Chen, H., Qu, J. and Dai, L. Recent Advances in Carbon-based Metal-free Electrocatalysts. *Advanced Materials* **31**, 1806403 (2019).
21. Zhou, W. *et al.* Recent Developments of Carbon-based Electrocatalysts for Hydrogen Evolution Reaction. *Nano Energy* **28**, 29-43 (2016).
22. Tiwari, J. N. *et al.* Multicomponent Electrocatalyst with Ultralow Pt Loading and High Hydrogen Evolution Activity. *Nature Energy* **3**, 773-782 (2018).
23. Mahmood, J. *et al.* An Efficient and pH-universal Ruthenium-based Catalyst for the Hydrogen Evolution Reaction. *Nature Nanotechnology* **12**, 441-446 (2017).
24. Zheng, Y. *et al.* Hydrogen Evolution by a Metal-free Electrocatalyst. *Nature Communications* **5**, 1-8 (2014).
25. Huang, X., Zhao, Y., Ao, Z. and Wang, G. Micelle-template Synthesis of Nitrogen-doped Mesoporous Graphene as an Efficient Metal-free Electrocatalyst for Hydrogen Production. *Scientific Reports* **4**, 1-6 (2014).
26. Singh, D. K., Jenjeti, R. N., Sampath, S. and Eswaramoorthy, M. Two in One: N-doped Tubular Carbon Nanostructure as an Efficient Metal-free Dual Electrocatalyst for Hydrogen Evolution and Oxygen Reduction Reactions. *Journal of Materials Chemistry A* **5**, 6025-6031 (2017).
27. Lin, Z. *et al.* Dual Graphitic-N Doping in a Six-Membered C-Ring of Graphene-Analogous Particles Enables an Efficient Electrocatalyst for the Hydrogen Evolution Reaction. *Angewandte Chemie International Edition* **58**, 16973-16980 (2019).
28. Maity, S. *et al.* High Surface Area NiCoP Nanostructure as Efficient Water Splitting Electrocatalyst for the Oxygen Evolution Reaction. *Materials Research Bulletin* **140**, 111312, doi:https://doi.org/10.1016/j.materresbull.2021.111312 (2021).
29. Xie, X. *et al.* Oxygen Evolution Reaction in Alkaline Environment: Material Challenges and Solutions. *Advanced Functional Materials* **32**, 2110036, doi:https://doi.org/10.1002/adfm.202110036 (2022).

30. Chakraborty, S. *et al.* Improvement in Oxygen Evolution Performance of NiFe Layered Double Hydroxide Grown in the Presence of 1T-Rich MoS2. *ACS Applied Materials & Interfaces* **14**, 31951-31961, doi:10.1021/acsami.2c06210 (2022).
31. Zhang, J., Zhao, Z., Xia, Z. and Dai, L. A Metal-free Bifunctional Electrocatalyst for Oxygen Reduction and Oxygen Evolution Reactions. *Nature Nanotechnology* **10**, 444-452, doi:10.1038/nnano.2015.48 (2015).
32. Hu, C. and Dai, L. Multifunctional Carbon-based Metal-free Electrocatalysts for Simultaneous Oxygen Reduction, Oxygen Evolution, and Hydrogen Evolution. *Advanced Materials* **29**, 1604942 (2017).
33. Yi, Y. *et al.* Electrochemical Degradation of Multiwall Carbon Nanotubes at High Anodic Potential for Oxygen Evolution in Acidic Media. *ChemElectroChem* **2**, 1929-1937 (2015).
34. Macauley, N. *et al.* Carbon Corrosion in PEM Fuel Cells and the Development of Accelerated Stress Tests. *Journal of the Electrochemical Society* **165**, F3148 (2018).
35. Tüysüz, H., Hwang, Y. J., Khan, S. B., Asiri, A. M. and Yang, P. Mesoporous Co3O4 as an Electrocatalyst for Water Oxidation. *Nano Research* **6**, 47-54 (2013).
36. Lu, X., Ng, Y. H. and Zhao, C. Gold Nanoparticles Embedded within Mesoporous Cobalt Oxide Enhance Electrochemical Oxygen Evolution. *ChemSusChem* **7**, 82-86, doi:https://doi.org/10.1002/cssc.201300975 (2014).
37. Singh, D. K., Chakraborty, S., Saha, A., Sampath, S. and Eswaramoorthy, M. Pick a Wick: A Simple, Ultrafast Combustion Synthesis of Co3O4 Dispersed Carbon for Enhanced Oxygen Evolution Kinetics. *ACS Applied Energy Materials* **1**, 4448-4452, doi:10.1021/acsaem.8b00906 (2018).
38. Clark, S., Latz, A. and Horstmann, B. A Review of Model-Based Design Tools for Metal-Air Batteries. *Batteries* **4**, 5 (2018).
39. Lee, J. S. *et al.* Metal–air Batteries with High Energy Density: Li–air versus Zn–air. *Advanced Energy Materials* **1**, 34-50 (2011).
40. Li, Y. and Dai, H. Recent Advances in Zinc–air Batteries. *Chemical Society Reviews* **43**, 5257-5275 (2014).
41. Fu, J., Cano, Z. P., Park, M. G., Yu, A., Fowler, M. and Chen, Z., Electrically Rechargeable Zinc–Air Batteries: Progress, Challenges, and Perspectives. *Advanced Materials* **29**, 1604685 (2017).
42. Li, Y. *et al.* Advanced Zinc-air Batteries Based on High-performance Hybrid Electrocatalysts. *Nature Communications* **4**, 1-7 (2013).
43. Jiang, Q. *et al.* Two-dimensional Metal–organic Framework Nanosheets: Synthetic Methodologies and Electrocatalytic Applications. *Journal of Materials Chemistry A* **8**, 15271-15301 (2020).
44. Xue, Y., Zheng, S., Xue, H. and Pang, H. Metal–organic Framework Composites and Their Electrochemical Applications. *Journal of Materials Chemistry A* **7**, 7301-7327 (2019).

45. Shinde, S. S. *et al.* Unveiling Dual-linkage 3D Hexaiminobenzene Metal–organic Frameworks Towards Long-lasting Advanced Reversible Zn–air Batteries. *Energy & Environmental Science* **12**, 727-738 (2019).
46. Lian, Y. *et al.* Unpaired 3d Electrons on Atomically Dispersed Cobalt Centres in Coordination Polymers Regulate Both Oxygen Reduction Reaction (ORR) Activity and Selectivity for Use in Zinc–air Batteries. *Angewandte Chemie International Edition* **59**, 286-294 (2020).
47. Xia, Z. *et al.* Pt Nanoparticles Embedded Metal-organic Framework Nanosheets: A Synergistic Strategy Towards Bifunctional Oxygen Electrocatalysis. *Applied Catalysis B: Environmental* **245**, 389-398 (2019).
48. Zheng, X. *et al.* Bimetallic Metal–organic-framework/reduced Graphene Oxide Composites as Bifunctional Electrocatalysts for Rechargeable Zn–air Batteries. *ACS Applied Materials & Interfaces* **11**, 15662-15669 (2019).
49. Ahn, S. H., Yu, X. and Manthiram, A. "Wiring" Fe-Nx-Embedded Porous Carbon Framework onto 1D Nanotubes for Efficient Oxygen Reduction Reaction in Alkaline and Acidic Media. *Advanced Materials* **29**, 1606534 (2017).
50. Lu, X. F., Xia, B. Y., Zang, S. Q. and Lou, X. W. Metal–organic Frameworks Based Electrocatalysts for the Oxygen Reduction Reaction. *Angewandte Chemie* **132**, 4662-4678 (2020).
51. Tian, G. L. *et al.* Toward Full Exposure of "Active Sites": Nanocarbon Electrocatalyst with Surface Enriched Nitrogen for Superior Oxygen Reduction and Evolution Reactivity. *Advanced Functional Materials* **24**, 5956-5961 (2014).
52. Tian, G. L. *et al.* Nitrogen-doped Graphene/carbon Nanotube Hybrids: In Situ Formation on Bifunctional Catalysts and Their Superior Electrocatalytic Activity for Oxygen Evolution/reduction Reaction. *Small* **10**, 2251-2259 (2014).
53. Yang, H. B. *et al.* Identification of Catalytic Sites for Oxygen Reduction and Oxygen Evolution in N-doped Graphene Materials: Development of Highly Efficient Metal-free Bifunctional Electrocatalyst. *Science Advances* **2**, e1501122 (2016).
54. Lu, Z. *et al.* Construction of 3D Carbon Network with N, B, F-tridoping for Efficient Oxygen Reduction Reaction Electrocatalysis and High Performance Zinc Air Battery. *Applied Surface Science* **507**, 145154 (2020).
55. Qu, K., Zheng, Y., Dai, S. and Qiao, S. Z. Graphene Oxide-polydopamine Derived N, S-codoped Carbon Nanosheets as Superior Bifunctional Electrocatalysts for Oxygen Reduction and Evolution. *Nano Energy* **19**, 373-381 (2016).
56. Lee, J.-S., Lee, T., Song, H.-K., Cho, J. and Kim, B.-S. Ionic Liquid Modified Graphene Nanosheets Anchoring Manganese Oxide Nanoparticles as Efficient Electrocatalysts for Zn–air Batteries. *Energy & Environmental Science* **4**, 4148-4154, doi:10.1039/C1EE01942B (2011).
57. Singh, T. *et al.* MOF Derived Co3O4@Co/NCNT Nanocomposite for Electrochemical Hydrogen Evolution, Flexible Zinc-Air Batteries, and Overall Water Splitting. *Inorganic Chemistry* **59**, 3160-3170, doi:10.1021/acs.inorgchem.9b03516 (2020).

58. He, X. *et al.* Long-range Interconnected Nanoporous Co/Ni/C Composites as Bifunctional Electrocatalysts for Long-life Rechargeable Zinc-air Batteries. *Electrochimica Acta* **413**, 140183, doi:https://doi.org/10.1016/j.electacta.2022.140183 (2022).
59. Monastersky, R. Global Carbon Dioxide Levels near Worrisome Milestone. *Nature* **497**, 13-14, doi:10.1038/497013a (2013).
60. Mukherjee, S., Kumar, A. and Zaworotko, M. J. in *Metal-Organic Frameworks (MOFs) for Environmental Applications*, ed. Sujit K. Ghosh 5-61 (Elsevier, 2019).
61. Ding, M., Flaig, R. W., Jiang, H.-L. and Yaghi, O. M. Carbon Capture and Conversion using Metal–organic Frameworks and MOF-based Materials. *Chemical Society Reviews* **48**, 2783-2828, doi:10.1039/C8CS00829A (2019).
62. Yu, J. *et al.* CO2 Capture and Separations Using MOFs: Computational and Experimental Studies. *Chemical Reviews* **117**, 9674-9754, doi:10.1021/acs.chemrev.6b00626 (2017).
63. Li, H., Davis, C. E., Groy, T. L., Kelley, D. G. and Yaghi, O. M. Coordinatively Unsaturated Metal Centers in the Extended Porous Framework of Zn3(BDC)3·6CH3OH (BDC = 1,4-Benzenedicarboxylate). *Journal of the American Chemical Society* **120**, 2186-2187, doi:10.1021/ja974172g (1998).
64. Millward, A. R. and Yaghi, O. M. Metal−Organic Frameworks with Exceptionally High Capacity for Storage of Carbon Dioxide at Room Temperature. *Journal of the American Chemical Society* **127**, 17998-17999, doi:10.1021/ja0570032 (2005).
65. Mason, J. A., Sumida, K., Herm, Z. R., Krishna, R. and Long, J. R. Evaluating Metal–organic Frameworks for Post-combustion Carbon Dioxide Capture via Temperature Swing Adsorption. *Energy & Environmental Science* **4**, 3030-3040, doi:10.1039/C1EE01720A (2011).
66. Zhou, W., Wu, H. and Yildirim, T. Enhanced H2 Adsorption in Isostructural Metal−Organic Frameworks with Open Metal Sites: Strong Dependence of the Binding Strength on Metal Ions. *Journal of the American Chemical Society* **130**, 15268-15269, doi:10.1021/ja807023q (2008).
67. Caskey, S. R., Wong-Foy, A. G. and Matzger, A. J. Dramatic Tuning of Carbon Dioxide Uptake via Metal Substitution in a Coordination Polymer with Cylindrical Pores. *Journal of the American Chemical Society* **130**, 10870-10871, doi:10.1021/ja8036096 (2008).
68. Roy, S., Laha, S. and Maji, T. K. in *Metal-Organic Frameworks (MOFs) for Environmental Applications*, ed. Sujit K. Ghosh 319-354 (Elsevier, 2019).
69. Flaig, R. W. *et al.* The Chemistry of CO2 Capture in an Amine-Functionalized Metal–Organic Framework under Dry and Humid Conditions. *Journal of the American Chemical Society* **139**, 12125-12128, doi:10.1021/jacs.7b06382 (2017).
70. Fracaroli, A. M. *et al.* Metal–Organic Frameworks with Precisely Designed Interior for Carbon Dioxide Capture in the Presence of Water. *Journal of the American Chemical Society* **136**, 8863-8866, doi:10.1021/ja503296c (2014).

71. Sadiq, M. M. *et al.* A Pilot-Scale Demonstration of Mobile Direct Air Capture Using Metal-Organic Frameworks. *Advanced Sustainable Systems* **4**, 2000101, doi:https://doi.org/10.1002/adsu.202000101 (2020).
72. Kumar, A. *et al.* Direct Air Capture of CO2 by Physisorbent Materials. *Angewandte Chemie International Edition* **54**, 14372-14377, doi:https://doi.org/10.1002/anie.201506952 (2015).
73. Chakraborty, A., Sutar, P., Yadav, P., Eswaramoorthy, M. and Maji, T. K. Charge-Assisted Self-Assembly of ZIF-8 and Laponite Clay toward a Functional Hydrogel Nanocomposite. *Inorg. Chem.* **57**, 14480-14483, doi:10.1021/acs.inorgchem.8b02545 (2018).
74. Chakraborty, A., Achari, A., Eswaramoorthy, M. and Maji, T. K. MOF–aminoclay Composites for Superior CO2 Capture, Separation and Enhanced Catalytic Activity in Chemical Fixation of CO2. *Chemical Communications* **52**, 11378-11381, doi:10.1039/C6CC05289D (2016).
75. Chakraborty, A., Roy, S., Eswaramoorthy, M. and Maji, T. K. Flexible MOF–aminoclay Nanocomposites Showing Tunable Stepwise/gated Sorption for C2H2, CO2 and Separation for CO2/N2 and CO2/CH4. *Journal of Materials Chemistry A* **5**, 8423-8430, doi:10.1039/C6TA09886J (2017).
76. Chakraborty, A. *et al.* In Situ Growth of Self-Assembled ZIF-8–Aminoclay Nanocomposites with Enhanced Surface Area and CO2 Uptake. *Inorg. Chem.* **56**, 9426-9435, doi:10.1021/acs.inorgchem.7b01601 (2017).
77. Kumar, R., Jayaramulu, K., Maji, T. K. and Rao, C. N. R. Hybrid Nanocomposites of ZIF-8 with Graphene Oxide Exhibiting Tunable Morphology, Significant CO2 Uptake and Other Novel Properties. *Chemical Communications* **49**, 4947-4949, doi:10.1039/C3CC00136A (2013).
78. Kumar, R., Jayaramulu, K., Maji, T. K. and Rao, C. N. R. Growth of 2D Sheets of a MOF on Graphene Surfaces to Yield Composites with Novel Gas Adsorption Characteristics. *Dalton Transactions* **43**, 7383-7386, doi:10.1039/C3DT53133C (2014).
79. Downes, C. A. and Marinescu, S. C. Electrocatalytic Metal–Organic Frameworks for Energy Applications. *ChemSusChem* **10**, 4374-4392, doi:https://doi.org/10.1002/cssc.201701420 (2017).
80. Singh, G. *et al.* Emerging Trends in Porous Materials for CO2 Capture and Conversion. *Chemical Society Reviews* **49**, 4360-4404, doi:10.1039/D0CS00075B (2020).
81. Kajiwara, T. *et al.* Photochemical Reduction of Low Concentrations of CO2 in a Porous Coordination Polymer with a Ruthenium(II)–CO Complex. **55**, 2697-2700, doi:https://doi.org/10.1002/anie.201508941 (2016).
82. Karmakar, S., Barman, S., Rahimi, F. A. and Maji, T. K. Covalent Grafting of Molecular Photosensitizer and Catalyst on MOF-808: Effect of Pore Confinement Toward Visible Light-driven CO2 Reduction in Water. *Energy & Environmental Science* **14**, 2429-2440, doi:10.1039/D0EE03643A (2021).

83. He, X. and Wang, W.-N. MOF-based Ternary Nanocomposites for Better CO2 Photoreduction: Roles of Heterojunctions and Coordinatively Unsaturated Metal Sites. *Journal of Materials Chemistry A* **6**, 932-940, doi:10.1039/C7TA09192C (2018).
84. Barman, S., Singh, A., Rahimi, F. A. and Maji, T. K. Metal-Free Catalysis: A Redox-Active Donor–Acceptor Conjugated Microporous Polymer for Selective Visible-Light-Driven CO2 Reduction to CH4. *Journal of the American Chemical Society* **143**, 16284-16292, doi:10.1021/jacs.1c07916 (2021).
85. Song, Y. *et al.* Metal-free Nitrogen-doped Mesoporous Carbon for Electroreduction of CO2 to Ethanol. *Angewandte Chemie* **129**, 10980-10984 (2017).
86. Sholl, D. S. and Lively, R. P. Seven Chemical Separations to Change the World. *Nature* **532**, 435-437, doi:10.1038/532435a (2016).
87. Li, J.-R., Kuppler, R. J. and Zhou, H.-C. Selective Gas Adsorption and Separation in Metal–organic Frameworks. *Chemical Society Reviews* **38**, 1477-1504, doi:10.1039/B802426J (2009).
88. Rosi, N. L. *et al.* Rod Packings and Metal−Organic Frameworks Constructed from Rod-Shaped Secondary Building Units. *Journal of the American Chemical Society* **127**, 1504-1518, doi:10.1021/ja045123o (2005).
89. Dietzel, P. D. C., Morita, Y., Blom, R. and Fjellvåg, H. An In Situ High-Temperature Single-Crystal Investigation of a Dehydrated Metal–Organic Framework Compound and Field-Induced Magnetization of One-Dimensional Metal–Oxygen Chains. **44**, 6354-6358, doi:https://doi.org/10.1002/anie.200501508 (2005).
90. Bao, Z. *et al.* Adsorption of Ethane, Ethylene, Propane, and Propylene on a Magnesium-Based Metal–Organic Framework. *Langmuir* **27**, 13554-13562, doi:10.1021/la2030473 (2011).
91. Bloch, E. D. *et al.* Hydrocarbon Separations in a Metal-Organic Framework with Open Iron(II) Coordination Sites. **335**, 1606-1610, doi:10.1126/science.1217544 %J Science (2012).
92. Li, L. *et al.* Ethane/ethylene Separation in a Metal-organic Framework with Iron-peroxo Sites. *Science* **362**, 443, doi:10.1126/science.aat0586 (2018).
93. Chen, K.-J. *et al.* Synergistic Sorbent Separation for One-step Ethylene Purification from a Four-component Mixture. *Science* **366**, 241, doi:10.1126/science.aax8666 (2019).
94. Laha, S. *et al.* Tailoring a Robust Al-MOF for Trapping C2H6 and C2H2 Towards Efficient C2H4 Purification from Quaternary Mixtures. *Chemical Science* **13**, 7172-7180, doi:10.1039/D2SC01180H (2022).
95. Achari, A. and Eswaramoorthy, M. Casting Molecular Channels Through Domain Formation: High Performance Graphene Oxide Membranes for H2/CO2 Separation. *Journal of Materials Chemistry A* **4**, 7560-7564, doi:10.1039/C6TA00296J (2016).
96. Qiu, S., Xue, M. and Zhu, G. Metal–organic Framework Membranes: From Synthesis to Separation Application. *Chemical Society Reviews* **43**, 6116-6140, doi:10.1039/C4CS00159A (2014).

97. Qian, Q. *et al.* MOF-Based Membranes for Gas Separations. *Chemical Reviews* **120**, 8161-8266, doi:10.1021/acs.chemrev.0c00119 (2020).
98. Mi, L. *et al.* Bioinspired in Vitro Lung Airway Model for Inflammatory Analysis via Hydrophobic Nanochannel Membrane with Joint Three-Phase Interface. *Analytical Chemistry* **91**, 15804-15810, doi:10.1021/acs.analchem.9b04114 (2019).
99. Laha, S. *et al.* A Dynamic Chemical Clip in Supramolecular Framework for Sorting Alkylaromatic Isomers using Thermodynamic and Kinetic Preferences. *Angewandte Chemie International Edition* **60**, 19921-19927, doi:https://doi.org/10.1002/anie.202106784 (2021).
100. Alaerts, L. *et al.* Selective Adsorption and Separation of Xylene Isomers and Ethylbenzene with the Microporous Vanadium(IV) Terephthalate MIL-47. *Angewandte Chemie International Edition* **46**, 4293-4297, doi:https://doi.org/10.1002/anie.200700056 (2007).
101. Alaerts, L. *et al.* Selective Adsorption and Separation of ortho-Substituted Alkylaromatics with the Microporous Aluminum Terephthalate MIL-53. *Journal of the American Chemical Society* **130**, 14170-14178, doi:10.1021/ja802761z (2008).
102. Moreira, M. A. *et al.* Reverse Shape Selectivity in the Liquid-Phase Adsorption of Xylene Isomers in Zirconium Terephthalate MOF UiO-66. *Langmuir* **28**, 5715-5723, doi:10.1021/la3004118 (2012).

https://doi.org/10.1142/9789811270956_0010

# Chapter 10

## Piezo and Triboelectric Energy Materials and Devices

Indrajit Mondal and Giridhar U. Kulkarni*

*Chemistry & Physics of Materials Unit, JNCASR, Bangalore 560064*
**kulkarni@jncasr.ac.in*

Piezo and triboelectric phenomena are known to mankind for the past several centuries, if not millenniums. It is only in the last few decades that these are exploited in the context of energy maneuvering and harvesting. While the physical understanding of the mechanisms involved are yet to be completely deciphered, empirical insights gained over time have served well towards the technological advancements. Thus, the piezoelectric materials have found applications as mechanical transducers in scanning probe microscopy, medical devices, and standalone sensors. Tribo has also earned equal interest as an energy scavenger at places starting from sports shoes to the windmills. There have been substantial efforts in making hybrid devices as well to enhance the transducing efficiency rendering them to be technologically more viable. Thus, the present chapter deals with the recent advancements in these areas.

## 1. Introduction

The earliest efforts of generation or interconversion of various forms of energy dealt with simple mechanical devices taking ground from basic thermodynamic principles. It dates back to 600 BC, when Thales, an ancient Greek thinker, demonstrated that amber resin, if rubbed against fur, could attract dust.[1] For inventors of yester era, a simple mechanical action like rubbing two material surfaces against each other had so many surprises to offer, which eventually led to a deeper understanding of electricity. The real discovery of frictional electricity, which is now known to be triboelectricity, was made by Francis Hauksbee in the year 1675 when he published his paper on "barometric light" demonstrating the production of intermittent flashes in a barometer when shaken, due to the accumulation of electrostatic charges.[2] The mechanical energy ruled the world through the industrial revolution during the 18th and 19th centuries.[3] Thus is the birth of triboelectricity, 'tribo' standing for 'to rub' and its cousin, piezoelectricity, 'piezo' for 'push'. Thus, the two fields have been visited every now and then, be it to understand deeper in terms of physics and specifically, the mechanisms involved, or be it the material structure, composition, and other chemical aspects that influence electricity production, importantly, in the context of harvesting energy.[4] In this sense, they are truly mechanical to electrical energy converters and in the case of piezo, electrical to mechanical converter as well.

In the last half century, there is renewed interest in piezo and tribo phenomena particularly because of the realization that the electric phenomena are intricately related to the crystal lattice. For instance, in the piezoelectric process, a minute change in crystal dimension can produce a measurable voltage; which also means that with reasonable operating voltages, one is able to cause crystal deformation in a controlled way even down to the angstrom level. This has opened a plethora of applications, ranging from wristwatches to micro-thrusters in satellites, and the past decades have witnessed the developments.[5] Similarly, the triboelectric aspect has been thought of in various contexts, from on-the-go energy generation devices to the killing of unwanted germs on surfaces.

The recent energy crisis has led to the quest for energy-saving and harvesting technologies such as solar cells, mechanical energy transducers from natural resources, etc.[6-9] The latter is the consideration of the present chapter, which focuses on various forms of nanomaterials and device architectures for tribo, and piezo-based nanogenerators which are not only used in energy devices,[10-12] but also exploited in various aspects ranging from standalone sensors and transistors[11,13] to healthcare applications,[14-16] and artificial intelligence (Figure 1).[17,18]

## 2. Piezoelectricity

The Greek word "piezein" means to "squeeze or press" and therefore, the word piezoelectricity represents the production of electricity through push or through squeeze. The first demonstration of the piezoelectric effect was

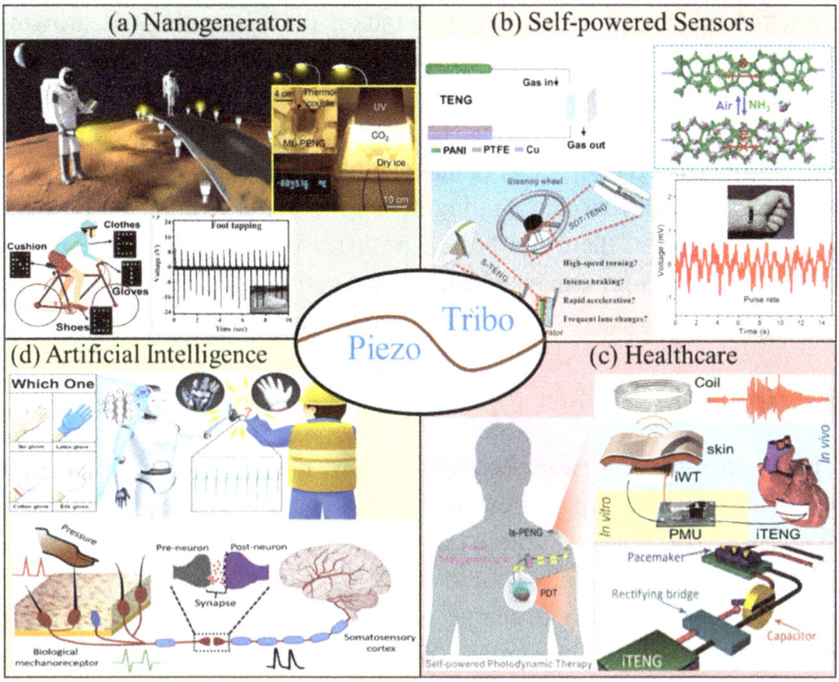


Figure. 1. Diverse applications of piezo and triboelectric devices in the field of (a) mechanical energy harvesting,[10-12] (b) self-powered sensors,[11,13] (c) stand-alone biomedical devices in healthcare,[14-16] and (d) artificial intelligence including neuromorphic computing, field effect transistors, and robotics.[17,18] Reproduced with permission from Refs. 10–18.

given in the year 1880 by the two Curie brothers, Pierre and Jacques, the former of the Pierre-Marie-Curie couple fame. Although many crystals of materials such as tourmaline, topaz and even cane sugar were shown to exhibit piezoelectricity, it was quartz, which was considered to be the best piezoelectric material those days. Later, it was found that most pyroelectric materials show piezoelectricity and are also ferroelectric. Further, it was realized that, besides generation of electricity, the piezo crystals can be elongated or contracted by applying an electrical signal, a finding which led to the very first application of the phenomenon in sonar devices during World War I. In the present day, piezoelectricity finds application in multitude of modern systems including scanning tunneling microscope (STM), atomic force microscope (AFM), optics focusing assembly, sound detectors, electronic frequency generators, etc. Recently, the growing interest in green energy requirements has driven the application toward energy generation, which in turn has led to the development of numerous piezoelectrically active materials. Examples of most efficient materials are barium titanate (BTO),[19] lead titanate and lead zirconate titanate (PZT).[20] Nevertheless, with lead-containing devices being environmentally toxic, in the recent past, there has been a push for designing lead-free piezoelectric materials for enhanced energy generation and stand-alone device fabrications; whose aspects are discussed in the later sections.

### 2.1. *Mechanism*

The mechanism for piezoelectricity is very meticulously related to the induction of electric dipole moment in crystalline solids having no inversion symmetry. For instance, zinc oxide (ZnO) lattice has a wurtzite crystal structure with tetrahedrally coordinated $Zn^{2+}$ and $O^{2-}$ (see Figure 2a).[21] At the undisturbed state, the charge centers of the $Zn^{2+}$ and $O^{2-}$ overlap, showing zero net dipole moment. Whenever the lattice is deformed by a compressive or tensile strain, the charge centers get separated, producing a finite dipole moment or piezo potential (Figures 2a and 2b).[22] At this state, if a load is connected, there will be a current flow through the external circuit to reach a new equilibrium with zero potential difference. A subsequent release of the strain breaks the equilibrium and results in an opposite current flow as depicted in Figure 2b. Thus, the piezoelectric materials work as a transducer to harvest mechanical energy into electrical energy.

A similar principle is used in Piezotronics to fabricate low energy-consuming field effect transistors (FETs) where the piezo potential acts as the gate voltage to control the source-drain current. The same is schematically demonstrated in Figures 2c and 2d, which show that the height of the Schottky barrier formed between the metal electrode and the piezo-active semiconductor can be controlled by applied strain.[23]

### 2.2. *Materials*

Piezoelectric materials are mostly ferroelectric transducers, which produce a potential difference when experiencing pressure or strain. The quality

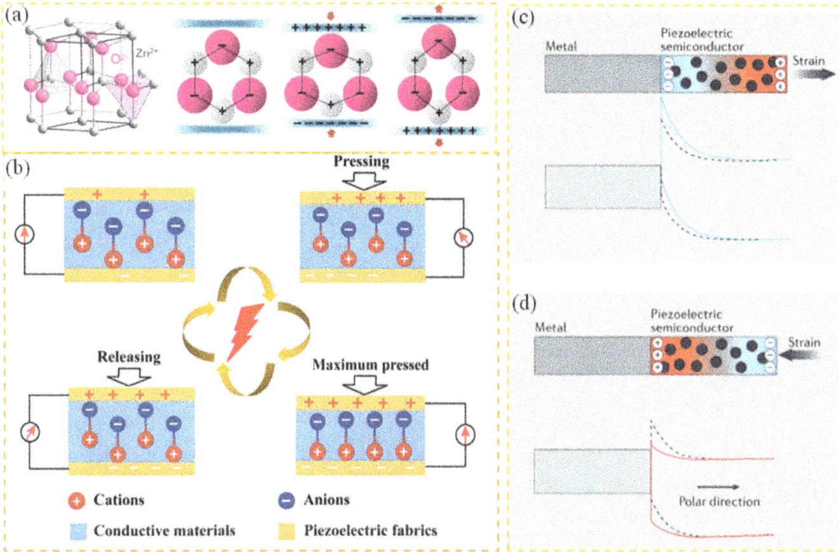


Figure 2. Mechanism of piezoelectric energy generation and piezotronics: (a) wurtzite structure of ZnO lattice having overlapped positive and negative charge centers. The compressive or tensile strain on the material can cause separation between the two centers resulting in a net non-zero dipole moment.[21] (b) When connected to an external load, the resulting piezo-potential drives electrons through the circuit and reaches equilibrium.[22] Piezotronics: (c) increment or (d) decrement in the Schottky barrier height depending upon the strain direction. Such piezo-potential controlled barriers are used in piezotronic transistors.[23] Reproduced with permission from Refs. 21–23.

factor or the figure of merit (FoM) of a piezoelectric material depends on two factors: charge constant (d) and voltage constant (g).[19,20] d is defined as the polarization experienced per unit external stress, whereas the ratio of the resultant electric field and the applied mechanical stress defines the voltage constant, g. Both d and g are direction-dependent since piezo-electric materials are non-centrosymmetric and hence, anisotropic. Therefore, for defining the FoM, we consider the multiplication of only $d_{33}$ and $g_{33}$, which represent the case when both the applied stress and the resultant strain are along the same direction (conventionally along Z-axis). Furthermore, the charge constant d is proportional to the dielectric constant of a material and hence, materials with a higher dielectric constant are desirable for piezoelectricity.

Based on recent developments, piezoelectric materials are categorized into four classes, namely zero-dimensional (0D), one-dimensional (1D), two-dimensional (2D), three-dimensional (3D), and composites. In the present chapter, emphasis has been given to those materials which are non-toxic, lightweight, flexible, inexpensive, chemically inert, and easily fabricable at a large scale. It is worthwhile that the nanostructures show superior piezoelectric behavior due to the better flexibility and pressure tolerance at the nano regime, and the high surface-to-volume ratio producing large interfaces for interaction.

### 2.2.1. *0D materials*

0D materials, being the building blocks of many other nano-microstructures, have received immense interest in various fields, explicitly solar cells, photocatalysts, energy generation, and storage. Presently, the most extensively studied 0D materials are the perovskite quantum dots (PQDs) which have been used in various high-efficient energy generating devices.[24] However, it is very lately that the PQDs are exploited in the field of piezoelectric energy generation.[25,26] Recently, Li *et al.* have reported an excellent ferroelectric behavior in an all-halide orthorhombic phase ($Pna2_1$) of $CsPbBr_3$ PQDs (Figure 3a) with a saturation polarization value of 0.25 $\mu C/cm^2$, which is attributed to the separated positive and negative charge centers appeared due to the distortion of octahedral $[PbBr_6]^{4-}$ units and the off-centered $Cs^+$.[27] Similarly, formamidinium lead halide PQDs (see the TEM image in Figure 3b) also show ferroelectric characteristics and therefore, are piezoelectrically active.[28] Other widely studied 0D materials include nanoparticles of ZnO, $BaTiO_3$, $BiFeO_3$, and so on.[29-31] Nevertheless, devices made of 0D materials suffer from low output current due to the large electron scattering at the inter-particle interfaces.

### 2.2.2. *1D materials*

To date, one-dimensional nanostructures such as nanowires, fibers, rods, nanobelts, nanotubes, and so on are the most extensively investigated geometry for piezoelectric materials due to the advantage of producing relatively higher output current ascribed to the easy electron percolation

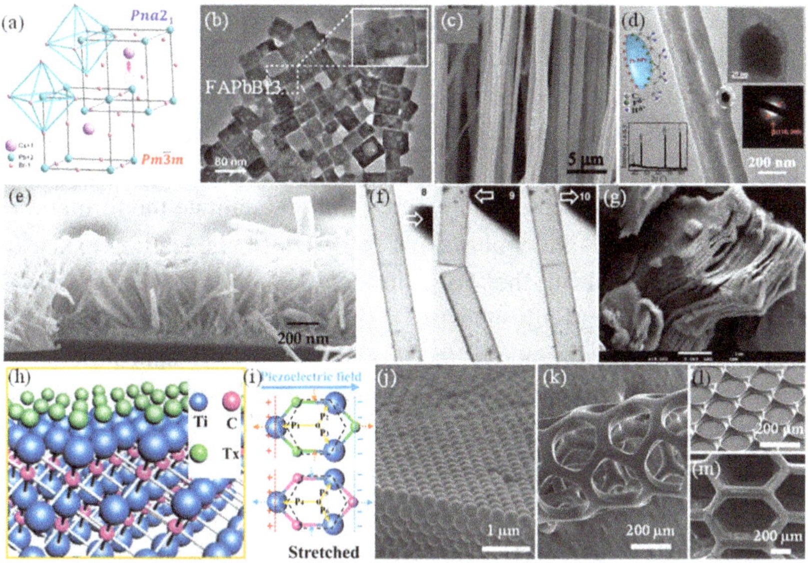

Figure 3. Piezoelectric materials: 0D: (a) cubic ($Pm\bar{3}m$) and orthorhombic phase ($Pna2_1$) of $CsPbBr_3$ quantum dots. The orthorhombic phase is more stable and can produce ferroelectricity due to non-centrosymmetry.[27] (b) TEM image of perovskite ($FAPbBr_3$) nanoparticles which are ferroelectric.[28] 1D: (c) FESEM image of Pt nanoparticle interfaced PVDF nanofibers (Pt-PVDF NFs). (d) High-resolution TEM image of a Pt-PVDF NF.[32] The magnified dotted circle and XRD pattern confirm the Pt decoration over the nanofibers. (e) Cross-sectional SEM image of 1D ZnO nanostructures.[36] (f) Polarized optical microscopy image of single-crystalline 3,3′,5,5′-tetramethyl-4,4′-bipyrazole nanofiber having self-healing property.[38] 2D: (g) FESEM image of MXene ($Ti_3C_2T_x$) multilayers.[39] Schematic representing (h) crystal structure of MXene and (i) structure deformation leading to piezoelectric polarization.[40] (j) SEM image of hollow 2D $MoS_2$ shells.[42] 3D: FESEM images of additive manufactured 3D structure of (k)-(l) barium titanate (BTO) nanoparticles and polyethylene glycol diacrylate composite,[19] and (m) piezoelectric complex of lead zirconate titanate (PZT).[20] Reproduced with permission from Refs. 19, 20, 27, 28, 32, 36, 38, 39, 40, 42.

through the 1D wires. Besides, the improved interaction due to the higher surface area of the nanostructure leads to the generation of enhanced piezo potential. A plethora of 1D materials including organic polymers, namely poly(vinylidene fluoride) (PVDF), polyacrylonitrile (PAN), poly(vinylidene fluoride-co-trifluoroethylene) (PVDF-TrFE), poly(L-lactic acid) (PLLA) etc.,[32-35] and inorganic non-centrosymmetric materials such as ZnO, BTO, PZT and so on have been developed and investigated in the recent past for the fabrication of piezoelectric nanogenerators (PENGs).[36,37] Figure 3c displays the FESEM image of aligned PVDF nanofibers formed by the electrospinning method.[32] These nanofibers were interfaced with Pt nanoparticles (see the high-resolution TEM image in Figure 3d and the insets) for the enhancement of the output power. The alignment of the fibers and the Pt incorporation facilitate the formation of pure polar phases ($\beta$- and $\gamma$-phases) in Pt-PVDF fibers, leading to nearly a five-time increment in the FoM in comparison to the pure PVDF nanofibers which partially consist of the nonpolar $\alpha$-phase. In another study, Mahmud *et al.* reported the growth of 1D ZnO nanostructure for PENG application (FESEM image in Figure 3e) by a simple hydrothermal method where AZO was used as the seed layer for controlling the orientation.[36] Recently, Bhunia *et al.* have identified that the piezoelectric compound 3,3′,5,5′-tetramethyl-4,4′-bipyrazole shows cyclic self-healing (see the polarized optical microscopy image in Figure 3f), which is analogous to the mechanism of wound healing in living tissues that use stress-induced opposite electrical charges for the healing.[38] While the biocompatibility of such materials can be advantageous in the biomedical field, exploiting the self-healing property in wearable device fabrication can be beneficial for technological advancement.

### 2.2.3. *2D materials*

Since the discovery of graphene in 2004, a perfect 2D material, materials with 2D geometry have been explored enormously in almost every field of research and PENGs are also not an exclusion. 2D materials having large areas of exposed polar basal planes and higher surface roughness due to the grain boundaries and wrinkles are beneficial specifically in the out-of-plane MIM (metal-insulator-metal) PENG device architecture.

MXene family, despite being studied vastly in different electronic devices, remains nearly unexplored as the piezoelectric materials. Auliya *et al.* recently disclosed that the 2D layered MXene ($Ti_3C_2T_x$) (FESEM image in Figure 3g) can be an excellent piezoelectric material since it possesses no inversion symmetry (see Figures 3h and 3i).[39,40] Tan *et al.* have reported the first experimental study of piezoelectricity in a monolayer $Ti_3C_2T_x$ MXene, which is mechanistically portrayed in the schematic in Figure 3i, where it is demonstrated that the lattice deformation due to the applied strain produces a non-zero dipole moment in the lattice.[40] 2D transition metal dichalcogenides such as $MoS_2$, $MoSe_2$, $WS_2$, $WSe_2$, and so on, along with graphene, 2D ZnO, InN, etc., are also explored parallelly for their outstanding piezoelectric properties.[36,41-44] Figure 3j depicts a 2D hollow shell structure of $MoS_2$ sheets formed by ammonium thiomolybdate coating on a superlattice of polystyrene (PS) beads, followed by high-temperature annealing.[42] The as-grown 2D structure could produce a peak output voltage of ~1.2 V at a minimal pressure of 4.2 kPa. Nonetheless, the commercial feasibility is still far from practice due to the less developments in automated manufacturing.

#### 2.2.4. *3D materials*

3D structuring of piezoelectric materials has the potential to bring performance enhancement to the device by introducing better flexibility, compactness, and multilayer geometry. Contextually, as a method of fast and automated product development, additive manufacturing or 3D printing brought a revolution recently in the traditional manufacturing industry for effectively fabricating complex 3D structures. Therefore, adopting the same into the piezoelectric field may substantially increase the production rate and reduce the fabrication complexity and cost.

In the literature, a few attempts have been recently found in this direction, and are depicted in Figures 3k, 3l and 3m, which display FESEM images of 3D structures of BTO and PZT, respectively.[19,20] The nanostructures portrayed in Figures 3k and 3l were fabricated by Kim *et al.* through a 3D printing technique termed digital projection printing (DPP), where the composite of BTO and a UV light-sensitive polymer, polyethylene glycol diacrylate (PEGDA) were first coated onto a substrate and subsequently irradiated with UV light through a dynamically alterable digital optical mask for selective area polymerization.[19] The best advantage of this process is that, unlike contact printing, large-area desired 3D patterns could be developed within a few seconds. Following a similar method, Yao *et al.* fabricated a honeycomb structure (see Figure 3m) of the PZT complex for force sensing application.[20] Albeit having been such a vast development in piezoelectric nanomaterials, the significantly poor power output due to the slow dynamics of charge separation in the high resistive active layer has hindered the commercialization of most of the piezo materials.

#### 2.2.5. *Composites*

In the context of achieving higher output energy, compositing materials has certainly proved to be a promising strategy owing to various synergistic effects leading to higher electron mobility, increased polarizability, and enhancement in dielectric constant. In a study by Karumuthil *et al.*, it was demonstrated that a composite of ZnO nanoparticles, polydimethylsiloxane (PDMS), graphene oxide (GO) nanoflakes, and multiwalled carbon nanotubes (MWCNTs) portrays much-improved voltage and current output.[45] Here, the GO nanoflakes act as a dispersing agent to reduce the agglomeration of ZnO in PDMS matrix, resulting in the increment in interacting surface area and hence, the output voltage. Besides, the conducting MWCNTs improve electron mobility, leading to the current enhancement. Similarly, as discussed earlier, the incorporation of Pt nanoparticles reduces the formation of the nonpolar α-phase of PVDF and hence improves the overall piezoelectric performance.[32]

### 2.3. *Applications*

Piezoelectric materials are known to be the remarkable transducers that convert mechanical energy into electrical energy and vice versa, which are exploited in a wide variety of applications. While the direct piezoelectric effect is extensively employed in fabricating PENGs as mechanical energy harvesters, the converse effect is explored in various other areas including actuators, frequency generators, and so on.

The basic device geometry of a PENG is very similar to a parallel plate capacitor where the dielectric piezo-active material is sandwiched between two parallel conducting electrodes as the charge collectors (see Figures 4a and 4b).[36] As explained previously, mechanical deformation of the piezoelectric layer produces an output voltage typically of the order of a few tens of volt and a maximum power of a few $\mu W/cm^2$ (see Table 1). The device (shown in Figure 4b) comprising of 1D/2D ZnO in poly(methyl

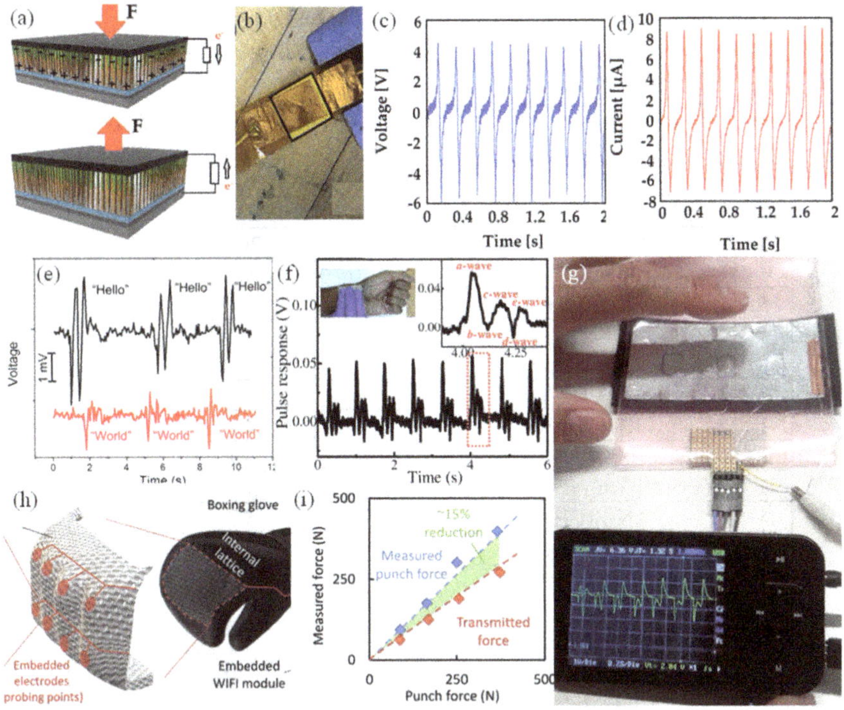


Figure 4. Applications: (a) schematic and (b) digital photograph of a typical metal-insulator-metal (MIM) geometry of a piezoelectric nanogenerator (PENG) and its (c) open-circuit voltage and (d) short-circuit current.[36] Diverse applications as self-powered sensors: (e) speech-recognition,[46] (f) pulse-monitoring,[32] (g) touch,[47] and (h)–(i) force sensors.[20] Reproduced with permission from Refs. 20, 32, 36, 46, 47.

methacrylate) (PMMA) matrix sandwiched between Al doped zinc oxide (AZO) and Al electrodes generates a peak-to-peak open circuit voltage ($V_{oc}$) of ~10 V and short circuit current ($I_{sc}$) of ~16 μA for an applied force of 5N (Figures 4c and 4d).[36] Lately, piezoelectric materials have also gained tremendous popularity in the field of sensors not only due to their high sensitivity to pressure but also for facilitating the production of self-powered stand-alone devices, which may bring revolution in wearable and implantable equipment by eliminating the requirement of external power sources. A few such instances are illustrated in Figures 4e–4g.[32,46,47] As depicted in Figure 4e, an in-plane sensor made of metal dichalcogenides $MoSe_2$ could produce unique electrical signals for voicing different words

Table 1. Comparison of open circuit voltage ($V_{oc}$), short circuit current ($J_{sc}$), and maximum output power ($P_{max}$) of piezo, tribo, and hybrid nanogenerators made of materials of different dimensionality (D).

| | Sl. No | Active materials | (D) | $V_{oc}$ (V) | $J_{sc}$ (μA/cm²) | $P_{max}$ (μW/cm²) | Ref. |
|---|---|---|---|---|---|---|---|
| Piezoelectric | 1 | $FAPbBr_3$-PDMS | 0 | 8.50 | 3.4 | 12 | 28 |
| | 2 | $FAPbBr_3$-PVDF | 0 | 30 | 6.2 | 27.4 | 26 |
| | 3 | ZnO-PMMA | 1 | 25 | 20 | 8.4 | 36 |
| | 4 | Pt-PVDF | 1 | 30 | 6000 | 22 | 32 |
| | 5 | $Ti_3C_2$-PVDF | 2 | 17 | 1.3 | 14 | 39 |
| | 6 | $Bi_{0.84}Sm_{0.16}Fe_{0.84}Ti_{0.16}O_3$-Silicone rubber | 3 | 16 | 0.62 | 3.11 | 48 |
| | 7 | $BaTiO_3$-P(VDF-TrFE) | 3 | 6 | 2 | 1.4 | 49 |
| Triboelectric | 1 | PET-$C_{60}$/Al | 0 | 1600 | 4 | 380 | 60 |
| | 2 | $Bi_2WO_6$:P(VDF-TrFE) | 0/1 | 205 | 1.19 | - | 59 |
| | 3 | Steel Yarn-PVDF | 1 | 175 | 0.03 | 4.4 | 17 |
| | 4 | PAN-$B(OH)_3$ | 1 | 200 | 5 | 667 | 61 |
| | 5 | Polylactic-co-glycolic acid(PLGA)-AgNWs-PLGA | 1 | 90 | 0.05 | 13 | 65 |
| | 6 | PA66-MWCNTs/PVDF | 1 | 142 | 4 | 130 | 63 |
| | 7 | Hexagonal boron nitride-PET | 2 | 200 | 0.048 | 14 | 66 |
| | 8 | Graphene-fluorinated ethylene propylene-polyimide | 2 | 210 | - | 4750 | 67 |
| | 9 | PTFE/Cu | 3 | 2360 | - | 670 | 73 |
| Hybrid | 1 | ZnO-PVDF | 2 | 78 | 0.46 | 24.5 | 79 |
| | 2 | Poly-L-lactic acid-PVDF | 0/1 | 35.7 | - | 52.5 | 80 |
| | 3 | Y-doped ZnO-PDMS | - | 247 | - | 600 | 81 |

when attached to the neck.[46] Such sensors could be used for speech recognition in robotics and human-machine interfacing. Another sensor consisting of Pt-decorated PVDF nanofibers could produce the distinguished blood pulse signals consisting of three characteristic waves: percussion (a), tidal (c), and dicrotic (e), which can be analyzed for the detection of blood flow abnormality in biomedical applications.[32] Such devices can also be utilized as a touch sensor as demonstrated in Figure 4g where the pixelated thin film of alkaline niobate $K_{0.485}Na_{0.485}Li_{0.03}NbO_3$ (KNLN) dispersed in a PDMS matrix could detect mild finger tapping.[47]

Among many other applications, Yao *et al.* demonstrated an interesting application of a 3D printed PZT-based piezoelectric material in devising a wireless self-sensing boxing glove to measure the transmitted force as a function of applied punch force (see Figures 4h and 4i).[20] Several other emerging applications are discussed later.

## 3. Triboelectricity

The word “triboelectricity” (TE) refers to the type of electricity that is produced from mechanical friction between two or more materials. This is due to dissimilar electronegativity values resulting in the electron flow from one to the other, maintaining the balance in electrostatic potential. Triboelectricity is commonly encountered in day-to-day life, for example, a sharp spark experienced while getting up from a chair. The process relates to instant plasma formation in presence of a high electric field. Triboelectricity can also be exemplified as the production of electrostatic charge while rubbing fur on a plastic rod (see Figure 5a) and something which is taught in the middle school physics even these days.[50] Despite TE being known for more than 2000 years, it remained neglected and as unwanted in the scientific community as well as technology sectors, until the very first demonstration of a triboelectric-based nanogenerator (TENG) by Wang *et al.* in 2012.[51] Since then, enormous progress is observed in this direction for not only harvesting mechanical energy from natural resources such as raindrops, seawater waves, wind, and so on but also fabricating energy-efficient sensors for artificial intelligence (AI). However, the basic mechanism still is not completely understood and has remained a topic of debate.

### 3.1. *Mechanism*

Even though triboelectricity is one among basic phenomena, the exact mechanism of charge separation is still not completely understood due to the simultaneous involvement of several complex chemical and physical occurrences relating to tribology. It, in some cases, may be driven by the electron transferring due to a difference in electronegativity, whereas in other cases it is possibly due to ion or material exchange at the interfaces.

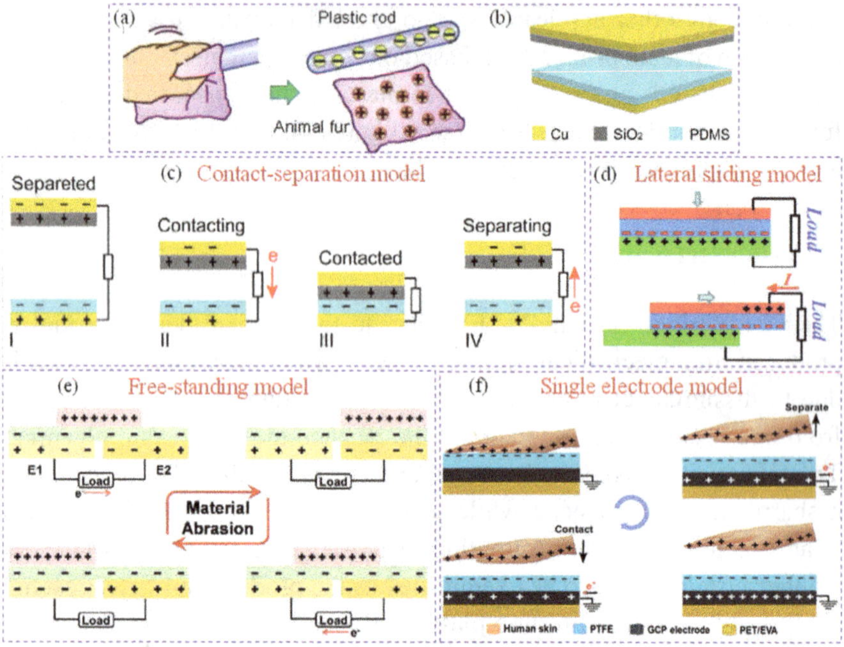


Figure 5. (a) Schematic showing the basic experiment of producing triboelectricity by rubbing fur against a plastic rod. Here, the tribo-positive material gets positively charged whereas the other gets negatively, producing a potential difference.[50] (b) Device architecture of a typical TENG where two triboelectric materials, $SiO_2$ and PDMS attached to Cu electrodes are separated by a finite gap.[52] (c)–(f) Schematic representing the mechanism of producing triboelectricity in various modes including contact-separation, lateral-sliding, free-standing, and single-electrode.[52-55] Reproduced with permission from Refs. 50, 52–55.

Albeit having a lack of complete understanding of the mechanism, there has been enormous progress in TENG technology since its inception.

Based on the device geometry and mode of energy scavenging, TENGs are categorized into primarily four classes, viz. contact-separation model, lateral sliding model, free-standing model, and single-electrode model. Figure 5b illustrates the simplest and the most basic architecture of a TENG that contains top and bottom electrodes and two strategically chosen materials, respectively attached to the electrodes.[52] In the contact-separation model (Figure 5c), the device experiences a repeated tapping force resulting in periodic contact and separation between the upper and lower parts.[52] During the contact, one part acquires positive charges while the other gets negative because of contact electrification. Since the process is spontaneously irreversible, during the separation, an electrostatic potential builds up driving electrons through the external circuit. The subsequent contact reduces the potential to zero resulting in an opposite electron flow. Thus, the iterated mechanical force gets transduced into an AC output power. Such a model can be incorporated into the automobile and footwear industries for the scavenging of energy, however tiny it may be.

The lateral sliding model is designed for implanting at those places where rotational or translational motions are involved. Unlike the contact-separation model, in this case, the two parts remain connected during motion. As demonstrated in Figure 5d, the relative sliding decreases the overlapped area, producing a potential difference, which in the next step of opposite motion, disappears.[53] Accordingly, the lateral sliding mode also produces AC outputs. The free-standing model is majorly based on a lateral geometry where a single triboelectric layer is coated on both electrodes.[54] Such a model is developed mostly to harvest the natural mechanical energy relating to raindrops, sea waves, wind, and so on. When such forces apply dynamic friction over the tribo-layer, it produces electrical energy (see Figure 5e).[54] The simplified type of all the other three is the single-electrode model where the device works only with one piezo layer attached to a single electrode (Figure 5f).[55] Such a model is suitable for the installation over railway tracks, touch screens, wind turbines, roadways, and so on.

### 3.2. *Materials*

The output power of a TENG depends on many factors ranging from electrode selection to the property of the tribo-materials and their position in triboelectric series. The materials which are at the top of the tribo series are donors and exhibit a higher tendency to donate electrons, whereas those at the bottom are the good electron acceptors. Thus, in a tribo-pair of a TENG, the materials are selected from the opposite ends of the series for higher power output. The same is depicted in Figure 6a which shows a strategic network to consider materials for TENG devices.[56] Similar to the piezoelectric section, the triboelectric materials are also classified into four categories discussed below.

### 3.2.1. *0D materials*

Triboelectricity is majorly a surface phenomenon. Therefore, for having a higher triboelectric output, materials with a higher surface area are advantageous mainly due to the increased friction. Zero-dimensional materials are thus beneficial in this context. In a recent study, Jiang *et al.* used Ag nanoparticles (AgNPs) densely coated on photographic paper as the positive electrodes sandwiching the negative counterpart- fluorinated ethylene propylene (FEP).[57] With the enhanced friction due to the rough surface of AgNP film, this TENG could produce an admirable output voltage of ~ 150 V and a current of ~ 23 μA. For a similar reason, Xu *et al.*, in their recent study on self-powered electronic skin, decorated the smooth surface of Ag nanowires (AGNWs) with ultra-lightweight graphene QDs (Figure 6b) that not only enhanced the power output but also the pressure sensitivity.[58] In a similar study, Bharti *et al.* demonstrated the synergistic effect between $Bi_2WO_6$ NPs and P(VDF-TrFE) fibers (Figure 6c) for boosting the output power.[59] As explained by them, the $(Bi_2O_2)_n^{2+}$ and $(WO_4)_n^{2-}$ lone-pair offers nucleation sites to the linker-P(VDF-TrFE) for the formation of hydrogen bonds that further dislocate the O and H atoms giving rise to higher non-centrosymmetry, which in turn increases the performance of the TENG. Fullerene, $C_{60}$, has also received a considerable attention in this direction due to its tendency of forming π-bonds on the surface to increase electron affinity.[60]

### 3.2.2. *1D materials*

The triboelectric nanogenerators are largely apposite in the portable and wearable technology which demands them to be light-weight and flexible alongside the high-power output. Films containing 0D materials are prone to form cracks upon bending and twisting and therefore, are less likely to be appropriate for wearable applications. Contextually, nano 1D materials possess outstanding flexibility owing to the high surface-to-volume ratio that helps in the easy release of accumulated stress during flexing. Consequently, 1D nanostructure, specifically electrospun nanofibers of

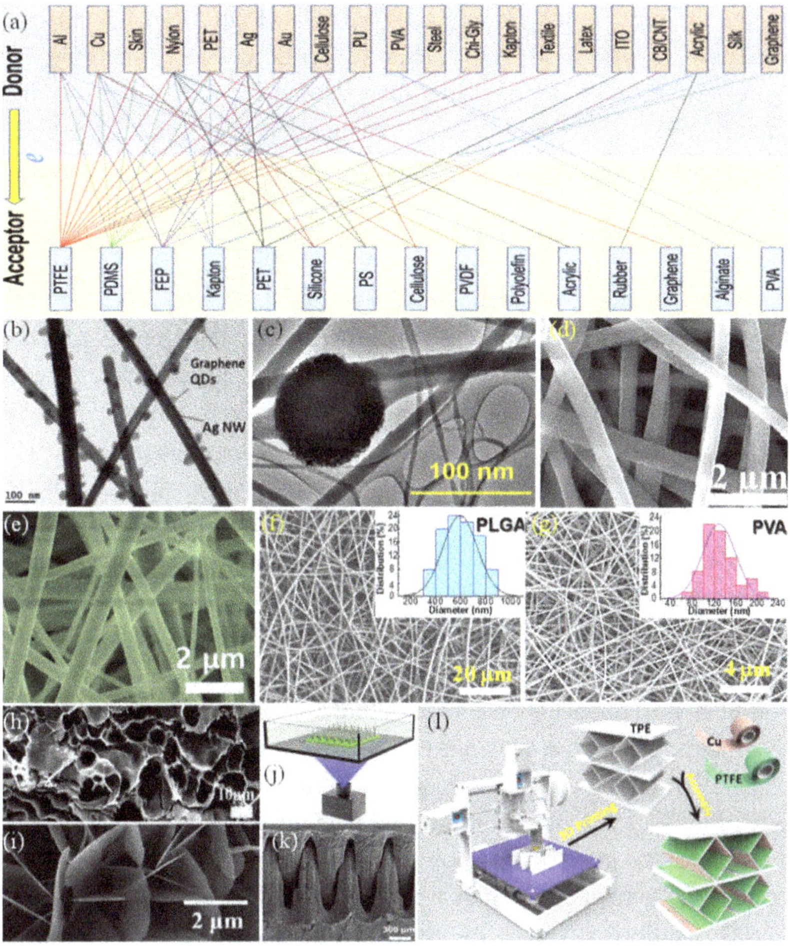


Figure 6. (a) A network showing the strategy to consider materials for triboelectricity. Generally, the most tribo-positive donor is chosen as the counterpart for the most tribo-negative acceptor.[56] 0D: Low-magnification TEM images of (b) graphene quantum dots,[58] and (c) $Bi_2WO_6$ nanoparticles incorporated in Ag nanowires and P(VDF-TrFE) nanofibers.[59] 1D: SEM images of (d) $B(OH)_3$ contained PAN nanofibers,[61] (e) (PVDF)/$C_{60}$ composite nanofibers,[62] (f) polylactic-co-glycolic acid (PLGA),[65] and (g) polyvinyl alcohol (PVA)[65] nanofibers used for TENG fabrications. 2D: SEM image of (h) laser-induced graphene sheets,[68] and (i) polyaniline-polypyrrole nanoflakes.[69] 3D: (j) Schematic demonstrating the additive manufacturing of polyurethane (PU) and fluorinated-PU (F-PU) 3D cones shown in (k) the SEM image.[72] (l) 3D printing of elastic triboelectric nanogenerator (PE-TENG) consisting of four-fold units of thermoplastic elastomer (TPE) structure, Cu, and PTFE.[73] Reproduced with permission from Refs. 56, 58, 59, 61, 62, 65, 68, 69, 72, 73.

polymers, namely PAN (Figure 6d),[61] PVDF (Figure 6e),[62,63] polytetrafluoroethylene (PTFE),[64] PLGA (Figure 6f),[65] polyvinyl alcohol (PVA) (Figure 6g),[65] and so on, has been investigated the most. In this niche, a special remark is ascribed to the electrospinning instrument, which albeit being simple and low-cost, can produce unbroken sub-micron nanofibers at the industrial scale, making TENG commercially more viable.

### 3.2.3. *2D materials*

2D materials, due to offering the layer thickness at the atomistic level are ideal for TENG application since they have a higher transverse area with most of the atoms occupying the surface. Besides, these materials possess ultra-high flexibility, mechanical stability, and even transparency, which meet the criteria of next-generation transparent and wearable devices. Nonetheless, they are comparatively less studied due to a lack of deeper understanding of the mechanism. In a recent study by Bhavya *et al.*, a contact-separation mode operable TENG was devised, where liquid-phase exfoliated hexagonal boron nitride (hBN) 2D sheets were coated on biaxially-oriented polyethylene terephthalate as the triboelectric layer to accomplish a 70 times higher power output compared to the device without hBN.[66] Graphene as a celebrated 2D material, has also contributed substantially to the TENG industry.[67] Figure 6h portrays the FESEM image of a spiky graphene film formed by a direct laser writing technique for the fabrication of a self-powered touch sensor where the boosted friction due to finger touching act as the sensing stimulus.[68] Other 2D materials in the framework of TENG include self-healable polyaniline-polypyrrole (Figure 6i),[69] MXene,[70] transition metal dichalcogenides such as $MoSe_2$, $MoS_2$, $WSe_2$, $WS_2$, and so on.[71]

### 3.2.4. *3D materials*

User-defined 3D nano-structuring of tribo layers by additive manufacturing facilitates increasing effective surface area, which is pivotal for enhanced electrical output. Recently, Feng *et al.* utilized direct image lithography (DIL) for the fabrication of a polyurethane (PU) layer consisting of nanocone structures (see Figure 6j).[72] DIL is advantageous compared to other known techniques due to its ability of maskless fast printing based on the spatial and temporal control of UV illumination to polymerize a pre-polymer selectively. In the device configuration, untreated PU was used as the tribo-positive layer, whereas the fluorinated PU layer was used as the bottom counterpart (see the FESEM image in Figure 6k).[72] The tactical structuring could increase the output current by five times compared to the device configured with flat films. In another instance, a fused deposition modeling (FDM) or a fused filament fabrication (FFF) printer was used for constructing a kirigami-inspired four-fold structure of a thermoplastic elastomer (TPE) which was further assembled with PTFE and Cu tape to produce a maximum power output of 6.7 $W/m^2$ (Figure 6l).[73]

### 3.3. *Applications*

In the era of green energy, TENG devices are most sought after. It is probably the simplest and most cost-effective scavenger of mechanical energy to transduce into electrical energy without releasing any greenhouse gases. Various resources of sustainable mechanical energy are seawater waves, wind energy, physical movement, automobile, raindrops, and so on, which can be easily harvested by TENGs to reduce the requirement for fossil fuels. In the context of portable and wearable applications, getting rid of the conventional battery and capacitor with a lightweight self-sustained TENG is always welcomed. In a recent development, Jeong *et al.* have demonstrated one such example, where a micro-brush array of nylon is designed to work in contact-separation and lateral sliding mode over numerous textile materials such as wool, silk, cotton, and so on to generate electrical energy (see Figures 7a, 7b, and 7c).[74] The advantages of the brush-like structure over the conventional monolithic design are that it has a higher area of contact, less contact resistance, less heat generation, and high abrasion resistance, which are the primary criteria for a device to work in a harsh environment like seawater. In another study by Xie *et al.*, a model rotary TENG (R-TENG) is discussed for harvesting wind energy.[53] The device geometry of the R-TENG shown schematically in Figure 7d comprises of PET sheet and PTFE nano-pillars as the tribo-negative layers coupled with Al as the counter material. As shown in the figure, the device

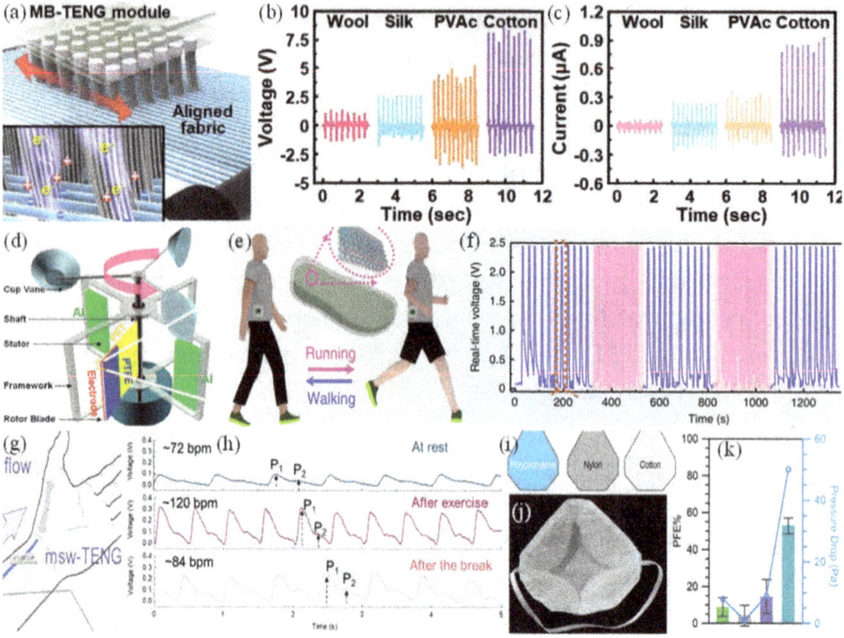


Figure 7. (a) Schematic illustration of the energy generation by nylon micro-brush TENG (MB-TENG) in sliding mode on aligned fabrics where the interlocking of micro-brush causes enhanced energy generation. Corresponding (b) voltage and (c) current output for various aligned fabrics.[74] (d) Model of a triboelectric windmill consisting of a rotary TENG (R-TENG) made of PTFE and Al.[53] (e) An intelligent footwear system for human motion monitoring and (f) voltage generation (the blue and pink peaks are due to walking and running, respectively).[75] (g) Schematic showing a TENG-based sensor placed over the wrist to monitor (h) blood pulse signals before and after exercising.[76] (i) Design of a face mask comprising polypropylene, nylon, and cotton. It filters particles via triboelectricity produced during facial muscle movements. (j) Photograph of the mask. (k) Comparison of particle filtration efficiency (PFE) and pressure drop across the mask for the individual layers (polypropylene, nylon, cotton), and the panel (consisting of all three layers) (from left to right, respectively).[77] Reproduced with permission from Refs. 53, 74–77.

is designed such that during the rotation, the PTFE-Al pair works in contact-separation mode and PET-Al operates in the lateral-sliding mode.

Along with energy generation, such devices can also function as sensors in diverse fields. For instance (Figure 7e), a TENG was fixed to footwear for transducing mechanical energy from physical movements into electrical energy as shown in Figure 7f, where the blue spikes are due to walking whereas the pink ones are generated while running.[75] As depicted, these electrical signals can further be investigated to understand the walking/running patterns and speed for healthcare applications. Additionally, such devices can be fixed to the wrist to monitor the heartbeat rate and blood pulse pattern before and after exercise (see Figures 7g and 7h).[76] In another instance related to healthcare, Santra *et al.* showed that the electrostatic potential, produced during the friction between a tribo-pair, has the ability to disintegrate virus/bacteria and filter sub-10-micron particulates.[77] In this work, a mask panel is designed by combining polypropylene, nylon, and cotton as tribo-layers, which enhances the particulate filtration efficiency by 18% while maintaining no pressure drop across the mask, leading to prolonged wearability (see Figures 7i, 7j, and 7k). The mask drew worldwide attention during the Covid pandemic.

## 4. Piezo-tribo Hybrid

Even though there has been a vast effort observed in developing materials for PENGs and TENGs to improve efficiency, much advancement remains unrealizable due to an inherent drawback of poor output power density. In this backdrop, hybridizing the two, tribo and piezo, has proved to be effective. One such structure of the hybrid piezo-tribo nanogenerator (PTENG) is shown in the schematic in Figure 8a and the photograph in Figure 8b.[78] As shown, the top section of the device is made of two Ni electrodes and a piezo-active layer, while the bottom part consisting of Cu electron and a nitro-cellulose nanofibril (CNF) work as a contact-separation mode tribo-layer in combination with Ni. The cyclic working mechanism of the hybrid device is schematically exemplified in Figure 8c(I-IV).[78] Initially, both the tribo and piezo parts remain at zero potential. As soon as a force is applied over the device (Figure 8c(I)), the piezo-layer experiences a tensile strain, thereby producing a positive piezo-potential. In the contact mode (Figure 8c(II)), the nitro-CNF film being more electronegative acquires electrons from the Ni electrode and delivers a triboelectric current. At this stage, the hybrid device affords the maximum power output due to the combined effect. In the next step (Figure 8c(III)), when the force is released, electrons flow back to the Ni electrode through the external circuit, while a negative piezo potential appears simultaneously at the upper layer. At the

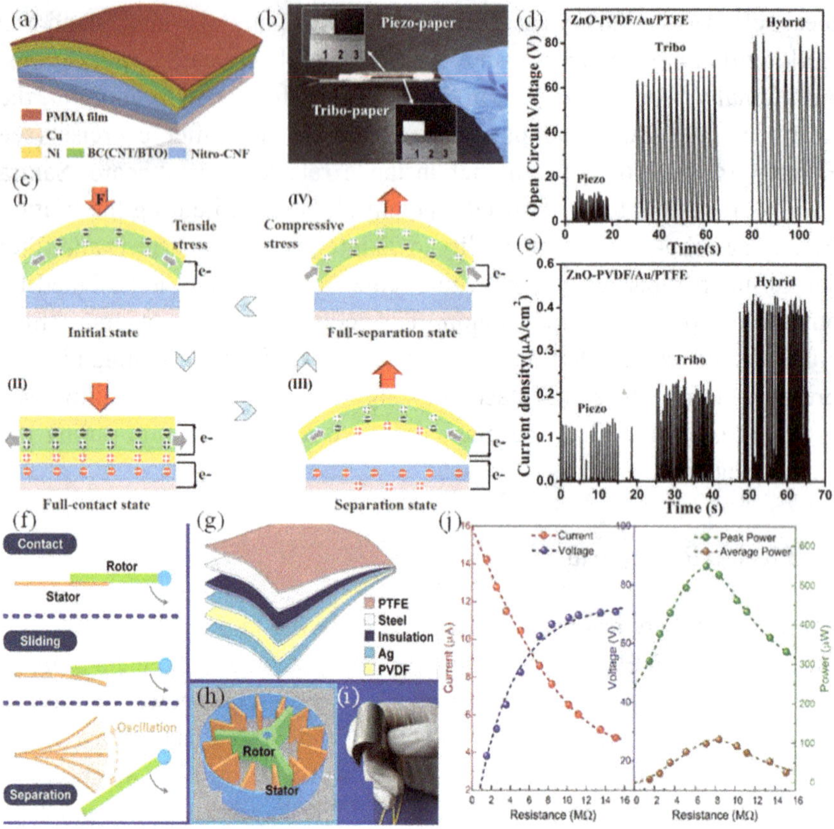


Figure 8. (a) Schematic portraying the structure of a piezo-tribo hybrid nanogenerator comprising top piezoelectric layers and bottom tribo layers. (b) A photograph of the hybrid device. (c) Working mechanism: the tensile and compressive strain on the piezo layer produces piezoelectricity, whereas contact and separation between Ni electrode and the tribo layer generate triboelectricity.[78] (d) Voltage and (e) current output comparison of a hybrid device with the individual PENG and TENG.[79] (f) Working mechanism of another hybrid device comprising of a rotor and a stator, (g) schematic of the flexible stator combined to a (h) rotor for energy generation, and (i) a photograph of the stator. The sliding of the rotor over the stator generates triboelectricity, whereas the afterward oscillation of the stator produces piezoelectricity. (j) Impedance-dependent output power characteristics of the hybrid generator.[82] Reproduced with permission from Refs. 78, 79, 82.

fourth state (Figure 8c(IV)), the potential due to triboelectricity reaches zero, and the piezo-layer experiences maximum compressive strain producing the maximum piezoelectric potential. Thus, due to the combined effect, devices with hybrid geometry show an enhanced voltage and current output compared to the individual counterparts, as displayed in Figures 8d and 8e.[79-81] For instance, Bu *et al.* devised a hybrid nanogenerator comprised of a stator and a rotor (see Figure 8h) that produced watt-level energy in a low-frequency rotation.[82] Figures 8g and 8i depict a laminar assembly of the stator comprised of six layers including PTFE, steel, double-sided insulating tape, Ag, and PVDF. The steel and PTFE layers combinedly constitute the tribo-part, whereas the bottom Ag/PVDF/Ag films act as the piezoelectric layers. As shown in the mechanism in Figure 8f, the tribo-power is produced during the rotation of the rotor by contact and sliding electrification, while the subsequent vibration of the stator produces piezo component in the PVDF layer. Accordingly, the hybrid device offers an improved output power (see Figure 8j).[82] Nonetheless, innovative device architecture and appropriate material selection still remain challenging for achieving better energy efficiency.

## 5. Emerging Applications

In recent years, implantable devices have received growing interest in the therapeutic field and biomedical engineering. The traditional energy source for these devices is batteries. However, such energy sources are generally bulky, having limited energy density and can be cytotoxic. Contextually, substantial progress has been observed in the literature in developing edible or biodegradable PENGs and TENGs which are lightweight and produce energy continuously from the heart, lung and muscle movements, eradicating the requirement of external batteries.[83-86] In this niche, chitosan (see Figure 9a) and cellulose nanofibers have shown significant potential for being non-toxic, biodegradable, comestible, and antimicrobial.[83] Recently, Yan *et al.* fabricated one such TENG made of all-biocompatible materials including carboxymethyl chitosan (CMCS), carboxymethyl cellulose sodium (CMC-Na), and Au foil (see Figure 9b) for *in vivo*, *in vitro* and baby care applications.[83] The load-dependent output power shown in Figure 9d portrays a maximum power density of 120 $mW/m^2$. In another illustration, chitosan/glycerol, PTFE, and Pt films were assembled to form a mouthguard-shaped TENG (see Figure 9c) that was utilized to produce electricity for antibacterial activity.[84] Similarly,

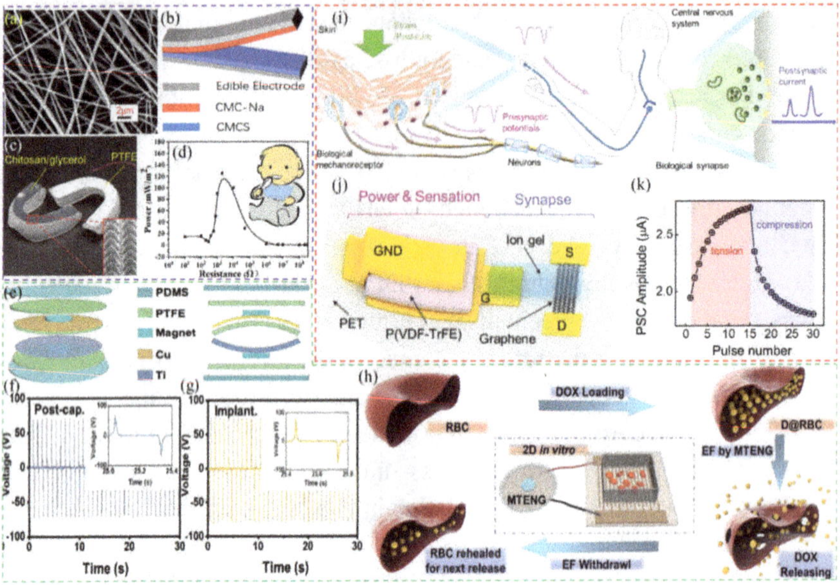


Figure 9. Edible electronics: (a) SEM image of bio-degradable chitosan nanofibers used for implantable nanogenerators. (b) Schematic of a nanogenerator made of all-biocompatible materials for *in vivo* and *in vitro* device applications.[83] (c) Photograph of a mouthguard TENG which can produce energy while speaking.[84] (d) The load-dependent output power of a chitosan-based TENG useful in childcare applications.[83] Biomedical applications: (e) Schematic of a magnet incorporated TENG (MTENG) for *in vivo* cancer therapy. (f), (g) Voltage output of the device after encapsulation and implantation in streaky pork, respectively. (h) Schematic illustrating the Doxorubicin (DOX) loading in RBC and subsequently its controlled release triggered by MTENG for anti-tumor drug delivery.[85] Artificial synapse: (i) Schematic of a biological neuron system where the mechanoreceptor in skin senses pressure, touch, or strain and produces electrical signals, which reach the central nervous system through several intermediate neurons and synapse. (j) An analogous artificial sensory synaptic device based on piezotronics. (k) Its post-synaptic current response as a function of the number of strain pulses.[86] Reproduced with permission from Refs. 83–86.

Zhao *et al.* utilized an implantable TENG incorporated with two magnets for *in vivo* cancer therapy application (Figure 9e).[85] The magnets were fixed to the tribo-layers such that the layers could trigger the contact and separation cycles, effectively increasing the output voltage even after encapsulation and implantation (see Figures 9f and 9g). Figure 9i demonstrates schematically the steps of *in vivo* cancer therapy where the doxorubicin (DOX) loaded in red blood cells (RBCs) was used as the chemotherapeutic drug which was electrically stimulated by the TENG for its fast release.

While most literature reports deal with direct energy harvesting by the PENGs and TENGs, there is emerging research making inroads to nanogenerators into the field of artificial intelligence. Present electronic components and digital computers are entirely based on von Neumann architecture. These devices are designed to be accurate for logic-based scientific calculations. However, when it comes to complex and deep calculations, such architectures fail miserably. Moreover, the computation is energy intensive and time-consuming. Contrastingly, the human brain handles much more complicated computations and learnings in a fast and easier way while consuming extremely small energy. Therefore, the growing requirement for vast data processing, fast computation, and the shortage of fossil fuels demand the computer to have energy-efficient brain-like functionality. In this context, perceptual computing emulating the human somatosensory system (Figure 9i) has gained considerable interest in robotics for the profound human-machine interactions.[86] In a recent study by Chen *et al.*, one such artificial sensory synapse is developed using an ion gel-gated piezotronic graphene transistor connected to a P(VDF-TrFE) PENG as the skin-like sensation and power component (see Figure 9j).[86] Similar to the human system, here the pressure-pulse or strain-pulse could produce the presynaptic current modulating the synaptic weight of the graphene transistor to transfer the information as the postsynaptic signals. Interestingly, the artificial system could sense the external tension/compression pulses applied on the TENG to emulate the pair pulse facilitation (PPF)/pair pulse depression (PPD) (see Figure 9k).

It is evident from the above discussion that both PENG and TENG could be interchangeably used in diverse applications. Nonetheless, they are tactically selected in applications based on their output values. As shown in Table T1, TENGs as compared to PENGs possess relatively higher power output and therefore, are more promising in energy scavenging applications, whereas PENGs are preferable when there is a requirement of higher current.

## 6. Outlook

The revolutionary inventions aimed to exploit the electromagnetic field have led the present-day generation and applications of electricity, in DC or in AC form. Can piezo or triboelectricity ever reach this level of technological maturity? Some researchers think so. According to Z. L. Wang, a pioneer of TENG, the 'blue energy' from tidal waves can be scavenged by transducing the mechanical energy of the waves. He has envisaged that a network of TENGs spread over the ocean surface across a few thousand square kilometers, can produce energy comparable to present-day generators.[87] Jeff Tollefson, the senior US correspondent of Nature Magazine, opined, "The sea could turn out to be an even more benign source of power than wind."[88] However, realization of such ideas into regular practice is still far-fetched. There are many challenges to deal with pertaining to the harsh conditions of the ocean environment, let alone other environmental concerns. Nonetheless, integrating TENG devices into life-style applications such as in sport shoes, textiles etc., has reached enough maturity for commercial launching. Future efforts would be towards developing methods to safely store the TENG-generated energy, or otherwise as in some cases, direct integration of intrinsically generated voltage spikes may destroy the attached electronics. On the other hand, piezoelectric devices which offer better control have already found applications in ultra-fine actuation of telescopic lenses, precise maneuvering of parts in satellites and sophisticated scientific instruments. The future demand for piezoelectric materials would be in artificial intelligence, specifically in developing sensory units for complex neuromorphic systems involving multiple stimuli such as electric, chemical, and optical pulses. Further developments would also focus on biocompatibility particularly in the context of in-vivo energy generation in implantable devices.

## References

1. P. Iversen and D.J. Lacks, A Life of its Own: The Tenuous Connection Between Thales of Miletus and the Study of Electrostatic Charging, *Journal of Electrostatics* **70**(3), 309-311 (2012).

2. D. Roller, D.H. Roller and F. Hauksbee, *Scientific American* **189**(2), 64-69 (1953).
3. E.A. Wrigley, Energy and the English Industrial Revolution, *Philosophical Transactions of the Royal Society A: Mathematical, Physical and Engineering Sciences* **371**(1986), 20110568 (2013).
4. S.M.A.Z. Shawon, A.X. Sun, V.S. Vega, B.D. Chowdhury, P. Tran, Z.D. Carballo, J.A. Tolentino, J. Li, M.S. Rafaqut and S. Danti, Piezo-tribo Dual Effect Hybrid Nanogenerators for Health Monitoring, *Nano Energy* **82**, 105691 (2021).
5. A. Eastman, J. Kiefer and M. Kimber, Thrust Measurements and Flow Field Analysis of a Piezoelectrically Actuated Oscillating Cantilever, *Experiments in fluids* **53**(5), 1533-1543 (2012).
6. N. Gupta, K. Rao, R. Gupta, F.C. Krebs and G.U. Kulkarni, Highly Conformal Ni Micromesh as a Current Collecting Front Electrode for Reduced Cost Si Solar Cell, *ACS Applied Materials & Interfaces* **9**(10), 8634-8640 (2017).
7. I. Mondal, S. Kiruthika, M.K. Ganesha, M. Baral, A. Kumar, S. Vimala, P.L. Madhuri, G.G. Nair, S.K. Prasad and A.K. Singh, ITO-free Large Area PDLC Smart Windows: A Cost-effective Fabrication Using Spray Coated $SnO_2$ on an Invisible Al Mesh, *Journal of Materials Chemistry A* **9**(40), 23157-23168 (2021).
8. A.K. Singh, S. Kiruthika, I. Mondal and G.U. Kulkarni, Fabrication of Solar and Electrically Adjustable Large Area Smart Windows for Indoor Light and Heat Modulation, *Journal of Materials Chemistry C* **5**(24), 5917-5922 (2017).
9. L.M. Zhang, C.B. Han, T. Jiang, T. Zhou, X.H. Li, C. Zhang and Z.L. Wang, Multilayer Wavy-structured Robust Triboelectric Nanogenerator for Harvesting Water Wave Energy, *Nano Energy* **22**, 87-94 (2016).
10. D.W. Jin, Y.J. Ko, C.W. Ahn, S. Hur, T.K. Lee, D.G. Jeong, M. Lee, C.Y. Kang and J.H. Jung, Polarization-and Electrode-Optimized Polyvinylidene Fluoride Films for Harsh Environmental Piezoelectric Nanogenerator Applications, *Small* **17**(14), 2007289 (2021).
11. Y. Liu, Y. Zheng, Z. Wu, L. Zhang, W. Sun, T. Li, D. Wang and F. Zhou, Conductive Elastic Sponge-based Triboelectric Nanogenerator (TENG) for Effective Random Mechanical Energy Harvesting and Ammonia Sensing, *Nano Energy* **79**, 105422 (2021).
12. C. Yoon, S. Ippili, V. Jella, A.M. Thomas, J.-S. Jung, Y. Han, T.-Y. Yang, S.-G. Yoon and G. Yoon, Synergistic Contribution of Flexoelectricity and Piezoelectricity Towards a Stretchable Robust Nanogenerator for Wearable Electronics, *Nano Energy* **91**, 106691 (2022).
13. X. Lu, H. Zhang, X. Zhao, H. Yang, L. Zheng, W. Wang and C. Sun, Triboelectric Nanogenerator Based Self-powered Sensor with a Turnable Sector Structure for Monitoring Driving Behavior, *Nano Energy* **89**, 106352 (2021).
14. Z. Liu, L. Xu, Q. Zheng, Y. Kang, B. Shi, D. Jiang, H. Li, X. Qu, Y. Fan and Z.L. Wang, Human Motion Driven Self-powered Photodynamic System for Long-term Autonomous Cancer Therapy, *ACS nano* **14**(7), 8074-8083 (2020).

15. Q. Zheng, H. Zhang, B. Shi, X. Xue, Z. Liu, Y. Jin, Y. Ma, Y. Zou, X. Wang and Z. An, In Vivo Self-powered Wireless Cardiac Monitoring via Implantable Triboelectric Nanogenerator, *ACS Nano* **10**(7), 6510-6518 (2016).
16. Q. Zheng, B. Shi, F. Fan, X. Wang, L. Yan, W. Yuan, S. Wang, H. Liu, Z. Li and Z.L. Wang, In Vivo Powering of Pacemaker by Breathing-driven Implanted Triboelectric Nanogenerator, *Advanced Materials* **26**(33), 5851-5856 (2014).
17. C. Ye, S. Yang, J. Ren, S. Dong, L. Cao, Y. Pei and S. Ling, Electroassisted Core-Spun Triboelectric Nanogenerator Fabrics for IntelliSense and Artificial Intelligence Perception, *ACS Nano* **16**(3), 4415-4425 (2022).
18. M. Kumar, R. Singh, H. Kang, S. Kim and H. Seo, An Artificial Piezotronic Synapse for Tactile Perception, *Nano Energy* **73**, 104756 (2020).
19. K. Kim, W. Zhu, X. Qu, C. Aaronson, W.R. McCall, S. Chen and D.J. Sirbuly, 3D Optical Printing of Piezoelectric Nanoparticle–polymer Composite Materials, *ACS Nano* **8**(10), 9799-9806 (2014).
20. D. Yao, H. Cui, R. Hensleigh, P. Smith, S. Alford, D. Bernero, S. Bush, K. Mann, H.F. Wu and M. Chin-Nieh, Achieving the Upper Bound of Piezoelectric Response in Tunable, Wearable 3D Printed Nanocomposites, *Advanced Functional Materials* **29**(42), 1903866 (2019).
21. Q. Zheng, B. Shi, Z. Li and Z.L. Wang, Recent Progress on Piezoelectric and Triboelectric Energy Harvesters in Biomedical Systems, *Advanced Science* **4**(7), 1700029 (2017).
22. C. Zhang, W. Fan, S. Wang, Q. Wang, Y. Zhang and K. Dong, Recent Progress of Wearable Piezoelectric Nanogenerators, *ACS Applied Electronic Materials* **3**(6), 2449-2467 (2021).
23. H. Yuan, T. Lei, Y. Qin and R. Yang, Flexible Electronic Skins Based on Piezoelectric Nanogenerators and Piezotronics, *Nano Energy* **59**, 84-90 (2019).
24. Q. Zhao, A. Hazarika, X. Chen, S.P. Harvey, B.W. Larson, G.R. Teeter, J. Liu, T. Song, C. Xiao and L. Shaw, High Efficiency Perovskite Quantum Dot Solar Cells with Charge Separating Heterostructure, *Nature Communications* **10**(1), 1-8 (2019).
25. J. Nie, L. Zhu, W. Zhai, A. Berbille, L. Li and Z.L. Wang, Flexible Piezoelectric Nanogenerators Based on P (VDF-TrFE)/$CsPbBr_3$ Quantum Dot Composite Films, *ACS Applied Electronic Materials* **3**(5), 2136-2144 (2021).
26. R. Ding, X. Zhang, G. Chen, H. Wang, R. Kishor, J. Xiao, F. Gao, K. Zeng, X. Chen and X.W. Sun, High-performance Piezoelectric Nanogenerators Composed of Formamidinium Lead Halide Perovskite Nanoparticles and Poly (vinylidene fluoride), *Nano Energy* **37**, 126-135 (2017).
27. X. Li, S. Chen, P.-F. Liu, Y. Zhang, Y. Chen, H.-L. Wang, H. Yuan and S. Feng, Evidence for Ferroelectricity of All-inorganic Perovskite $CsPbBr_3$ Quantum Dots, *Journal of the American Chemical Society* **142**(7), 3316-3320 (2020).
28. R. Ding, H. Liu, X. Zhang, J. Xiao, R. Kishor, H. Sun, B. Zhu, G. Chen, F. Gao and X. Feng, Flexible Piezoelectric Nanocomposite Generators Based on Formamidinium

Lead Halide Perovskite Nanoparticles, *Advanced Functional Materials* **26**(42), 7708-7716 (2016).

29. H. Yadav, N. Sinha, S. Goel and B. Kumar, Eu-doped ZnO Nanoparticles for Dielectric, Ferroelectric and Piezoelectric Applications, *Journal of Alloys and Compounds* **689**, (2016) 333-341.
30. S.M. Villa, V.M. Mazzola, T. Santaniello, E. Locatelli, M. Maturi, L. Migliorini, I. Monaco, C. Lenardi, M. Comes Franchini and P. Milani, Soft Piezoionic/piezoelectric Nanocomposites Based on Ionogel/$BaTiO_3$ Nanoparticles for Low Frequency and Directional Discriminative Pressure Sensing, *ACS Macro Letters* **8**(4), 414-420 (2019).
31. X. Ren, H. Fan, Y. Zhao and Z. Liu, Flexible Lead-free $BiFeO_3$/PDMS-based Nanogenerator as Piezoelectric Energy Harvester, *ACS Applied Materials & Interfaces* **8**(39), 26190-26197 (2016).
32. S.K. Ghosh and D. Mandal, Synergistically Enhanced Piezoelectric Output in Highly Aligned 1D Polymer Nanofibers Integrated All-fiber Nanogenerator for Wearable Nano-tactile Sensor, *Nano Energy* **53**, 245-257 (2018).
33. W. Wang, Y. Zheng, X. Jin, Y. Sun, B. Lu, H. Wang, J. Fang, H. Shao and T. Lin, Unexpectedly High Piezoelectricity of Electrospun Polyacrylonitrile Nanofiber Membranes, *Nano Energy* **56**, 588-594 (2019).
34. K. Shi, B. Chai, H. Zou, P. Shen, B. Sun, P. Jiang, Z. Shi and X. Huang, Interface Induced Performance Enhancement in Flexible $BaTiO_3$/PVDF-TrFE Based Piezoelectric Nanogenerators, *Nano Energy* **80**, 105515 (2021).
35. N.T. Cuong, S. Barrau, M. Dufay, N. Tabary, A. Da Costa, A. Ferri, R. Lazzaroni, J.-M. Raquez and P. Leclère, On the Nanoscale Mapping of the Mechanical and Piezoelectric Properties of Poly (L-lactic acid) Electrospun Nanofibers, *Applied Sciences* **10**(2), 652 (2020).
36. A. Mahmud, A.A. Khan, P. Voss, T. Das, E. Abdel-Rahman and D. Ban, A High Performance and Consolidated Piezoelectric Energy Harvester Based on 1D/2D Hybrid Zinc Oxide Nanostructures, *Advanced Materials Interfaces* **5**(23), 1801167 (2018).
37. J. Chang, M. Dommer, C. Chang and L. Lin, Piezoelectric Nanofibers for Energy Scavenging Applications, *Nano Energy* **1**(3), 356-371 (2012).
38. S. Bhunia, S. Chandel, S.K. Karan, S. Dey, A. Tiwari, S. Das, N. Kumar, R. Chowdhury, S. Mondal and I. Ghosh, Autonomous Self-repair in Piezoelectric Molecular Crystals, *Science* **373**(6552), 321-327 (2021).
39. R.Z. Auliya, P.C. Ooi, R. Sadri, N.A. Talik, Z.Y. Yau, M.A.S. Mohammad Haniff, B.T. Goh, C.F. Dee, N. Aslfattahi and S. Al-Bati, Exploration of 2D $Ti_3C_2$ MXene for All Solution Processed Piezoelectric Nanogenerator Applications, *Scientific Reports* **11**(1), 1-13 (2021).
40. D. Tan, C. Jiang, N. Sun, J. Huang, Z. Zhang, Q. Zhang, J. Bu, S. Bi, Q. Guo and J. Song, Piezoelectricity in Monolayer MXene for Nanogenerators and Piezotronics, *Nano Energy* **90**, 106528 (2021).

41. J. Kou, Y. Liu, Y. Zhu and J. Zhai, Progress in Piezotronics of Transition-metal Dichalcogenides, *Journal of Physics D: Applied Physics* **51**(49), 493002 (2018).
42. J.K. Han, S. Kim, S. Jang, Y.R. Lim, S.-W. Kim, H. Chang, W. Song, S.S. Lee, J. Lim and K.-S. An, Tunable Piezoelectric Nanogenerators Using Flexoelectricity of Well-ordered Hollow 2D $MoS_2$ Shells Arrays for Energy Harvesting, *Nano Energy* **61**, 471-477 (2019).
43. P.C. Sherrell, M. Fronzi, N.A. Shepelin, A. Corletto, D.A. Winkler, M. Ford, J.G. Shapter and A.V. Ellis, A Bright Future for Engineering Piezoelectric 2D Crystals, *Chemical Society Reviews* **51**, 650-671 (2022).
44. M. Islam, M. Zamil, M. Mojumder, R. Hasan, C. Stampfl and J. Park, Strong Tribo-Piezoelectric Effect in Bilayer Indium Nitride (InN), *Scientific Reports* **11**(1), 1-12 (2021).
45. S.C. Karumuthil, S.P. Rajeev and S. Varghese, Piezo-tribo Nanoenergy Harvester Using Hybrid Polydimethyl Siloxane Based Nanocomposite, *Nano Energy* **40**, 487-494 (2017).
46. P. Li and Z. Zhang, Self-powered 2D Material-based pH Sensor and Photodetector Driven by Monolayer $MoSe_2$ Piezoelectric Nanogenerator, *ACS Applied Materials & Interfaces* **12**(52), 58132-58139 (2020).
47. D.B. Deutz, N.T. Mascarenhas, J.B.J. Schelen, D.M. de Leeuw, S. van der Zwaag and P. Groen, Flexible Piezoelectric Touch Sensor by Alignment of Lead-free Alkaline Niobate Microcubes in PDMS, *Advanced Functional Materials* **27**(24), 1700728 (2017).
48. Y. Zhang, M. Wu, Q. Zhu, F. Wang, H. Su, H. Li, C. Diao, H. Zheng, Y. Wu and Z.L. Wang, Performance Enhancement of Flexible Piezoelectric Nanogenerator via Doping and Rational 3D Structure Design for Self-powered Mechanosensational System, *Advanced Functional Materials* **29**(42), 1904259 (2019).
49. X. Zhou, K. Parida, O. Halevi, Y. Liu, J. Xiong, S. Magdassi and P.S. Lee, All 3D-printed Stretchable Piezoelectric Nanogenerator with Non-protruding Kirigami Structure, *Nano Energy* **72**, 104676 (2020).
50. Z.L. Wang, From Contact Electrification to Triboelectric Nanogenerators, *Reports on Progress in Physics* **84**(9), 096502 (2021).
51. F.-R. Fan, Z.-Q. Tian and Z.L. Wang, Flexible Triboelectric Generator, *Nano Energy* **1**(2), 328-334 (2012).
52. Y. Liu, G. Liu, T. Bu and C. Zhang, Effects of Interfacial Acid–base on the Performance of Contact–separation Mode Triboelectric Nanogenerator, *Materials Today Energy* **20**, 100686 (2021).
53. Y. Xie, S. Wang, L. Lin, Q. Jing, Z.-H. Lin, S. Niu, Z. Wu and Z.L. Wang, Rotary Triboelectric Nanogenerator Based on a Hybridized Mechanism for Harvesting Wind Energy, *ACS Nano* **7**(8), 7119-7125 (2013).
54. J. Chen, H. Guo, C. Hu and Z.L. Wang, Robust Triboelectric Nanogenerator Achieved by Centrifugal Force Induced Automatic Working Mode Transition, *Advanced Energy Materials* **10**(23), 2000886 (2020).

55. S.A. Shankaregowda, R.F.S.M. Ahmed, C.B. Nanjegowda, J. Wang, S. Guan, M. Puttaswamy, A. Amini, Y. Zhang, D. Kong and K. Sannathammegowda, Single-electrode Triboelectric Nanogenerator Based on Economical Graphite Coated Paper for Harvesting Waste Environmental Energy, *Nano Energy* **66**, 104141 (2019).
56. R. Zhang and H. Olin, Material Choices for Triboelectric Nanogenerators: A Critical Review, *EcoMat* **2**(4), e12062 (2020).
57. Q. Jiang, B. Chen, K. Zhang and Y. Yang, Ag Nanoparticle-based Triboelectric Nanogenerator to Scavenge Wind Energy for a Self-charging Power Unit, *ACS Applied Materials & Interfaces* **9**(50), 43716-43723 (2017).
58. Z. Xu, C. Wu, F. Li, W. Chen, T. Guo and T.W. Kim, Triboelectric Electronic-skin Based on Graphene Quantum Dots for Application in Self-powered, Smart, Artificial Fingers, *Nano Energy* **49**, 274-282 (2018).
59. D.K. Bharti, S. Veeralingam and S. Badhulika, An Ultra High Performance, Lead-free $Bi_2WO_6$: P(VDF-TrFE)-based Triboelectric Nanogenerator for Self-powered Sensors and Smart Electronic Applications, *Materials Horizons* **9**(2), 663-674 (2022).
60. P. Parajuli, B. Sharma, H. Behlow and A.M. Rao, Fullerene-Enhanced Triboelectric Nanogenerators, *Advanced Materials Technologies* **5**(8), 2000295 (2020).
61. A. Yar, A. Karabiber, A. Ozen, F. Ozel and S. Coskun, Flexible Nanofiber Based Triboelectric Nanogenerators with High Power Conversion, *Renewable Energy* **162**, 1428-1437 (2020).
62. D. J. Sim, G. J. Choi, S. H. Sohn and I. K. Park, Electronegative Polyvinylidene Fluoride/C60 Composite Nanofibers for Performance Enhancement of Triboelectric Nanogenerators, *Journal of Alloys and Compounds* **898**, 162805 (2022).
63. N. Sun, G.-G. Wang, H.-X. Zhao, Y.-W. Cai, J.-Z. Li, G.-Z. Li, X.-N. Zhang, B.-L. Wang, J.-C. Han and Y. Wang, Waterproof, Breathable and Washable Triboelectric Nanogenerator Based on Electrospun Nanofiber Films for Wearable Electronics, *Nano Energy* **90**, 106639 (2021).
64. P. Zhao, N. Soin, K. Prashanthi, J. Chen, S. Dong, E. Zhou, Z. Zhu, A.A. Narasimulu, C.D. Montemagno and L. Yu, Emulsion Electrospinning of Polytetrafluoroethylene (PTFE) Nanofibrous Membranes for High-performance Triboelectric Nanogenerators, *ACS Applied Materials & Interfaces* **10**(6), 5880-5891 (2018).
65. X. Peng, K. Dong, C. Ye, Y. Jiang, S. Zhai, R. Cheng, D. Liu, X. Gao, J. Wang and Z.L. Wang, A Breathable, Biodegradable, Antibacterial, and Self-powered Electronic Skin Based on All-nanofiber Triboelectric Nanogenerators, *Science Advances* **6**(26) eaba9624 (2020).
66. A.S. Bhavya, H. Varghese, A. Chandran and K.P. Surendran, Massive Enhancement in Power Output of BoPET-paper Triboelectric Nanogenerator Using 2D-hexagonal Boron Nitride Nanosheets, *Nano Energy* **90**, 106628 (2021).
67. Y. Chen, B. Xie, J. Long, Y. Kuang, X. Chen, M. Hou, J. Gao, S. Zhou, B. Fan and Y. He, Interfacial Laser-Induced Graphene Enabling High-Performance Liquid–Solid Triboelectric Nanogenerator, *Advanced Materials* **33**(44), 2104290 (2021).

68. Z. Yan, L. Wang, Y. Xia, R. Qiu, W. Liu, M. Wu, Y. Zhu, S. Zhu, C. Jia and M. Zhu, Flexible High-resolution Triboelectric Sensor Array Based on Patterned Laser-induced Graphene for Self-powered Real-time Tactile Sensing, *Advanced Functional Materials* **31**(23), 2100709 (2021).
69. S. Singh, R.K. Tripathi, M.K. Gupta, G.I. Dzhardimalieva, I.E. Uflyand and B. Yadav, 2-D Self-healable Polyaniline-polypyrrole Nanoflakes Based Triboelectric Nanogenerator for Self-powered Solar Light Photo Detector with DFT Study, *Journal of Colloid and Interface Science* **600**, 572-585 (2021).
70. Y. Dong, S.S.K. Mallineni, K. Maleski, H. Behlow, V.N. Mochalin, A.M. Rao, Y. Gogotsi and R. Podila, Metallic MXenes: A New Family of Materials for Flexible Triboelectric Nanogenerators, *Nano Energy* **44**, 103-110 (2018).
71. S.A. Han, J. Lee, J. Lin, S.-W. Kim and J.H. Kim, Piezo/triboelectric Nanogenerators Based on 2-dimensional Layered Structure Materials, *Nano Energy* **57**, 680-691 (2019).
72. H. Feng, H. Li, J. Xu, Y. Yin, J. Cao, R. Yu, B. Wang, R. Li and G. Zhu, Triboelectric Nanogenerator Based on Direct Image Lithography and Surface Fluorination for Biomechanical Energy Harvesting and Self-powered Sterilization, *Nano Energy* **98**, 107279 (2022).
73. S. Gao, Y. Zhu, Y. Chen, M. Tian, Y. Yang, T. Jiang and Z.L. Wang, Self-power Electroreduction of $N_2$ into $NH_3$ by 3D Printed Triboelectric Nanogenerators, *Materials Today* **28**, 17-24 (2019).
74. J. Jeong, S. Jeon, X. Ma, Y.W. Kwon, D.M. Shin and S.W. Hong, A Sustainable and Flexible Microbrush-Faced Triboelectric Generator for Portable/Wearable Applications, *Advanced Materials* **33**(39), 2102530 (2021).
75. K. Dong, X. Peng, J. An, A.C. Wang, J. Luo, B. Sun, J. Wang and Z.L. Wang, Shape Adaptable and Highly Resilient 3D Braided Triboelectric Nanogenerators as E-textiles for Power and Sensing, *Nature Communications* **11**(1), 1-11 (2020).
76. Y. Wu, Y. Li, Y. Zou, W. Rao, Y. Gai, J. Xue, L. Wu, X. Qu, Y. Liu and G. Xu, A Multi-mode Triboelectric Nanogenerator for Energy Harvesting and Biomedical Monitoring, *Nano Energy* **92**, 106715 (2022).
77. P.K. Santra, A.K. Singh, G.U. Kulkarni, S. Kundu, T.S. Rao and M.K. Ganesha, Aesthetically Acceptable, Breath Friendly Triboelectric Face Masks: Design, Fabrication, and Its Efficacy, *Energy Technology* **9**(12), 2100614 (2021).
78. M. Li, Y. Jie, L.-H. Shao, Y. Guo, X. Cao, N. Wang and Z.L. Wang, All-in-one Cellulose Based Hybrid Tribo/piezoelectric Nanogenerator, *Nano Research* **12**(8), 1831-1835 (2019).
79. H.H. Singh and N. Khare, Flexible ZnO-PVDF/PTFE Based Piezo-tribo Hybrid Nanogenerator, *Nano Energy* **51**, 216-222 (2018).
80. X. Li, W. Yu, X. Gao, H. Liu, N. Han and X. Zhang, PVDF Microspheres@ PLLA Nanofibers-based Hybrid Tribo/piezoelectric Nanogenerator with Excellent Electrical Output Properties, *Materials Advances* **2**(18), 6011-6019 (2021).

81. H. Patnam, S.A. Graham and J.S. Yu, Y-ZnO Microflowers Embedded Polymeric Composite Films to Enhance the Electrical Performance of Piezo/Tribo Hybrid Nanogenerators for Biomechanical Energy Harvesting and Sensing Applications, *ACS Sustainable Chemistry & Engineering* **9**(12), 4600-4610 (2021).
82. L. Bu, Z. Chen, Z. Chen, L. Qin, F. Yang, K. Xu, J. Han and X. Wang, Impact Induced Compound Method for Triboelectric-piezoelectric Hybrid Nanogenerators to Achieve Watt Level Average Power in Low Frequency Rotations, *Nano Energy* **70**, 104500 (2020).
83. K. Yan, X. Li, X.-X. Wang, M. Yu, Z. Fan, S. Ramakrishna, H. Hu and Y.-Z. Long, A Non-toxic Triboelectric Nanogenerator for Baby Care Applications, *Journal of Materials Chemistry A* **8**(43), 22745-22753 (2020).
84. Z. H. Lin, W. S. Hsu, A. Preet, L. H. Yeh, Y. H. Chen, Y. P. Pao, S. F. Lin, S. Lee, J. C. Fan and L. Wang, Ingestible Polysaccharide Battery Coupled with a Self-charging Nanogenerator for Controllable Disinfection System, *Nano Energy* **79**, 105440 (2021).
85. C. Zhao, H. Feng, L. Zhang, Z. Li, Y. Zou, P. Tan, H. Ouyang, D. Jiang, M. Yu and C. Wang, Highly Efficient in Vivo Cancer Therapy by an Implantable Magnet Triboelectric Nanogenerator, *Advanced Functional Materials* **29**(41), 1808640 (2019).
86. Y. Chen, G. Gao, J. Zhao, H. Zhang, J. Yu, X. Yang, Q. Zhang, W. Zhang, S. Xu and J. Sun, Piezotronic Graphene Artificial Sensory Synapse, *Advanced Functional Materials* **29**(41), 1900959 (2019).
87. Z.L. Wang, T. Jiang and L. Xu, Toward the Blue Energy Dream by Triboelectric Nanogenerator Networks, *Nano Energy* **39**, 9-23 (2017).
88. J. Tollefson, Blue Energy, *Nature* **508**(7496), 302 (2014).

https://doi.org/10.1142/9789811270956_0011

# Chapter 11

# Nuclear Energy Materials

Dheeraj Jain[1] and A.K. Tyagi[1,2,*]

[1]*Chemistry Division, Bhabha Atomic Research Centre, Trombay, Mumbai 400085, India*
[2]*Homi Bhabha National Institute, Anushaktinagar, Mumbai 400094, India*
[*]*aktyagi@barc.gov.in*

Amidst alarming climate deterioration and rising energy demands, the share of nuclear electricity, which is a proven clean and green energy source, to global energy mix is inevitable to increase from its present value (~10%). Commensurate evolution of advanced nuclear fuels is needed to implement efficient and safe nuclear energy systems. An overview of different types of nuclear fuels would therefore be useful to the scientific community engaged in the development of energy materials. With this objective, the present article briefly introduces different types of nuclear fuels; covers their basic attributes and the reactor systems in which they are usable or being used, and also outlines the potential areas where innovative advancement is desired/anticipated. Salient features of solid-state metallic, ceramic and composite fuels are explained and a brief introduction to the concept of accident tolerant fuels is provided. A glimpse of ongoing nuclear fuel development programs in India is also provided with the aim that information contained herein would make students and materials researchers aware about the subject and help then pursue further developments.

## 1. Introduction

Curiosity to understand observations around us has led to discoveries. Remarkable human ability to utilize discoveries in multiple ways more often results in inventions and innovations. Evolution seen around is therefore closely linked to impact created by such discoveries, innovations, and inventions. The area of nuclear energy technology is an example of one of the most intricate scientific and technological marvels; successfully and commercially practiced globally on a large scale. Energy continuously generated through fusion-driven nuclear transformations in the Sun, is a remarkable testimony of the inexhaustible and sustainable attribute of nuclear energy. While energy resourcefulness of Sun is an example of naturally available nuclear energy to the solar system, the man-crafted might of it, in the form of fission-based nuclear electricity, is serving about 10% of the global electricity requirements with 441 nuclear reactors under operation.[1] Contribution of clean and green nuclear energy to global energy mix is all set to increase with 51 new reactors presently under construction,[2] and numerous additional units under various stages of planning. In addition, with joint efforts of 35 countries, progressive implementation of world's largest fusion-based experimental reactor; the International Thermonuclear Experimental Reactor (ITER) is set to test the might of controlled nuclear fusion for power generation.[3] Genesis of nuclear energy can be attributed to the fundamental discoveries of natural radioactivity (in 1896), nuclear fusion (in 1920), neutron (in 1932) and nuclear fission (in 1938).[4] Phenomenological understanding of these discoveries led to identification of neutron-induced nuclear chain reaction and the remarkable demonstration of sustained fission chain reaction.[5] Unprecedented destructive power of uncontrolled nuclear fission was also witnessed by the world through atomic bombs explosion at Hiroshima and Nagasaki, which led to the end of the World War–II. Amidst these nuclear events that occurred in the first half of the Twentieth century, and with the vision of Dr. Homi Jahangir Bhabha; the father of Indian Nuclear Program, India took a timely decision of indigenously developing nuclear technology for both, national prosperity as well as security.[6] In retro-spection, discoveries like radioactivity, neutron as a sub-atomic particle and neutron-induced fission in heavy elements (like uranium) can be owed to the basic human instinct of understanding material behavior as a function of external perturbation(s). Progressive clarity on material behavior under various conditions catalyzes the advent of applications arising from a system of engineered material(s). In the context of nuclear energy technologies, effect of ionizing radiations ($\alpha$, $\beta$, $\gamma$–rays, neutrons, X–rays, *etc.*) on materials is an additional factor; apart from perturbations such as temperature, pressure, non-ionizing radiations (UV, visible, IR, microwaves, radio waves, *etc.*) and chemical environment that govern their behavior. As a result, qualification criteria for material(s) utilization in nuclear energy systems are much more stringent and good radiation stability is an essential requirement. This is applicable not only to nuclear fuel materials that undergo nuclear fission/fusion to produce energy, but also to all peripheral materials; continuously exposed to ionizing radiations during and after reactor's operational lifetime. These peripheral materials include fuel clad (immediate enclosure of fuel), coolant, coolant flow system (tubes, vessels), moderator (for thermal neutron spectrum reactors), neutron reflectors, neutron absorbers, reactor control instrumentation systems, reactor vessel and even the reactor containment structure. For safe and economically viable operation of nuclear reactors over many decades, design and development of nuclear-grade materials is very important. The term 'nuclear-grade' has distinct qualification criterion for different materials, which essentially depends on their utilization purpose in a given nuclear energy system. For example, 'nuclear-grade uranium (NGU)' is chemically pure enough with respect to strong neutron absorbers (B, Gd, Cd, Co, *etc.*) so that neutron density inside reactor's core is not affected adversely and controlled nuclear chain reaction progresses as desired. Similarly, qualification criterion for 'nuclear-grade heavy water ($D_2O$)' include high isotopic purity ($^2H$ fraction = 99.82 to 99.91 wt. %),[7] so that the ratio of neutron moderation to neutron capture can be maintained for reactivity control in a heavy water moderated/cooled reactor. Additionally, control of both inorganic and organic impurities in coolant/moderator ($H_2O$, $D_2O$) is essential to minimize corrosion in fluid circulation system and avoid undesired products formed by reactions of $\gamma$–ray radiolysis products of water (free radicals, ions, aquated electrons, *etc.*) with organic impurities, respectively. In contrast with NGU and $D_2O$, specifications for 'nuclear-grade boron' depend on the type of nuclear application. For instance, elemental boron or boron trifluoride ($BF_3$) gas, which is used in neutron detection application, must have high isotopic purity for $^{10}B$ (> 90%). On the other hand, boron carbide ($B_4C$) used as a control rod material in fast neutron spectrum reactors requires variable $^{10}B$ isotopic purity for FBTR (> 90%) and PFBR (> 65%).[8] For zirconium-based alloys and/or stainless steel, which are used as fuel-clad and/or structural materials, minimization of impurities (e.g., hafnium in zirconium, cobalt in steel, *etc.*) that have high neutron absorption cross sections and may convert to long-lived radioactive activation products, are essential chemical qualification criteria. For clad materials (Zircaloy, Aluminum-alloys, Steel, *etc.*), apart from neutron interaction properties of clad constituents and impurities therein, nature and extent of thermo-chemical interaction with given fuel forms (metallic, oxide, carbide, molten salts, *etc.*) as well as coolant (water, liquid metal, gas, molten salts, *etc.*) also govern the material selection. From above discussion and given that fission-based nuclear energy technology could be fully industrialized in a span of few decades from the discovery of nuclear fission, it is evident that material issues in respect to nuclear technology have been enormously researched to arrive at the industrial scale maturity. Further, the nuclear fraternity is continuously engaged in development of superior nuclear materials for safer and better performance of nuclear energy systems. For readers interested in study of material aspects in nuclear technology, the subject of 'Nuclear Materials' is so vast to bring out full treatises on a regular interval.[9] The present chapter however confines to a brief overview of various nuclear fuels; materials from which nuclear energy is harnessed.

## 2. Classification of Nuclear Fuels

### 2.1. *Classification based on nuclear process*

Based on the elementary nuclear processes, through which energy is harnessed from nuclear fuels, they are categorized into two groups namely, fission-based nuclear fuels and fusion-based nuclear fuels. Naturally occurring fission-fuel elements include Uranium and Thorium, whereas other fission-fuel elements (Plutonium and minor actinides) are artificially produced in a nuclear reactor. For fusion-based nuclear fuels, although different isotopes of light elements can be used to achieve fusion reaction, the Deuterium-Tritium (D–T) fusion is the most efficient for fusion devices. These two heavier isotopes of hydrogen namely, Deuterium ($^2$H or D) and Tritium ($^3$H or T) are therefore the major fusion-based nuclear fuels.

Naturally occurring Uranium is composed of three isotopes namely, $^{234}$U (< 0.005%; half-life ($t_{1/2}$) = $2.47 \times 10^5$ years), $^{235}$U (~0.72 %, $t_{1/2} = 7.1 \times 10^8$ years) and $^{238}$U (~99.28% $t_{1/2} = 4.51 \times 10^9$ years) while three times more abundant natural Thorium exists exclusively as $^{232}$Th (100%; $t_{1/2}$ ~

$1.4 \times 10^{10}$ years).[10] Among all above isotopes, only $^{235}U$ has high probability of undergoing nuclear fission upon capture of thermal neutrons (fission cross section ($\sigma_{fission-th}$) ~585.1 barn for 0.0253 eV neutrons; 1 barn or $1 b = 10^{-24} cm^2$) and referred as fissile. Other two most abundant isotopes namely, $^{238}U$ and $^{232}Th$ have extremely small thermal neutron fission cross sections ($\sigma_{fission-th}(^{238}U)$ ~16.8 μb; $\sigma_{fission-th}(^{232}Th)$ ~53.7 μb). $^{238}U$ and $^{232}Th$ however capture thermal neutrons ($\sigma_{capture-th}(^{238}U) = 2.683$ b; $\sigma_{capture-th}(^{232}Th) = 7.338$ b) to produce $^{239}U$ and $^{233}Th$, which subsequently undergo two successive beta particle emissions to finally yield long lived $^{239}Pu$ ($\sigma_{fission-th} = 747.4$ b; $t_{1/2} = 2.41 \times 10^4$ years) and $^{233}U$ ($\sigma_{fission-th} = 531.3$ b; $t_{1/2} = 1.6 \times 10^5$ years), respectively. $^{239}Pu$ and $^{233}U$ are two major artificially produced fissile isotopes while their precursors namely, $^{238}U$ and $^{232}Th$, respectively, are called fertile isotopes. Cross sections of nuclear reactions including that of fission typically decrease with increasing neutron energy. Apart from producing two (sometimes three) fission fragments, each fission event is also associated with release of multiple high energy neutrons. These additional neutrons are the main resource to establish controlled fission chain reaction inside an operating nuclear reactor. For reactors operating in thermal energy spectrum, fission neutrons are slowed down by use of moderators that are materials composed of low atomic number elements ($H_2O$, $D_2O$, graphite, *etc.*). Reactors designed to operate with high energy neutrons (fast neutrons) do not use moderator and are called Fast Breeder Reactors (FBRs). By fast neutrons capture in fertile nuclei ($^{238}U$ and $^{232}Th$), FBRs not only produce their own fuel for subsequent fuel loading cycles but also create additional fissile inventory; a concept known as breeding. FBRs are therefore excellent means to enhance the overall utilization of fission-based fuel materials; especially for countries like India where indigenous uranium reserves are limited. With successful and continuing operation of Fast Breeder Test Reactor (FBTR; 40 $MW_{th}$) for more than 36 years and steadily advancing towards the commissioning of indigenous Prototype Fast Breeder Reactor (PFBR; 500 $MW_{el}$), India is close to enter into 2nd stage of its nuclear power program. During prolonged neutron irradiation of Uranium-based fuels, apart from Plutonium; other long-lived higher actinides such as Neptunium (Np), Americium (Am) and Curium (Cm) are also produced in small amounts due to successive neutron capture by

Plutonium. These higher actinides; also known as minor actinides (MAs) or transuranics (TRUs), are also good fissile materials in fast neutron spectrum. If chemically separated from irradiated nuclear fuel (also known as spent nuclear fuel; SNF), these TRUs can be put back preferably into FBRs and fissioned to produce nuclear energy as well as get rid of their long-term safe storage requirement liability for several thousands of years; which otherwise is essential. With closed fuel cycle approach since the formulation of three stage nuclear power program,[6] India has made notable R&D advances in this direction and successfully demonstrated the separation of TRU elements from spent nuclear fuel.[11]

Between two major fusion fuels namely, Deuterium ($^2H$) and Tritium ($^3H$), the former is naturally available along with Hydrogen in tiny amount ($^2H$ fraction = 0.0153%) and is routinely extracted at industrial scale. Concurrent with the evolution of 1$^{st}$ stage of Indigenous nuclear power program; largely based on Pressurized Heavy Water Reactors (PHWRs), India has mastered deuterium extraction technologies and is currently the largest global producer of heavy water ($D_2O$). It is noteworthy to mention that fusion energy that can theoretically be produced by Deuterium present in one litre of water is equivalent to that obtained from combustion of 300 litres of oil.[12] It implies that fusion-energy; when realized industrially for power production, would alone be effectively sufficient to meet all global energy requirements for millions of years and serve as truly sustainable source of energy. Tritium (T or $^3H$); the other major fusion-based fuel, is a beta-emitting radioactive element ($t_{1/2}$ = 12.3 years) that occurs in trace quantities in nature. For nuclear fusion experiments, it is artificially produced by neutron capture in Deuterium containing compounds (c.a., $D_2O$) by $^2D\ (n,\gamma)\ ^3T$ reaction. Since Indian PHWRs use heavy water ($D_2O$) as moderator as well as primary heat transport media, Tritium activity is produced in operating PHWRs. Tritium-containing compounds can be obtained in liquid (c.a., DTO), gas (c.a., DT, $T_2$) as well as solid (c.a., $TiT_x$) phase. Storage of Tritium in solid phase as metal tritides ($TiT_x$, $PdT_y$, $TiCrVT_z$, *etc.*) is safer as well as cost effective than in gas ($T_2$) form.[13] Other than neutron capture by Deuterium, Tritium can also be produced by neutron capture reactions of Lithium such as $^6Li(n_{thermal},\alpha)^3T$ and $^7Li(n_{fast},\alpha n)^3T$. Tritium production from abundant $^7Li$ isotope (natural abundance ~92.6%) is favoured with high energy (fast) neutrons while that from $^{6}Li$ (natural abundance ~7.4%) is facilitated with slower (thermal) neutrons. One of the attractive ways therefore to *in situ* breed Tritium within fusion reactors (also known as Tokamak) is to place Lithium containing materials in the blanket region of Tokamak so that fast neutrons escaping the fusion plasma react with Lithium and produce Tritium.

## 2.2. *Classification based on chemical state of nuclear fuels*

Nuclear fission was discovered through neutron irradiation experiments on uranyl nitrate $\{UO_2(NO_3)_2\}$ crystals.[4] Since then, enormous advancement has been made to develop fission-based energy technologies that utilize nuclear fuels in various chemical states such as (i) metals, alloys and intermetallics, (ii) ceramics such as oxides, carbides, nitrides and hydrides, (iii) liquid halides such as molten fluorides, *etc.* In terms of structural phase uniformity, nuclear fuels can be broadly classified as single-phase and multi-phase fuels. In the former, fuel has uniform crystal structure while the later form is composed of more than one structural/ phase forms. Examples of single-phase fuels include (i) pure uranium metal, U–Mo cubic alloys, Pu-based cubic alloys, *etc.*, which are mainly used as research reactors fuels and special nuclear materials, (ii) cubic fluorite phase oxide ($MO_2$) fuels (M = U, Pu, Th, $U_{1-x}Pu_x$, $U_{1-y}Th_y$, $Pu_{1-z}Th_z$, $U_{1-n}Gd_n$, *etc.*) that are largely used in commercial nuclear power reactors, (iii) intermetallic fuels such as $U_3Si_2$, $UAl_x$ (x = 2, 3, 4), *etc.*, which are also used in research and naval reactors and (iv) homogenous liquid fuels that are composed of one or more solid component but are utilized in single molten phase fuel form (e.g., molten $ThF_4$–$UF_4$, $UF_4$–$ThF_4$–LiF, *etc.*). In case of multi-phase composites fuels comprising of more than one phase, all constituents need not essentially consist of fissile/fertile materials. Dispersion nuclear fuel is one such example, wherein fuel phase is uniformly dispersed in a suitable matrix such that the neutronic behavior is governed by fuel phase while mechanical, thermal, and chemical behavior of fuel system is governed by the dispersion matrix. Uranium metal or $UAl_x$ intermetallics dispersed in Al matrix, U–Mo alloy particles dispersed in Aluminum alloys, $U_3Si_2$ fuel dispersed in Aluminum-based alloys, $UZr_2$ dispersed in Aluminum matrix, *etc.* are a few examples of metallic dispersion fuels wherein matrix phase is neither fissile nor fertile. On the other hand, Uranium metal dispersed (up to ~10 wt.% U) in Thorium matrix, $UO_2/PuO_2$ particles dispersed in depleted Uranium metal, *etc.* are examples of dispersion fuels wherein the matrix usually consists of fertile materials. Another category in multi-phase nuclear fuels is composite nuclear fuels. Unlike dispersion fuels, homogenous and isolated distribution of fuel particles in the second phase (matrix) is not essential for composite fuels. U–Zr binary alloys, U–Pu binary alloys, Th–U binary alloys (U > 10 wt.%), U–Pu–Zr ternary alloys, U–Fs binary alloys (here Fs = fissium; a mixture consisting of 2.4 Mo–1.9 Ru–0.3 Rh–0.2 Pd–0.1 Zr–0.1 Nb (wt.%), where the composition is obtained by electro-refining of irradiated uranium metal[14]) are a few examples of composite metallic nuclear fuels. Carbide fuel (UC, $U_{1-x}Pu_xC$, *etc.*) is an interesting and unique example of composite nuclear fuels wherein body centered cubic sesquicarbide phase(s) ($UC_{1.5}$, $PuC_{1.5}$, $U_{1-x}Pu_xC_{1.5}$) are mostly formed along with the rock salt structured monocarbide phase(s) (UC, PuC, $U_{1-x}Pu_xC$) during carbide fuel fabrication stages. Although both monocarbide and sesquicarbide phases are composed of heavy metal(s) and carbon, from crystal structure point of view, they classify as composites fuels. Salient features, recent advances, and future R&D scope of different types of fission-based nuclear fuels are briefly discussed in the following sections.

## 3. Metallic Nuclear Fuel Materials

Fission-based nuclear fuel development all began with metals. Well-known physico-chemical properties of metals and alloys, ease of alloy formation and fuel fabrication, simple and compact design of metal-fuelled reactor cores and efficient recovery of artificial fissile materials ($^{239}Pu/^{233}U$) produced by neutron irradiated fertile targets ($^{238}U/^{232}Th$) were among the reasons for choice of metallic fuels.[14,15] Experimental Breeder Reactor–I (EBR–I); the first ever reactor to produce electricity for lighting application and demonstrating the breeding concept, was a liquid metal (Na, K, NaK) cooled, metal fuelled reactor.[16] Inherently safe performance of several metal/alloy fuels (U, U–Fs, U–Zr, U–Pu–Zr, *etc.*) up to high burn-ups (~19 at.%; 1 at.% ≈ 10000 MWd/Te) was established in the following EBR–II reactor.[14,17] Experience obtained in other reactors like

Fast Flux Test Facility (FFTF), Fermi nuclear power station, and Dounreay Fast Reactor (DFR) also demonstrated excellent performance of metal fuels, which proved their candidature among the best options for safe and rapid expansion of nuclear power. Beyond 1960s, actinide oxides ($UO_2$, $ThO_2$, $PuO_2$, $U_{1-x}Pu_xO_2$, $Th_{1-x}U_xO_2$, $Th_{1-x}Pu_xO_2$) dominated as the power reactor nuclear fuels; mainly due to their high temperature stability (up to 2800 K) and ease of large-scale production under near-atmospheric environment. Irradiation tests still continued on metallic fuels to validate their high burn-up potential (up to ~20 atom %) and safety features. Inherent advantages and challenges associated with metallic nuclear fuels have been reported by Jain *et al.*[18] Although global interest shifted towards oxide-fuelled commercial light water thermal reactors (BWR, PWR, PHWR), countries like Russia (BN–series reactors), France (Rapsodie, Phoenix, Super-Phoenix), Japan (Monju) and India (FBTR, PFBR) continued to develop fast breeder reactors (FBRs), wherein metallic fuel option provides best performance potential in terms of the shortest doubling time (measure of rate of production of excess fissile material in a breeder reactor) and the largest breeding ratio (rate of production of fissile isotopes divided by rate of destruction of fissile isotopes). Although a large number of metal-fuelled research reactors continue to provide safe and efficient means for transmutation of fertile actinides into fissile ones and radioisotope production, commercial metal fuelled FBRs for electricity production are yet to come. In India, research reactors CIRUS and Dhruva; both fuelled with aluminum-clad natural uranium metal, have served with impeccable safety for more than four decades to steer indigenous development of nuclear technology. Other research reactors namely Apsara, KAMINI and the latest Apsara–U are also fuelled with U–based alloys or intermetallic dispersion fuels and add to the operational experience gained over metallic fuel systems.[19] A global renaissance is around on shifting towards metal fuels again for future FBRs due to their excellent breeding potential, proven safety during normal as well as transient conditions, compact core designs and minor actinide transmutation capability. Metal fuelled, liquid-sodium-cooled FBRs are also being developed as one of the six generation–IV reactor systems.[20] In the second stage, India too plans to establish metal fuelled FBRs in future after a fleet of MOX ($U_{1-x}Pu_xO_2$, $Th_{1-y}Pu_yO_2$) fuelled FBRs are set up over the period of next two decades.[21] To achieve this goal, extensive research activities are presently underway on indigenous development of fuel cycle aspects of various metallic fuels (U–Zr, U–Pu–Zr, Th–U, Th–U–Zr, U–Mo alloys, *etc.*).[18,22-30] Some of the innovative approaches currently being pursued towards metallic fuel development program in India include mechanically bonded Zr–alloy lined alloys, Uranium metal particles uniformly dispersed in metallic thorium matrix,[22] T91 clad U–based metallic fuels,[31] *etc.*

Travelling Wave Reactor (TWR) is an emerging concept in metal fuelled FBRs, which works on *in situ* breeding of subcritical fuel (depleted uranium, natural uranium, LEU) into a new critical fuel so that a breed and burn wave propagates in the reactor. In TRWs, slightly higher enrichment in the starting reactor core can lead to once through utilisation of metallic fuels up to very high burn ups (~50 atom %), which has never been achieved till date.[32] Since metallic fuelled FBRs are one of the safest and most efficient options for rapid expansion of nuclear energy capacity with maximum possible heavy metal (U, Pu, Th) utilization potential, greater advancements in fuel development are expected.

## 4. Oxide Nuclear Fuel Materials

A majority of commercial nuclear power reactors operational worldwide use ceramic oxide nuclear fuels. In India too, $UO_2$ and its mixed oxides (MOX) with Plutonium or Thorium are utilized in power reactors, wherein the extent of $^{235}U$ enrichment is governed by reactor type. In the case of light water reactors (LWRs), fuel for boiling water reactors (BWR; Tarapur Atomic Power Station; TAPS–1&2), is ~2–3% $^{235}U$ enriched $UO_2$. For Pressurized Water Reactors (Kudankulam Nuclear Power Plant; KKNPP–1 & 2) however, enrichment is relatively higher (~3–5% $^{235}U$) in dioxide form of fuel. For all 18 indigenous Pressurized Heavy Water Reactors (PHWRs), $UO_2$ fuel comprising of natural Uranium (~0.7% $^{235}U$) is used. Studies have shown that if Uranium enrichment in PHWR fuel can be increased up to ~1.0 to 1.1% (also known as Slightly Enriched Uranium; SEU), average discharge burn up of SEU–$UO_2$ PHWR fuel can be enhanced from ~7000 MWD/Te (for natural $UO_2$) to in excess of 15000 MWD/Te.[33] Commensurate with India's vision to utilize its abundant indigenous thorium reserves for nuclear power production, Advanced Heavy Water Reactor(s) (AHWR and AHWR–LEU) has been designed. AHWR is one among the safest thermal nuclear reactor designs globally with several passive safety features and is designed to utilize $(Th_{1-x}U_x)O_2$ and/or $(Th_{1-y}Pu_y)O_2$ mixed oxide (MOX) fuels. While global understanding on bulk monolithic $UO_2$ fuels has reached great maturity, India is leading the development of advanced thorium-based oxide nuclear fuels. As a result of over two decades of extensive research, deeper understanding on adoption of thoria-based fuels in different reactor systems has been obtained.[34,35] Numerous irradiation campaigns on $ThO_2$ fuel have also been completed both in research reactors as well as Indian PHWRs followed by chemical reprocessing of spent $ThO_2$ fuel to obtain $^{233}U$, which has been utilized in $^{233}U$–fuelled KAMINI research reactor. Further development of oxide nuclear fuels focuses on (i) imparting enhanced thermo-chemical stability under transient conditions and (ii) developing bulk nanocrystalline oxide fuels that mimic high burn-up RIM structure of oxide fuels and show superior thermo-mechanical properties to impart improved fission gas retention and better radiation-tolerance as compared to conventional micron-sized bulk oxide fuels.[36]

## 5. Carbide Nuclear Fuel Materials

Actinide carbides; particularly Uranium-Plutonium mixed carbides are advanced fuels for both, liquid-metal-cooled FBRs and gas-cooled fast reactors. Both U and Pu form multiple carbide phases such as monocarbides (UC, PuC), sesquicarbides ($U_2C_3$, $Pu_2C_3$) and dicarbides ($UC_2$, $PuC_2$). Individual carbides dissolve well into each other to form wide range of U–Pu carbide solid-solutions. Monocarbide form is preferred fuel choice for advantages like higher heavy metal fraction, better thermo-physical properties, low carbon potential, *etc.* Other advantages offered by carbide fuels include high-thermal conductivity, high density, high temperature stable isotropic cubic structure (mono and sesquicarbide), *etc.* These advantages benefit in terms of superior breeding performance and high specific power operation. Although carbide fuel development had been globally initiated as early as in the 1960s, its utilization as a driver fuel in a reactor has not been implemented anywhere except in India.

Efforts undertaken by other countries like Japan, France, Russia, USA, and Germany have mostly been limited to small-scale test irradiation of carbide fuels and their post-irradiation studies. At present, India is the only country in the world to operate mixed carbide fuelled fast reactor (Fast Breeder Test Reactor; FBTR). FBTR (40 $MW_{th}$) is fuelled by indigenously developed Plutonium-rich U–Pu mixed carbide fuels (Mark–I: $Pu_{0.7}U_{0.3}C$ and Mark-II: $Pu_{0.55}U_{0.45}C$) with SS 316 clad, and has for the first time generated electricity from a carbide fuelled fast neutron spectrum reactor; thereby paving the way for our entry into the 2$^{nd}$ stage of Indian nuclear power program. High burn-up potential of U–Pu mixed carbide fuels (up to 16.5 atom %) and their safe behavior under postulated transient conditions has been amply demonstrated in FBTR.[37] Mixed carbide fuel for FBTR is fabricated through indigenously developed process flow sheet consisting of a sequence of vacuum carbothermic reduction of ($UO_2$ + $PuO_2$ + graphite) powder to form carbide clinkers → powder processing (crushing, milling and pelletization) of clinkers → sintering under controlled environment to obtain dense mixed carbide pellets → pellet inspection → encapsulation of pellets in SS 316 clad tube → pin welding → decontamination → Quality check and assurance (QC & QA). Challenges associated with both, carbide fuel fabrication technology and spent carbide fuel reprocessing, have been successfully overcome, which is a hallmark of Indian fast reactor development program that boosts the confidence to pursue it further for commercial power production. Although FBTR has provided extensive operational experience on carbide fuel cycle, the upcoming Prototype Fast Breeder Reactor (PFBR; 500 $MW_{el}$) employs U–Pu mixed oxides (MOX) driver fuel. Some of the challenges associated with both front and back end of carbide fuel cycle, which could affect the fuel cycle economics for a commercial FBR have been among the reasons for shifting to MOX fuel for the PFBR. A few of the material issues with carbide fuels are as follows: Oxygen and nitrogen usually remain as dissolved impurities in U–Pu monocarbides. Oxygen comes from starting actinide oxides while nitrogen is picked up during carburization in ($N_2$-$H_2$) atmosphere. These impurities adversely affect fuel performance by raising both, carbon potential of fuel and partial pressure of carbon monoxide, which in turn accelerate clad-carburization. Sesquicarbides of Uranium and Plutonium are also formed (up to 20 vol. %) during carbothermic reduction of oxides, which render multi-phase attribute to as-prepared fuel pellets and adversely affect fuel's physico-chemical properties. Lower solidus temperature, lower thermal conductivity, and pyrophoric nature of Pu-rich U–Pu carbides are also challenging while utilizing carbide fuels in a large-scale power generation fast reactor. During dissolution of Pu-rich spent carbide fuels in nitric acid medium, soluble organics are formed in dissolver solution, which pose challenges in separation of plutonium by solvent extraction and lead to its diversion into the raffinate waste stream. Also, Pu is stripped off from organic phase and stable emulsions too get formed. Use of electro-oxidative dissolution technique has been reported to eliminate soluble organics as well as oxalic acid from the dissolver solution, thereby facilitating carbide fuel reprocessing.[38] The operational experience being gained on mixed carbide fuel cycle in FBTR for many decades is immensely helpful in designing fuels for compact fast reactors in future. Database generated on both in-pile and out-of-pile fuel behavior of mixed carbide fuels would be invaluable input for design of advanced fuels such as actinide carbides ($AnC_x$) uniformly distributed in thermo-chemically stable, high thermal conductivity inert carbide matrix such as SiC.

## 6. Nitride Nuclear Fuel Materials

Similar to carbides, potential of actinide nitrides as nuclear fuels has also been investigated since the beginning of nuclear technology development. Compared to corresponding oxides, nitrides show superior thermal conductivity, higher density, comparable high temperature stability (under nitrogen pressure) and isotropic crystal structure(s). Important properties of nitride fuels have been reported by Bauer.[39] For their excellent high temperature physico-chemical properties, uranium nitrides are preferred fuel options for space power reactor applications.[40] Generally, these fuels are preferred as cubic structured mononitrides (UN, PuN, $U_{1-x}Pu_xN$, MN, $U_{1-x-y}Pu_xM_yN$; MA = Np, Am, Cm) and sesquinitrides ($U_2N_3$, $Pu_2N_3$, $(U_{1-x}Pu_x)_2N_3$). Nitride fuels can be fabricated in the form of steel-clad pins either filled with sintered pellets of actinide nitrides or those containing actinide nitrides in suitable inert matrix (c.a., $AnN_x$ in ZrN matrix). Nitride fuels are also suitable for fast reactors[41] and/or accelerator driven sub-critical systems (ADSS) as they support hard neutron spectrum.[42,43] In LWR conditions, U/Pu–nitrides react with water especially at high temperatures. Use of nitride fuels in LWRs is therefore proposed in the form of composites made in water-compatible secondary matrices ($U_3Si_5$, $U_3Si_2$, $U^{11}B_4$, $Y_2O_3$–stabilized $ZrO_2$, *etc.*) so that U/Pu-nitride phase remains shielded from water.[44] Inert nitride matrix fuels ($Pu_{1-y}Zr_yN$, $ZrN_x$–$AnN_x$, PuN–TiN, *etc.*) have been envisaged for incineration of plutonium and/or long-lived minor actinides present in stock-piled spent nuclear fuels. Actinide nitrides (UN, PuN, $U_{1-x}Pu_xN$) can be prepared by carbothermic nitridation of oxides in a narrow reaction temperature range, wherein exclusion of carbon and oxygen impurities is a yet to be resolved challenge. Significant deterioration in high temperature heat conductivity (up to 10–13%; over 1000–1500 K) of sintered UN and (U,Pu)N consisting ~1 wt.% oxygen impurity has been reported.[45] Results reported on fission gas release, swelling behavior, constituent redistribution and clad deformation for fast neutron irradiated Helium-bonded pins comprising of (U,Pu)N and uranium-free nitride fuels ((Pu,Zr)N, PuN+TiN) have indicated their suitability in fast reactors.[43,46] For special reactor systems, where burn-up optimization is crucial, fabrication of nitride fuels in specific shapes by direct coagulation casting or freeze drying is preferred over powder compaction route. Nitride fuels are also attractive for lead-bismuth eutectic (LBE) cooled long-life small modular reactors (SMRs), which are being designed for remote/isolated locations with a target refuelling (or fuel shuffling) duration over a decade.[47] While irradiation experience for nitride fuels is the least among metals, oxides, carbides and nitrides, there is a renewed interest in nitride fuels; mainly for their utilization in minor actinide transmutation application. With rapidly increasing use of computational chemistry and simulation studies tools for design and understanding of innovative nuclear fuels, development of inert matrix nitride fuels is expected to advance.

## 7. TRISO-Coated Particle Nuclear Fuels

TRISO-coated particle fuel concept emerged from fuel engineering developed for high temperatures gas cooled nuclear reactors (HTGRs). This fuel type is being used in most of the HTGRs and is also a candidate fuel for other high temperature reactors. For example, both The Compact High temperature Reactor (CHTR) and the Indian High Temperature Reactor (IHTR) designs, which are being developed in India for utilization of nuclear energy for large-scale hydrogen ($H_2$) production; apart from generation of nuclear electricity, are based on TRISO coated particle fuels.[48] The term 'TRISO' here represents 'Tri-isotropic' coatings that are provided over fuel particles. Fuel is taken as tiny (typically $\leq$ 500 μm) near-spherical particles and successively coated with (i) inner low density pyrolytic carbon (PyC) (ii) second layer of high density inner PyC, (iii) silicon carbide (SiC) interlayer and (iv) outer high density PyC. High thermal conductivity, high temperature stability and high heat capacity of SiC and graphitic PyC phase impart high heat retention capability and high temperature stability to TRISO coating. Depending upon reactor type, fuel particle (also known as kernel) could be in different forms such as oxides, carbides, nitrides, oxycarbides and oxynitrides of actinides (U, Th, Pu). Near-spherical micro fuel particles (oxides) are mostly prepared by sol-gel technique mostly following the internal gelation route.[49] Sol-gel synthesized particles (beads) are sintered to obtain high density bare fuel kernel. In order to make carbide, nitride and their oxy-derivatives fuel particles, carbon is incorporated in the particles during the sol-gel process. Subsequent high temperature treatment under controlled environment ($H_2$–$N_2$, $H_2$–Ar, *etc.*) enables carbothermic reduction or carbothermic nitridation to produce carbide/oxycarbide or nitride/oxy-nitride fuel particle, respectively. TRISO coatings over bare sintered kernel are successively provided by CVD route. A complete TRISO particle therefore behaves like a tiny reactor pressure vessel, wherein successive heat conducting and radiation resistant coatings ($PyC_{low\text{-}density}$–$PyC_{dense}$–SiC–$PyC_{dense}$; total coating thickness < 500 μm) allow for efficient removal of fission-induced heat from the fuel kernel but act as firm barrier for release of fission fragments, fission gases, alpha particles, *etc.* Coating technology over the fuel particle is therefore central to performance of TRISO fuels and safety of high temperature reactor systems. These TRISO particles are then suitably packed in fuel channels of given high temperature reactor. Advantages of TRISO-coated particle fuels include exceptional high temperature as well as radiation stability due to which fuel can be irradiated to very high burn ups in excess of 50 atom %. Fuel meltdown is virtually impossible in case of TRISO coated particle fuels as they can withstand very high temperatures well beyond the threshold of conventional bulk monolithic nuclear fuels. An example of further advancement in TRISO fuels evolution includes outermost ZrC coating of TRISO particles. ZrC layer acts as an oxygen getter and further strengthens fuel against failure mediated by fuel's oxygen potential that rises rapidly at higher burn-ups. Oxygen getter action of ZrC coating also delays over pressurization of particle and prevents kernel migration (known as Amoeba effect).[50] Instead of oxides, TRISO fuels comprising of UN kernels offer significantly higher heavy metal density and are proposed to be optimal for advanced reactor systems.[51]

## 8. Inert Matrix Nuclear Fuels

Commercial operation of hundreds of nuclear energy reactors over the last six decades has resulted into large quantities of safely stock-piled spent nuclear fuels (for countries adopting once through open fuel cycle) and reprocessed inventories of Plutonium and transuranic elements like Np, Am and Cm (for countries adopting closed fuel cycle). As such, these materials require safely engineered storage for several thousands of years; an approach that is rapidly being recognized as an unnecessary liability as well as proliferation threat. A largely accepted solution to stockpiled Pu and TRU inventory is transmutation where it is reused in reactors in a suitable fuel form to produces energy in an economically viable manner. Also, the spent fuel obtained after transmutation mostly contains short-lived isotopes, which pose neither proliferation concerns nor long-term storage requirements. Inert matrix fuels (IMFs) meet both these attributes and are among the potential candidates for transmutation fuels. IMF approach is based on replacement of monolithic bulk actinide fuel matrix ($UO_2$, $PuO_2$, U–Pu MOX, *etc.*) by suitable neutron transparent matrix; also known as Inert Matrix.[52,53] The inert matrix acts as host to actinides (Pu and TRU) such that the IMF is either a solid solution of fissile actinides in the matrix or a composite with fissile phase uniformly distributed in the matrix. Key attribute of inert matrix is its inability to produce any fissile isotope during neutron irradiation. Two approaches are being pursued towards development of IMFs. One approach is fission of Pu and TRUs in an inert matrix using a once through fuel cycle followed by engineered disposal of actinide-depleted spent fuel. The second approach considers closed fuel cycle starting with uranium-free fuel in such a way that while energy production is maximized by consumption of Pu and TRUs through multi-cycle reprocessing, additional generation of Pu/TRU liable for storage is minimized. Various IMF hosts have recently been reviewed by Nandi.[54] Potential oxide-based IMF hosts for development of solid-solution IMFs include cubic stabilized zirconia, $Al_2O_3$, MgO, $CeO_2$, pyrochlores, spinels ($MgAl_2O_4$), $Y_3Al_5O_{12}$, *etc.* Non-oxide inert matrices that are being explored for IMF development include nitrides (AlN, TiN, ZrN), carbides (SiC, TiC, ZrC) and even metals (Mg, Si, Zr, Nb, *etc.*). Multi-phase composite IMFs can be developed in the form of ceramic fuel dispersed in ceramic matrix (CERCER), ceramic fuel in metallic matrix (CERMET) and metallic fuel dispersed in metallic matrix (METMET). With continued evolution of IMF concept and progressive irradiation campaigns in research/power reactors, commercial implementation of inert matrix fuels for safe, economic, and proliferation-resistant utilization of stock-piled fissile inventories is inevitable.

## 9. Accident Tolerant Nuclear Fuels

Since the inception of civil nuclear energy technology, and over a cumulative period of 18000 reactor years of operation,[55] the world has witnessed exceptionally safe performance of nuclear power reactors except on three occasions where major accidents took place. These accidents namely, Three Mile Island (TMI, USA) in 1979, Chernobyl (USSR; now Ukraine) in 1986 and Fukushima (Japan) in 2011 highlighted the need to further strengthen nuclear safety and accelerated scientific efforts to further advance materials' development for safer performance of future nuclear reactors during both, postulated design basis (DB) as well as beyond design basis accident (BDB) scenarios. Concept of accident tolerant fuels (ATF) has its origin in this pursuit. The objective towards ATF development is to avoid/delay the physico-chemical consequences that a nuclear fuel can undergo with time from the onset of severe accident, such as loss of coolant accident (LOCA) or reactivity-initiated accidents (RIA). Severe physico-chemical interactions inside the reactor core may include clad hydriding/oxidation, Fuel-clad chemical interaction (FCCI) leading to clad failure by ballooning & burst process (melting/rupture/embrittlement/fragmentation), fuel interaction with steam/air, *etc.* Rates of these interactions bring to the consequent events like fuel pin collapse, fuel pellets relocation, pressure (hydrogen/steam) build-up in reactor vessel/containment, temperature surge and fuel-melting, *etc.* Two possible approaches to avoid the material degradation consequences post onset of sever accident are (i) incorporation of passive core cooling mechanisms and (ii) use of accident tolerant fuels. ATF development approach is fundamentally based on material/process modifications that can delay or disrupt the chain of core degradation events.[56] Importantly, any ATF strategy must not adversely affect the fuel performance during the normal operation conditions, as that would affect the overall economics of given nuclear energy system. Three general approaches are being presently explored towards ATF development. These include (i) improvement in elevated temperature oxidation resistance and strength of Zircaloy-based clad materials, (ii) adoption of non-zirconium alternative clad material with equal or superior oxidation resistance and high temperature strength and (iii) use of alternative fuels (other than monolithic $UO_2$) that have higher fission product retention capacity, superior thermal/mechanical properties and improved fuel-clad interaction resistance. One approach towards alternative fuel forms includes the design of high-density fuels that have higher thermal conductivity and lower heat capacity. The former facilitates heat extraction during transients while the latter minimizes rapid build-up of stored energy. Uniform dispersion of bulk oxide fuel particles in a thermo-chemically stable high thermal conductivity second phase (SiC) could be useful to develop such fuels.[57] Development of micro-encapsulated ceramics is another approach towards alternative fuel forms wherein oxide fuel particles are embedded within an inert oxide matrix.[58] Extensive research is required to address the issues related to reactor physics suitability, fabrication technology and cost, fuel enrichment optimization, behavior under normal irradiation as well as off-normal (LOCA, RIA, and BDB) scenarios.

## 10. Conclusions

In conclusion, a brief overview is presented on different types of nuclear fuels, which form the main class of 'nuclear energy materials'. It is aimed that readers familiar with the subject of nuclear energy technology would find this chapter as a starting point to explore the area of nuclear fuels from the perspective of their material forms and their properties. Cited references would help in furthering their efforts to study these materials and contribute to both deeper understanding of discussed fuel systems and evolution of futuristic advanced fuel materials.

## References

1. Nuclear Power in the World Today, Web-link: https://world-nuclear.org/information-library/current-and-future-generation/nuclear-power-in-the-world-today.aspx.
2. IAEA PRIS database (April 2022) Web-link: https://pris.iaea.org/PRIS/home.aspx.
3. International Thermonuclear Experimental Reactor (ITER), Web-link: https://www.iter.org/.
4. O. Hahn and F. Strassmann, Über den Nachweis und das Verhalten der bei der Bestrahlung des Urans mittels Neutronen entstehenden Erdalkalimetalle, *Die Naturwissenschaften* 27 (1939) 11–15.
5. Chicago pile commemorated, *Chem. Engg. News* 24(24) (1946) 3353.
6. S. Banerjee and H.P. Gupta, The Evolution of the Indian Nuclear Power Programme, *Prog. Nucl. Energy* 101 (2017) 4–18.
7. Nuclear Applications of Heavy Water, Heavy Water Board, India, Web-link: https://www.hwb.gov.in/nuclear-applications.
8. Boron in Nuclear Applications, Heavy Water Board, India, Web-link: https://www.hwb.gov.in/boron-0.
9. R. Konings and R. Stoller (eds.), Comprehensive Nuclear Materials, 2nd Edition, Elsevier (2020).
10. Depleted Uranium, International Atomic Energy Agency, Web-link: https://www.iaea.org/topics/spent-fuel-management/depleted-uranium.
11. P.K. Wattal, Back End of Indian Nuclear Fuel Cycle-A Road to Sustainability, *Prog. Nucl. Energy* 101 (2017) 133-145.
12. Nuclear Fusion Basics, International Atomic Energy Agency, Web-link: https://www.iaea.org/newscenter/news/nuclear-fusion-basics.
13. L.K. Heung, Titanium for Long Term Tritium Storage, *US DoE Report: WSRC–TR–94-0596* (1994) 1–14.

14. G.L. Hofman, L.C. Walters and T.H. Bauer, Metallic Fast Reactor Fuels, *Prog. Nucl. Energy* 31(1–2) (1997) 83–110.
15. W.J. Carmack, D.L. Porter, Y.I. Chang, S.L. Hayes, M.K. Meyer, D.E. Burkes, C.B. Lee, T. Mizuno, F. Delage and J. Somers, Metallic Fuels for Advanced Reactors, *J. Nucl. Mater.* 392 (2) (2009) 139–150.
16. Idaho National Laboratory, Experimental breeder reactor–I (EBR–I) (Web-link: https://inl.gov/experimental-breeder-reactor-i/).
17. L.C. Walters, Thirty Years of Fuels and Materials Information from EBR-II, *J. Nucl. Mater.* 270(1) (1999) 39–48.
18. Dheeraj Jain, Preparation, Characterization and Thermophysical Investigations on Thorium-based Nuclear Fuels and Related Systems, PhD Thesis, Homi Bhabha National Institute (2020) (Web-link: http://www.hbni.ac.in/thesis.html).
19. Research Reactors in BARC, (Web-link: https://www.barc.gov.in/reactor/).
20. G. Locatelli, M. Mancini and N. Todeschini, Generation IV Nuclear Reactors: Current Status and Future Prospects, *Energ. Policy* 61 (2013) 1503–1520.
21. S.C. Chetal, P. Chellapandi, P. Puthiyavinayagam, S. Raghupathy, V. Balasubramaniyan, P. Selvaraj, P. Mohanakrishnan and B. Raj, Current Status of Fast Reactors and Future Plans in India, *Energy Proceed.* 7 (2011) 64–73.
22. U. Kumar, S. Kaity, A. Arya and A. Banerjee, Characterization and Thermal Expansion of Th–10Zr Alloy, *J. Radioanal. Nucl. Chem.* 331(4) (2022) 1619–1627.
23. Dheeraj Jain, V. Sudarsan and A.K. Tyagi, Thorium-based Metallic Alloys as Nuclear Fuels: Present Status, Potential Advantages and Challenges, *SMC Bulletin* 4 (1) (2013) 27–40 (Web-link: https://www.smcindia.org/smc-bulletin.php).
24. S. Das, R. Kumar, S.B. Roy and A.K. Suri, Development Study on Metallic Thorium and Uranium-thorium Alloy, *BARC Newsletter-Founder's Day Special issue* (2011) 72–76.
25. S. Das, R. Kumar, S. Kaity, S. Neogy, K.N. Hareendran, S.B. Roy, G.K. Dey, B.S.S. Daniel and G.P. Chaudhari, Characterization of Microstructural, Mechanical and Thermal Properties and Ageing Study of Th–3 wt.% U Alloy, *Nucl. Eng. Des.* 282 (2015) 116–125.
26. S. Das, S. Kaity, R. Kumar, J. Banerjee, S.B. Roy, G.P. Chaudhari and B.S.S. Daniel, Characterization of Microstructural, Mechanical and Thermophysical Properties of Th–52 U Alloy, *J. Nucl. Mater.* 480 (2016) 223–234.
27. S. Das, S.B. Roy, G.P. Chaudhari and B.S.S. Daniel, Microstructural Evolution of As-cast Th–U Alloys, *Prog. Nucl. Energy* 88 (2016) 285–296.
28. R. Kumar, S. Das, S.B. Roy, R. Agarwal, J. Banerjee and S.K. Satpati, Effect of Mo Addition on the Microstructural Evolution and γ–U Stability in Th–U Alloys, *J. Nucl. Mater.* 539 (2020) 152317.
29. B. Sreenivasulu, S. Rajeswari, A. Suresh and N. Sivaraman, Demonstration of Aqueous Reprocessing of U–Zr and U–Pu–Zr Metallic Alloy Fuels Using an Ejector Mixer-settler with Tri-n-butyl Phosphate (TBP) as the Extractant, *Solvent Extraction Ion Exchange* 39(3) (2021) 271–289.

30. V.G. Mishra, S.K. Sali, D.J. Shah, U.K. Thakur, R.M. Sawant and B.S. Tomar, Studies on U-Zr and U-Pu-Zr Alloys for Determination of Cl and F Using Pyrohydrolysis, *Radiochim. Acta* 102(10) (2014) 895-901.
31. S. Kaity, T.R.G. Kutty, R. Agarwal, A. Laik and A. Kumar, Chemical Compatibility of Uranium-based Metallic Fuels with T91 Cladding, *Nucl. Engg. Design* 250 (2012) 267–276.
32. K. D. Weaver, J. Gilleland, C. Whitmer and G. Zimmerman, High Burn-up Fuels for Fast Reactors: Past Experience and Novel Applications, *International Congress on Advances in Nuclear Power Plants* (ICAPP 2009) (2009) 795.
33. H.P. Gupta, S.V.G. Menon and S. Banerjee, Advanced Fuel Cycles for Use in PHWRs, *J. Nucl. Mater.* 383 (2008) 54–62.
34. D. Das and S.R. Bharadwaj (eds.), Thoria-based Nuclear Fuels: Thermophysical and Thermodynamic Properties, Fabrication, Reprocessing, and Waste Management, Springer (2013) (DOI: 10.1007/978–1–4471–5589–8).
35. A.K. Nayak and B.R. Sehgal (eds.), Thorium-Energy for the Future, Springer (2019) (DOI: 10.1007/978–981–13–2658–5).
36. J. Spino, H.S. Cruz, R. Jovani-Abril, R. Birtcher and C. Ferrero, Bulk-nanocrystalline Oxide Nuclear Fuels — An Innovative Material Option for Increasing Fission Gas Retention, Plasticity and Radiation-tolerance, *J. Nucl. Mater.* 422 (1–3) (2012) 27–44.
37. Journey of FBTR - Reaching Further Heights: 50 Years Journey of FBTR, (2022) (Web-link: http://www.igcar.gov.in/document/fbtr_50_2022.pdf)
38. A. Palamalai, S.K. Rajan, A. Chinnusamy, M. Sampath, P.K. Varghese, T.N. Ravi, V.R. Raman and G.R. Balasubramanian, Development of an Electro-oxidative Dissolution Technique for Fast Reactor Carbide Fuels, *Radiochim. Acta* 55(1) (1991) 29–36.
39. A.A. Bauer, Nitride Fuels: Properties and Potentials, *Reactor Technol.* 15(2) (1972) 87–104.
40. R.B. Matthews, K.M. Chidester, C.W. Hoth, R.E. Mason and R.L. Petty, Fabrication and Testing of Uranium Nitride Fuel for Space Power Reactors, *J. Nucl. Mater.* 151(3) (1988) 334–344.
41. B. D. Rogozkin, N. M. Stepennova and A. A. Proshkin, Mononitride Fuel for Fast Reactors, *Atomic Energy* 95 (2003) 624–636.
42. M. Streit and F. Ingold, Nitrides as a Nuclear Fuel Option, *J. Eur. Ceram. Soc.* 25 (12) (2005) 2687–2692.
43. M. Akabori, M. Takano, Y. Arai, K. Nakajima, A. Itoh and T. Ogawa, Fabrication of Nitride Fuels for Transmutation of Minor Actinides *J. Nucl. Mater.* 320(1–2) (2003) 18–24.
44. N.R. Brown, A. Aronson, M. Todosow, R. Brito and K.J. McClellan, Neutronic Performance of Uranium Nitride Composite Fuels in a PWR, *Nucl. Engg. Design* 275 (2014) 393–407.

45. Y. Arai, M. Morihira and T. Ohmichi, The Effect of Oxygen Impurity on the Characteristics of Uranium and Uranium-plutonium Mixed Nitride Fuels, *J. Nucl. Mater.* 202(1–2) (1993) 70–78.
46. K. Tanaka, K. Maeda, K. Katsuyama, M. Inoue, T. Iwai and Y. Arai, Fission Gas Release and Swelling in Uranium–plutonium Mixed Nitride Fuels, *J. Nucl. Mater.* 327(2–3) (2004) 77–87.
47. H. Sekimoto and Z. Su'ud, Design Study of Lead- and Lead-bismuth-cooled Small Long-life Nuclear Power Reactors Using Metallic and Nitride Fuel, *Nucl. Tech.* 109 (1995) 307–313.
48. I.V. Dulera, R.K. Sinha, A. Rama Rao and R.J. Patel, High Temperature Reactor Technology Development in India, *Prog. Nucl. Energy* 101 (2017) 82–99.
49. V.N. Vaidya, Sol-gel Process for Ceramic Nuclear Fuels, *Trans. Indian Ceram. Soc.* 63(3) (2004) 163–167.
50. I.E. Porter, T.W. Knight, M.C. Dulude, E. Roberts and J. Hobbs, Design and Fabrication of an Advanced TRISO Fuel with ZrC Coating, *Nucl. Engg. Design* 259 (2013) 180–186.
51. K.A.Terrani, B.C. Jolly and J.M. Harp, Uranium Nitride Tristructural-isotropic Fuel Particle, *J. Nucl. Mater.* 531 (2020) 152034.
52. C. Degueldre and J. Paratte, Concepts for an Inert Matrix Fuel, an Overview, *J. Nucl. Mater.* 274 (1999) 1–6.
53. C. Degueldre and T. Yamashita, Inert Matrix Fuel Strategies in the Nuclear Fuel Cycle: The Status of the Initiative Efforts at the 8th Inert Matrix Fuel Workshop, *J. Nucl. Mater.* 319 (2003) 1–5.
54. C. Nandi, Oxide Materials for Hosting Minor Actinides: Investigation of Phase Relations and Thermophysical Properties, PhD Thesis, Homi Bhabha National Institute (2018) (Web-link:http://www.hbni.ac.in/thesis.html).
55. Nuclear Power in the World Today (July 2022), Web-link: https://world-nuclear.org/information-library/current-and-future-generation/nuclear-power-in-the-world-today.aspx.
56. S.J. Zinkle, K.A. Terrani, J.C. Gehin, L.J. Ott and L.L. Snead, Accident Tolerant Fuels for LWRs: A Perspective, *J. Nucl. Mater.* 448 (2014) 374–379.
57. S. Yeo, E. Mckenna, R. Baney, G. Subhash and J. Tulenko, Enhanced Thermal Conductivity of Uranium Dioxide–silicon Carbide Composite Fuel Pellets Prepared by Spark Plasma Sintering (SPS), *J. Nucl. Mater.* 433 (2013) 66–73.
58. K.A. Terrani, L.L. Snead and J.C. Gehin, Microencapsulated Fuel Technology for Commercial Light Water and Advanced Reactor Application, *J. Nucl. Mater.* 427 (2012) 209–224.

https://doi.org/10.1142/9789811270956_0012

# Chapter 12

## Emerging Supramolecular Strategies towards Energy Materials

Saikat Ghosh, Shagufi N. Ansari and Subi J. George*

*New Chemistry Unit and School of Advanced Materials,*
*Jawaharlal Nehru Centre for Advanced Scientific Research (JNCASR),*
*Jakkur, Bangalore 560064, India*
**george@jncasr.ac.in*

In plants, the light gets efficiently converted from solar energy to form chemical energy through the assembly and hierarchical organization of the light-harvesting protein machinery present in the green leaves. In artificial systems, noncovalent interactions-driven self-assembly provides an attractive strategy to mimic this functional correlation among molecules for the optimization of photocatalysis, energy storage and photovoltaics. Weak intermolecular forces are used to direct the organization of the supramolecular assemblies to construct complex architectures that act as efficient sensitizers by generating long-lived excitons with excellent charge separations. The emphasis of this chapter is on recent developments and progress in supramolecular energy materials, and architectures exhibiting effective energy conversion in sight of new applications in photocatalysis.

## 1. Introduction

Natural evolution has always been the intellectual source for the development of modern chemistry. The first man-made organic molecule was urea, prepared unexpectedly by Wöhler in the eighteenth century.[1] Till the late 1960s, chemists mainly dealt with the covalent and ionic interactions that formed strong bonds within molecules. In the last decades of the 20th century, a treasure of new possibilities has been accomplished in the supramolecular chemistry field, where relatively weak and non-covalent interactions play an important role in holding the functional molecular structures and materials.[2] Encouraged by the beauty and molecular architectures found in nature, the construction of amazingly complex supramolecular structures became possible by applying the concept of hierarchical self-assembly. These self-assembling materials are often referred to as supramolecular soft materials. A plethora of functional molecules and materials have been reported, which include self-assembled supramolecular objects,[3] specific host-guest complexes through molecular recognition,[4-6] hydrated gel scaffolds,[7-9] organogels,[10-12] crystalline assemblies such as metal-organic frameworks (MOFs), and covalent organic frameworks.[13-15] Due to their dynamic nature, these self-assembled structures are easily processable, self-healable, and can change and adopt various functionalities by responding to their environment. Most importantly, various functionalities can be harnessed *via* multicomponent assembly by using the properties of all components simultaneously. Over the last two decades, supramolecular nanostructures are being extensively exploited to make various functional energy materials such as organic photovoltaic devices (Figures 1b and 1c),[16-18] ferroelectric materials,[19] energy storage devices,[20-25] magneto-resistive devices[26], and most recently to use them as photocatalytic energy conversion systems.[27-29]

The real engrossment in the field of the functional supramolecular systems came with the synthesis of multicomponent supramolecular polymers.[30-38] The multicomponent systems flourished in the last decade because of the beauty of their ability to compile the properties of each component in a single system. Manners and coworkers, in their pioneering work, have successfully synthesized supramolecular block co-polymer using a crystallization-driven self-assembly approach of controlled length and architecture.[31] Aida and coworkers came up with a seeded growth approach to synthesize the supramolecular linear p-n heterostructures and had extended their heterostructures to show efficient electronic communications like exciton and charge transport through the junction (Figure 1a).[33] Later, Würthner and Meijer groups came up with the idea to synthesize supramolecular block co-polymer with precise control over the kinetics of the self-assembly process in a living supramolecular polymerization approach[34] and utilizing the thermodynamic parameters to get blocky heterostructures by a fine-tuning over hydrogen bonding, and mismatch penalty between monomer units, respectively.[35]

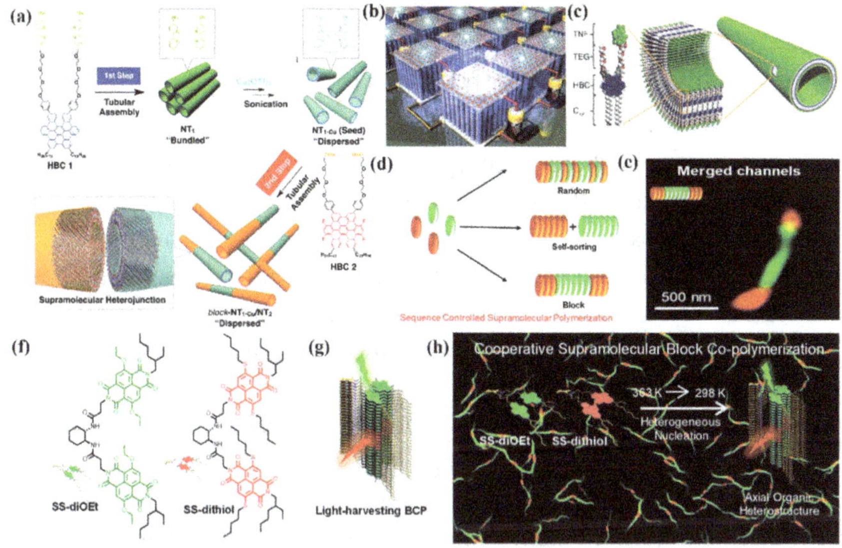


Figure 1. (a) Structural representation and schematic display of the synthesis of axial p-n heterojunction. (b) Demonstration of the visualization for photovoltaic devices holding big sets of well-arranged semiconducting supramolecular polymers with integrated p/n-heterojunctions. (c) Pictorial illustration of the monomer and the consequent supramolecular nanotube comprising of a coaxial p-n heterojunction. The donor component present in the monomer is Hexabenzocoronene (HBC) (blue), and the acceptor moiety extant is trinitrofluorenone (TNF, green); additional fragments of the monomer configuration contain triethylene glycol (TEG, red) and dodecyl (C12, white) chains designed for enhanced solubility. (d) Schematic representation for various possibilities of heterostructure formation, random, self-sorted and block using two different monomer units. (e) SIM image representing supramolecular BCPs via two different core substituted cholesterol appended Naphthalene diimide monomer units. (f) Molecular design along with a schematic for the synthesis of light harvesting axial BCPs. (g) Depiction of axial organic supramolecular BCP (light-harvesting heterostructures) with green-red-green segmented microstructure. (h) Image showing synthesis of axial heterostructure with charge transfer property. Background: SIM image of the block structures. Reproduced with permission from Refs. 33, 34b, 36, 37. Copyright 2011, Nature Research; 2012, Science; 2020, ACS; 2020, ACS respectively.

Many significant developments[30-35] have emerged in this aspect including contributions from our group, where we have achieved unprecedented control on the length, structure, and sequence of multicomponent supramolecular polymers using both the thermodynamic and kinetic approach to explore light-harvesting properties (Figures 1d and 1e).[36,37] In an attempt to use these supramolecular heterostructures as a functional materials, we have shown directional energy transfer through the axial block-co-polymer architecture (Figures 1f–1h).[37] For an advanced level, in the current scenario, higher hierarchical supramolecular structures are needed to use a plethora of morphology-dependent functions in device materials. We envisage that primary, secondary, and heterogeneous nucleation processes would be the way forward to manipulate the hierarchical structural complexity supramolecular functional nano-structures.[39-48] Our recent findings on this aspect open new possibilities to enhance the functional aspect of supramolecular systems in every segment of the functional supramolecular polymer field.[48]

An impulsive growth in research has been witnessed in the area of visible-light photocatalysis in the beginning of the 21st century. The power of the photosynthesis process amazingly revealed by Nature *via* the conversion of water and carbon dioxide into oxygen and carbohydrates is still unmatched by any man-made synthetic route.[49] In biological systems, the solar energy is absorbed mostly *via* chlorophylls from where energy is passed through a series of chromophores onto the reaction centers of Photosystems II (PSII) and I (PSI), in which charge separation is induced followed by consequent electron transfer through the central chlorophyll pigments. Such electron transfer follows a sequence of small-distance steps for the transportation of electrons within the redox mediators of the thylakoid membrane and protein scaffold. The transmission of electrons from the PSII to the plastoquinones (PQs), enclosed in the thylakoid membrane, further to the Cytochrome b6f complex (Cytb6f), and finally to an iron-sulfur protein (ferredoxin and ferredoxin) in the stroma of the chloroplasts happens synchronously in an effective manner.[49] Much efforts have been undertaken on the development of synthetic materials whose structure, properties or functions mimic natural photosynthetic systems e.g. efficient catalysts for water oxidation and proton reduction. Recently, many catalysts have been incorporated into light-absorbing silicon micro rods, CdSe quantum dots, and organic dyes to imitate photosynthesis artificially.[58-60] It has been quite recent that the synthesis and shaping of bioinspired soft materials into various forms to incorporate catalytic, charge-transport, and light-harvesting functions for the fabrication of solar cell devices as well as solar fuel-generating devices have begun to develop.[23,50-60] Due to the huge structural diversity presented by organic-based materials in comparison to inorganic composites, their tuning and optimization can be accomplished within a much broader range. With the current developments in visible-light-driven photo-catalysis, we aim in this chapter to highlight pioneering examples of the photocatalytic reduction of $CO_2$ and $H_2$ production from water using supramolecular materials.

## 2. Photocatalytic Water Splitting and $CO_2$ Reduction: Thermodynamics and Kinetic Aspects

The water-splitting reaction comprises of two parts. If we perform water-splitting reactions in an electrochemical cell in a cathode, hydronium ion ($H_3O^+$ or $H^+$) will consume 2 electrons from the cathode plate and will get reduced to generate the hydrogen gas (Reaction Scheme 1).[61-64] However, at the anode, water will release 4 electrons and get oxidized to produce the oxygen gas. To make this reaction feasible, external energy in the form of electrical voltage, is needed, i.e. a minimum of 1.23 V or more due to the overpotential. Inspired by nature, light can be utilized instead of electrical energy to do the same to produce hydrogen and oxygen gas. Organic molecules absorb visible light and generate electron-hole pairs in their excited state, known as Frenkel excitons.[65-68] In analogy to the inorganic semiconductors, for the organic molecules, the conduction band is the LUMO (Lowest Unoccupied Molecular Orbital) of the molecule containing the excited electrons and the generated holes reside in the HOMO (Highest Occupied Molecular Orbital) of the molecule which can be considered as the valence band. Subsequently, the excited electron gets transferred to a catalytic center where hydronium ion can consume it and get reduced to evolve hydrogen. On the other hand, the generated hole in the HOMO can take electrons directly from the water molecule, or *via* a catalytic center to produce oxygen. This whole process is called Overall water splitting (Figure 2). But due to four-electron transfer occurring, the oxygen evolution reaction requires high energy which is thermo-dynamically very unfavorable. Compared to the water oxidation process, sacrificial electron donors are easy to oxidize, and the reaction is kinetically faster due to the involvement of less electron transfer in the

$$2H^+(aq) + 2e^- \rightarrow H_2(g) \qquad E^{0'} = -0.41\ V$$

$$O_2(g) + 4H^+(aq) + 4e^- \rightarrow 2H_2O(l) \qquad E^{0'} = 1.23\ V$$

$$CO_2 + e^- \rightarrow CO_2^{\bullet -} \qquad E^{0'} = -1.93\ V$$

$$CO_2 + 2e^- + 2H^+ \rightarrow CO + H_2O \qquad E^{0'} = -0.53\ V$$

$$CO_2 + 2e^- + H^+ \rightarrow HCO_2^- \qquad E^{0'} = -0.39\ V$$

$$CO_2 + 4e^- + 4H^+ \rightarrow HCHO + H_2O \qquad E^{0'} = -0.48\ V$$

$$CO_2 + 6e^- + 6H^+ \rightarrow CH_3OH + H_2O \qquad E^{0'} = -0.38\ V$$

$$CO_2 + 8e^- + 8H^+ \rightarrow CH_4 + 2H_2O \qquad E^{0'} = -0.24\ V$$

Reaction Scheme 1. Redox potential of water splitting reactions and $CO_2$ reduction reaction vs NHE in pH 7.

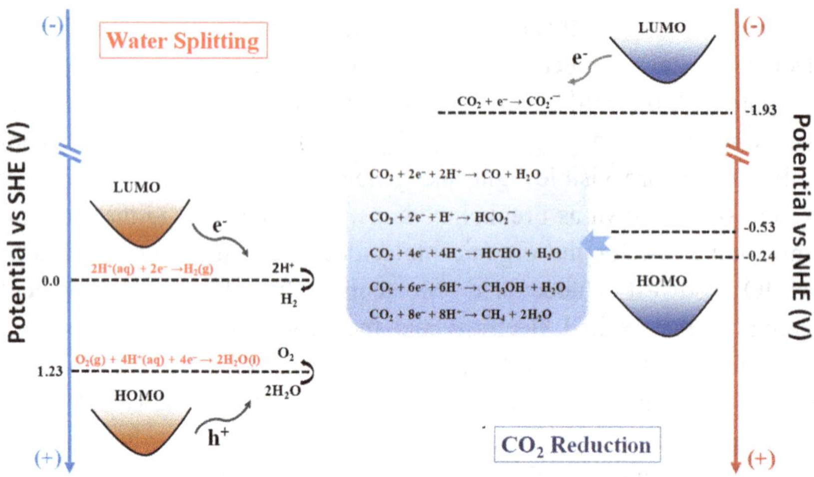


Figure 2. Schematic representation for the illustration of energy and potential requirements of the free charge carriers to achieve water splitting and $CO_2$ reduction.

whole process. Thus, to produce hydrogen *via* the photochemical process, the sacrificial agents are always preferably selected as a substitute rather than performing the entire water splitting process, and to fasten the hydrogen evolution reaction.

The $CO_2$ reduction reaction is rather more challenging to perform in terms of the selectivity of the product formation. $CO_2$ can take a series of electrons to form various reduced $C_1$ products in a very narrow range of external voltage (−0.53 to −0.24 V vs NHE in pH 7) (Reaction Scheme 1).[61-64] In addition to that, the $CO_2$ reduction reaction performed in an aqueous solution can also compete with the hydrogen evolution reactions as the reduction potential of hydronium ion to hydrogen production reactions has very close values to each other. Thus getting a particular $C_1$ product selectively from the pool of possibilities is still a very challenging task to work on. Selecting a perfect system that will match the prerequisite of performing a $CO_2$ reduction reaction is very crucial. In the subsequent sections, we will discuss the conditions and requirements to choose a photocatalytic system precisely in more detail.

## 3. Prerequisites for the Design of the Artificial Photocatalytic System

Carrying out a successful photocatalytic process needs three basic requirements: (a) A stable photosensitizer that can effectively absorb in the range of visible light. (b) Separations of generated excitons so that the electrons and holes can be channelized to the different catalytic centers without recombination for redox reactions. (c) Choosing proper sensitizers, catalytic centers and sacrificial electron donor molecules with suitable redox potential values of the water splitting or $CO_2$ reduction reaction, and also selecting them with necessary HOMO-LUMO energy match with each other for working synchronously (Figure 3).

Starting from Fujishima's simple $TiO_2$ electrode,[68] many developments have been made till now.[69-72] Along with inorganic semiconductors, additional organic photosensitizers have been incorporated for the better functional ability of the photosystem.[73-76] The effective charge separation of the inorganic semiconductors is achieved by the introduction of a co-catalyst or by doping the system. In an advanced Z-scheme method, organic photosensitizer, inorganic semiconductor material and an additional co-catalytic metal center are coupled simultaneously to get a very effective system as a whole for the photocatalytic reactions.[75,76] This increases the possibility of charge separation along with an increment in

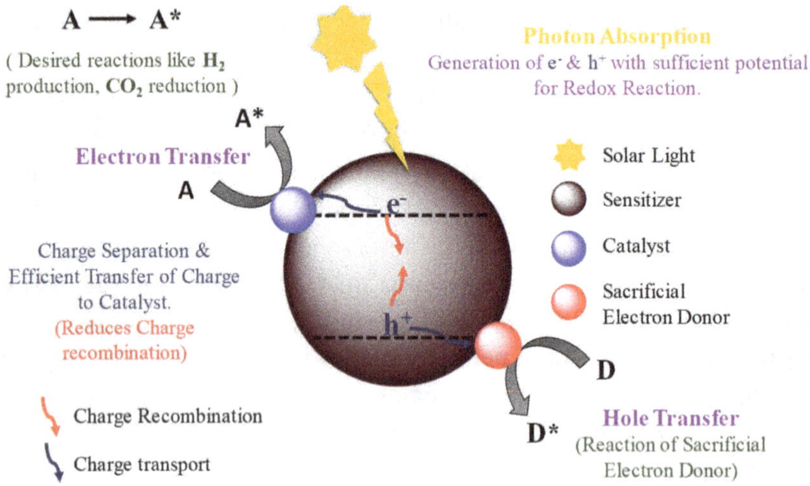


Figure 3. General schematic of a synthetic photocatalytic system.

the range of light absorption of the solar spectrum. Though high charge separation and sensitizing properties can be achieved through Z-scheme, as a whole, inorganic systems lack precise control over redox potential tunability and need extensive use of transition metals and expensive rare metal-based sensitizers. In comparison to inorganic systems, organic chromophores have better tunable absorption and the sensitizing capacity for photocatalysis.[77] The beauty of the organic sensitizers is, along with the broad absorption range, that they can be tuned to a vast range of redox potentials as per the requirements by changing simple structural motifs.[77] Exploration of common organic sensitizers has been widely performed using organic chromophore-metal catalytic center dyads.[78] Various sensitizer chromophore molecules, either coupled with molecular catalysts separately or attached to the metal center as a ligand, have been extensively studied for the last two to three decades.[78] These systems are excellent in terms of sensitizing, generating long-lived excitons, and tuning the redox potentials. As this high-performance ability is achieved through metal-to-ligand charge transfer transition, only a few purely organic sensitizer dyes are reported as efficient as metal-organic coordinate complexes.[79] The charge transfer state provides the metal-organic complexes to generate long-lived excited state, which earns enough time for the photocatalytic reactions to happen. For purely organic dyes, the very fast nature of exciton recombination processes *via* fluorescence or other thermal deactivation processes makes them a poor candidate to be used as photocatalytic sensitizers. As an outcome of this challenge, the use of the long-lived triplet state of the purely organic dyes has brought much interest in the research community.[80]

Also, the photobleaching of pure organic sensitizers often emerges as a major concern in this regard.[81,82] Unlike triplet sensitizers, pure organic singlet sensitizer dyes used in their monomeric state fail to generate long-lived excitons, which eventually destroys the performance of the photo-catalytic activity. Thus, use of small molecule-based purely organic dyes as an efficient sensitizer is highly desirable and simultaneously challenging as well. In recent years, scientists found supramolecular strategy to be an effective way out to use purely organic dyes for the development of environmentally friendly, renewable, robust, cost-effective and easily tunable synthetic photocatalytic systems.

## 4. Supramolecular Approach to Design Photocatalytic Systems: Self-assembly of Chromophores

Self-assembled chromophores are being extensively studied for their potential applications in energy-harvesting. Usually, π-conjugated chromophores are assembled *via* supramolecular interactions to make supramolecular nanostructures, which further augments their photo-physical properties too. From the pioneering work of Scheibe and Jelly with their cyanine dye system,[83,84] it has been established that frontier orbitals of the π-conjugated chromophores in their assembled state can interact with each other and thus can tune their energy levels in the assembled state. Excitons that generate photo excitations can delocalize through the assembly from one chromophore to the other. This leads to the generation of a long-lived charge-separated state. Thus, the use of a self-assembled chromophore as a photosensitizer can provide the desired long-lived charge-separated state without any additional requirement of a triplet state in the dye system and can be very effective towards photocatalytic activity. These aggregates are very robust and can be easily tuned by changing very minimal conditions based on the requirements. Thus, by only using a chromophore-based assembly and a molecular catalyst we can easily construct a simple but efficient photocatalytic system.

Upon irradiation by solar light, these self-assembled chromophore sensitizer molecules can generate the excitons and delocalize them through the stacks. Considering an overall reduction reaction to perform, a molecular catalyst co-assembled to it can take the excited electrons from the LUMO level of the stack and perform the photocatalytic reduction reactions. To quench the generated holes in the HOMO level, a sacrificial electron donor can be used which may give the electrons to the supra-molecular stacks of the sensitizer. Sacrificial electron donors favor the photocatalytic process by making it kinetically faster. Quenching of the excited states by sacrificial agents generally happens *via* two processes as depicted in Figure 4. One where excited state photosensitizer ($PS^*$) takes an electron first to get reduced and forms $PS^-$ which is called reductive quenching. Whereas in the other process, the $PS^*$ gets oxidized to $PS^+$ by

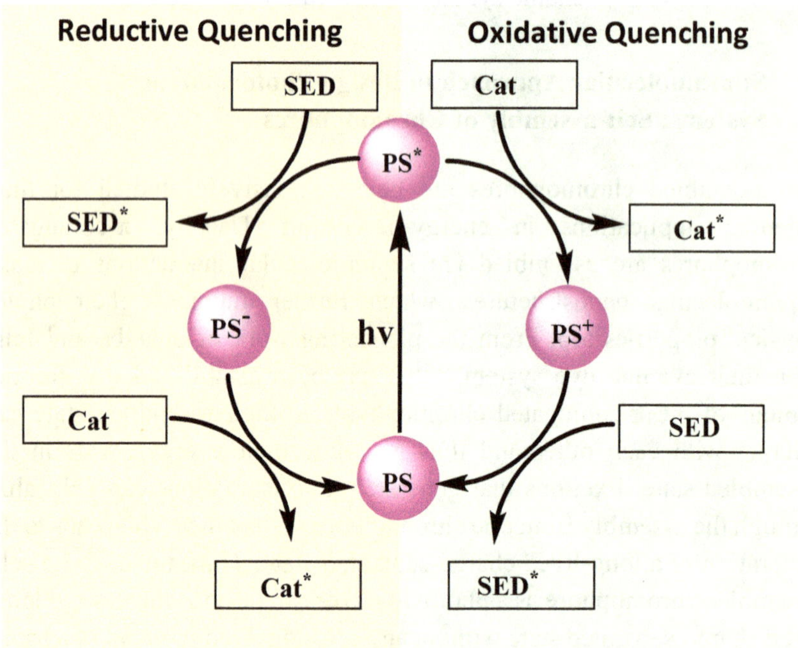


Figure 4. Schematic representation of oxidative and reductive quenching mechanism. SED, PS, and Cat represent Sacrificial Electron donors, Photosensitizer and Catalyst, respectively.

giving an electron to the catalyst and then takes an electron from the donor in the later stage. This process is called oxidative quenching. It is worth mentioning that the later process is not so common, as it requires strong dye systems to be stable in their higher oxidation state. Depending upon the photocatalytic reaction, whether the overall reduction or oxidation to perform, various sacrificial agents can be chosen; accordingly, either an electron donor or an electron acceptor, respectively.

For the performance of the self-assembled chromophores as an effective sensitizer, the nature of their aggregate plays a crucial role. In general, supramolecular self-assembled states form two types of aggregates with different molecular organization. One is the slip-stack type aggregate, which is called J-aggregate, and the other is the co-facial parallel aggregate which is known as H-aggregate (Figure 5a).[85,86] Detailed study on the coupling properties of the frontier orbitals reveals that the electronic

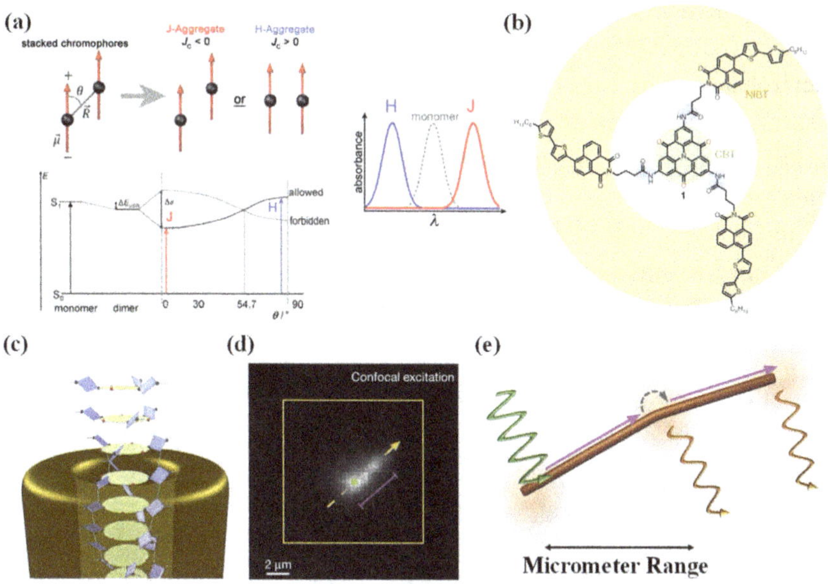


Figure 5. (a) Schematic representation for the theory of J- and H-aggregate alignment and coupling nature. J-aggregate shows red-shifted absorbance where H-aggregate shows blue-shifted absorbance. Supramolecular aggregate shows long-range exciton coupling. (b) molecular design (c) schematic of aggregated state (d) confocal image of exciton transport (e) schematic of micrometer range exciton transport. Reproduced with permission from Refs. 85 and 87. Copyright 2011, Wiley; 2015, Springer Nature respectively.

transition from the lower energy excited state to the ground state of an H-aggregate is forbidden in nature, whereas for the J aggregate it is allowed.[85] Thus generally, J-aggregates spontaneously emit light, and generated excitons get quenched readily. But for the H-aggregate, as the transition is not allowed, excitons are more accessible to couple with each other and can lead to the generation of long-lived excitons. Though this phenomenon comes out to be flexible in nature,[87] many systems are found to differ from the classical theory of exciton diffusion. Recent findings show that along with Frenkel exciton coupling, charge transfer coupling also can influence exciton diffusion. This leads to an efficient diffusion length for the charge carriers even in the J-aggregates. These types of aggregates are termed as hJ aggregates.[87] Thus, depending upon the aggregate and morphology of the supramolecular assembly, a plethora of possibilities exists to tune the exciton diffusion and lifetime of the excitons in a precise manner. Recent development from Schmidt and Hildner group shows an unprecedented long-range exciton transfer using self-assembly of tris(carbonyl)-N-heterotriangulene core (Figures 5b–5d).[88] This gives supramolecular self-assembled photosensitizers a new dimension, which essentially can be an effective approach toward constructing efficient, and tunable photo-catalytic systems.

In this chapter, we will discuss some of the supramolecular systems and their mechanisms in photocatalytic water splitting, oxygen evolution, and $CO_2$ reduction in the coming sections by taking some recent examples.

## 5. Photocatalytic Water Splitting

### 5.1. *Hydrogen evolution reaction (HER)*

In 2014, Stupp and coworkers in their pioneering work developed a biomimetic photocatalytic system with perylene monoimide derivative to perform hydrogen evolution reaction in the aqueous medium (Figure 6a).[23] Chromophoric perylene moiety appended with a charged carboxylate group separated by a five-member aliphatic hydrocarbon spacer makes the molecule amphiphilic, which at a higher concentration forms self-assembled nanoribbons in water (Figure 6b). In these nanoribbons, the intermolecular

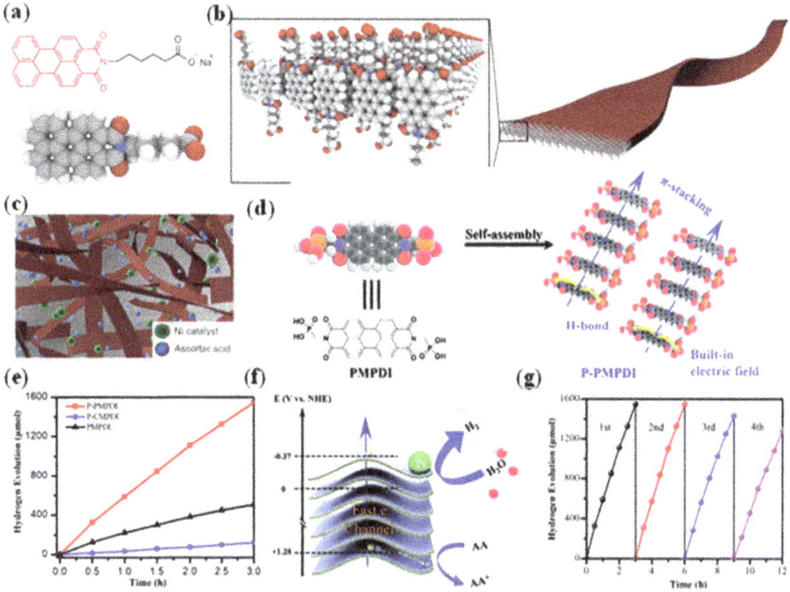


Figure 6. (a) Illustration of the molecular structure and space-filling model of PMI-based (perylene monoimide-based) chromophore amphiphile (CA), (b) CA molecules packed in an antiparallel direction forming an interlocked bilayer to generate supramolecular ribbons. a.u. (c) Demonstration of a 3D architecture of gel showing CA nanoribbons (in red) allows where the ascorbic acid and catalyst to be incorporated with CA ribbons for photoinduced $H_2$ production. (d) Schematic representation of the supramolecular synthesis of P-PMPDI. (e) Photocatalytic evolution of hydrogen over P-PMPDI and P-CMPDI through the time-dependent analysis. (f) Graphical illustration showing the possible mechanism of electron transfer in supramolecular P-PMPDI for producing $H_2$ in photocatalytic process. (g) Stability and performance of the photocatalytic system through multiple cycles. Reproduced with permission from Ref. 91. Copyright 2018, RSC.

interactions are weak due to the charge repulsion between the chromophores and thus fail to bind strongly with each other. As a result of the lack of long-range coupling in this system no photocatalytic activity was observed. A surface charge stabilization process with polyelectrolyte or by the addition of NaCl leads to stronger $\pi$- $\pi$ interactions which enhanced the exciton migration in the assembled state. Thus, upon addition of a polycationic electrolyte solution of ammonium polydiallyl-dimethylammonium chloride (PDDA), the system forms a hydrogel. The formation of hydrogel is essential here for the performance of hydrogen evolution because it helps the substrate to diffuse throughout the system and also to incorporate the catalysts (Figure 6c).

In 2018, same group extended their work with the same sensitizer motif of perylene monoamide chromophore amphiphiles synthesized by various components in the 9-position to probe the impact of self-assembled nanostructure and the chromophore organization in the photocatalytic efficiency.[89] Taking different systems, they first showed the molecular dipole strength dependence with the nature of the substituents of various electron-donating and electron-withdrawing groups. A compound with more electron-donating substituents shows stronger molecular dipoles, which favors the assembly to stabilize, whereas the electron-withdrawing substituents make the assembly weaker. This phenomenon dictates the morphology of the assembly significantly. As a consequence, the former forms crystalline nanoribbons and the latter forms nanofibers. However, a compound with moderate dipole moment shows the highest photocatalytic $H_2$ production owing to the appropriate molecular packing to access charge-transfer excitons.[90]

In 2019, Jianli Hua and coworkers compared various electron-withdrawing group substituted perylenedimide (PDI) to narrow the band gap of the photocatalyst for further broadening of the absorption range and exploiting the photocatalytic activity (Figure 6d).[91] The researchers found that phosphoric acid substituted perylene diimide (P-PMPDI) containing self-assembled supramolecular system displayed a higher HER rate of 11.7 mmol $h^{-1}$ $g^{-1}$ under visible light irradiation. The system exhibited a recorded apparent quantum yield (AQY) of 2.96% at 550 nm, which is so far the greatest AQY value reported for supramolecular photocatalysts (Figure 6e). Compared to carboxylic acid (P-CMPDI), phosphoric acid derivative (P-PMPDI) shows a lowering in the energy gap between the HOMO and LUMO, essentially facilitating the use of a broader absorption range. The synthesized P-PMPDI nano-belts elevate the uniformity of coplanar $\pi$–$\pi$ stacking through the assembly of perylene skeletons, further boosting the long-range $\pi$-electron delocalization and electron coupling effects. The improved light absorption range enhances the photocurrent density of P-PMPDI signifying a remarkable charge separation in the photocatalysts. It is worth to mention that the system also shows remarkable stability towards multicycle performances (Figures 6f and 6g).

In recent times, gels have been presented as a unique class of catalysts for solar energy conversion supporting the water splitting reaction in presence of light within a hydrated environment. Furthermore, modification in the metal center can readily regulate the properties of porphyrin-based gels. For example, in 2018, Cheng-Yong Su and coworkers assembled a series of metallo-porphyrin imine gels, and their photocatalytic performance was explored for visible-light-irradiated HER.[92] Sodium ascorbate was utilized as the sacrificial donor to accelerate the reaction rate in the existence of Pt catalyst for the proton reduction. Covalent imine gels with dynamic nature were prepared from polycondensation of bridging amines 5,10,15,20-tetrakis(p-aminophenyl)porphyrin ($H_2TAPP$) and its metalized forms (MTAPP, M = $SnCl_2$ and Pd) with 1,3,5-tris-(4-formyl-phenyl)triazine (TFPT) in a 1 : 1 molar ratio. The porphyrin-based materials can have tailored band gap energies through the selection of incorporated metals at the centers of porphyrin moieties. The PdTAPP–TFPT gel displayed a hierarchically porous structure which is beneficial for the loading of Pt cocatalyst. Furthermore, they show a high photocatalytic HER rate by uninterrupted $H_2$ production from aqueous solution of sodium ascorbate when illuminated with visible light $1.0744 \times 10^5$ μmol $g^{-1}$ for 120 h showing insignificant degradation.

In 2019, a unique strategy of electron transmission proceeded by two diverse pathways was reported by Tong-Bu Lu for the augmentation of artificial photosynthesis.[93] The scientists innovatively integrated Ir-complex with coumarins and boron dipyrromethene to maximize the utilization of solar energy and to develop BSVLA (broadband and strong visible-light-absorbing) photosensitizers. Ir-4 indicated the broadband covering of 50% of the visible light region. In this specific system (Ir-4), the Ccumarin ligand got excited by visible light and proficiently converts excitation energy into bodipy-confined triplet state *via* ISC cascade processes and intramolecular triplet-triplet energy transfer (TTET), subsequently producing 32 μmol of $H_2$ in 12 h. The reported turnover number of 115840, under optimized conditions, represents Ir-4 as the most active sensitizer among the molecular photocatalytic systems reported so far.

Among π-conjugated molecules, porphyrin-based nanomaterials exhibiting well-organized structural arrangements can work as energy transfer platforms. Their characteristics to stimulate the natural light-harvesting systems make porphyrins a perfect elementary unit for the synthesis of photoactive nanomaterials. In 2019, Hongyou Fan and coworkers developed an innovative example of self-assembled semiconductor porphyrin nanocrystals (In-TPP) *via* microemulsion-controlled nucleation and growth process.[94] These photoactive nanocrystals revealed an exceptionally efficient energy transmission progression for enhanced photocatalytic hydrogen production, retaining the crystallinity and structural integrity of nanorods. The rigid aggregate framework of In-TPP protects them from photobleaching or solvolysis by reactive species and resulted in hydrogen production activity up to 60 h under visible light irradiation showing a superior photostability in comparison to the original porphyrin monomers.

In 2020, Ian Manners's group prepared block copolymer assemblies with catalysts or photosensitizers implanted in the corona-forming segment of the block.[95] Tailored nanofibers assemblies were generated through these building blocks *via* the living crystallization-assisted self-assembly seeded growth method (Figure 7). In this method, the photosensitizers and catalysts were positioned adjacently to be used as a metal-free photocatalytic system for producing hydrogen from water. Visible light-sustained artificial assemblies were found to be extremely proficient, stable and recyclable at ambient temperature. Various controlled experiments showed the importance of the precise structural arrangement of sensitizer dye and the catalyst for efficient photocatalysis to occur. The best performance they observed was when the catalyst and the dye remained in the blended structure of the heterostructure.

In addition to these well-documented photocatalysts, an approach of designing new photocatalytic systems, assisted by charge transfer (CT) interactions, can bring about the modulation of the properties, and hybridization of the suitable energy levels to achieve desirable outcomes. HOMO–LUMO energy gaps can be reduced through the association of electron-deficient acceptor and electron-rich donor molecules. Stoichiometric CT complexation achieved via mechanically interlocked donor acceptors fosters the dynamicity of the system making it a potential contender to be used in adaptive catalytic processes. Weak interactions could probably be improved in a mechanically interlocked molecular system (MIMs) which facilitates their components to reside in

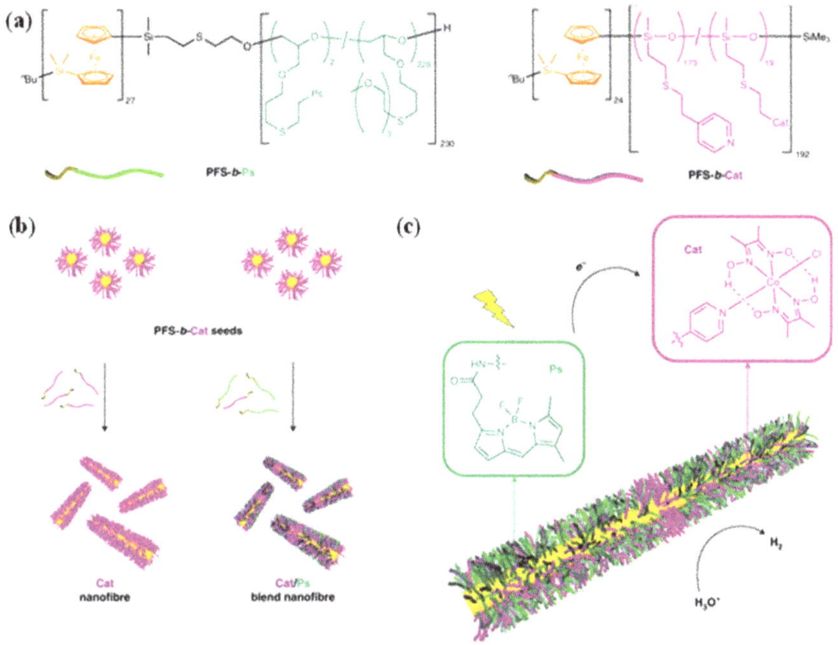


Figure 7. (a) Representation of the structures of the photosensitizer block copolymer PFS-b-Ps and the catalytic block copolymer PFS-b-Cat. (b) Demonstrating the construction of nanofibres by seeded growth. (c) Diagram showing the Cat/Ps blend nanofibers structure for HER with corona-forming segments bearing the catalyst (purple) and photosensitizer (green) and crystallizable PFS core-forming block (yellow). Reproduced with permission from Ref. 95. Copyright 2020, Springer Nature.

nanoconfined spaces. Very recently, J.F. Stoddart and coworkers employed a donor-acceptor catenane for the implementation of visible-light photocatalysis in the absence of any additional photosensitizers.[96] The intramolecular CT interactions between dioxynaphthalene units and bipyridinium dications provided enhanced absorption of the visible light and also enabled the photo-induced transfer of electrons to form a persistent bipyridinium radical cation. This photoinduced radical is very proficient in (i) oxidation of L-methionine in an aerobic environment through the production of highly reactive superoxide radical anion and (ii) proton reduction further producing hydrogen in water. However, the CT excited states exhibited a smaller lifetime and low coefficients of molar absorption which limits their application for practical uses.

Perylene-based molecular systems despite being a promising organic photocatalyst, are still not deliberated as excellent photocatalytic HER candidates due to their (i) insufficient reduction potential required for HER thermodynamics, and (ii) less hydrophilic sites leading to an inferior surface catalytic activity. However, appropriate structural modifications can be introduced to vary the molecular energy level, for example, the frontier orbital level can be lowered *via* the introduction of electron-withdrawing groups. Following the above design strategies, Yongfa Zhu and group synthesized supramolecular perylene tetracarboxylic acid (PTA) nanosheets as photocatalysts for HE.[98a] (Figure 8a). The carboxyl groups promoted the charge-carrier separation by amplifying the internal electric field intensity by 10.3 times. The decreased molecular conjugation effect in PTA upgraded the reduction potential for HER from −0.47 V to −1.13 V presenting a boosted photocatalytic HER activity with production rates of 118.9 mmol $g^{-1}$ $h^{-1}$ (Figures 8b and 8c).

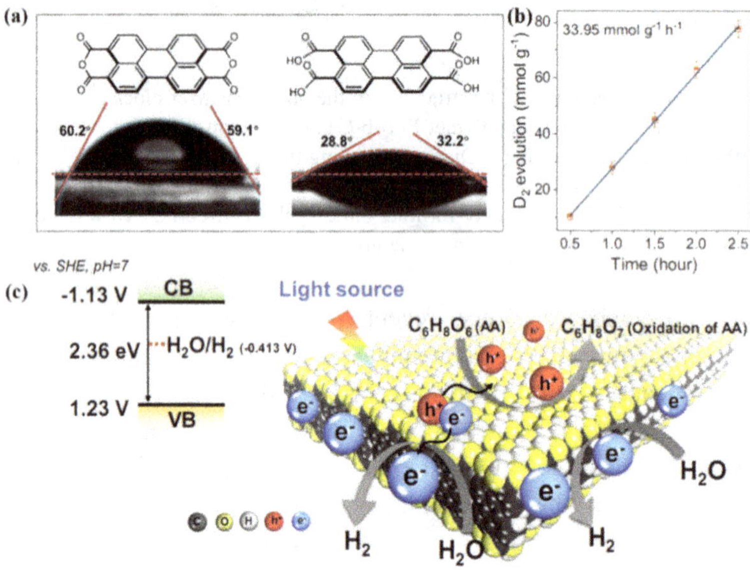


Figure 8. (a) Schematic illustration showing contact angle of PTA and PTCDA (Perylenetetracarboxylic dianhydride) with water, (b) Graphical outline presentation of $D_2$ evolution performance of PTA-split $D_2O$, (c) Representation of the mechanism of the photocatalytic HER by PTA nanosheets showing the band positions on the left and migration and reaction of photo-generated charges on the right side. Reproduced with permission from Ref. 98a. Copyright 2022, Springer Nature.

## 5.2. *Oxygen evolution reaction (OER)*

Perylene 3,4,9,10-tetracarboxylic diimide (PDI) possessing π–π conjugated-electrons structure displays outstanding thermal and photostability allowing strong visible light absorption. Their pronounced electron affinity and charge carrier mobility place them among n-type organic semiconductors. The above-mentioned properties of PDI roved scientists to explore it for photocatalysis using self-assembled nanostructures.[98b] In 2016, Duan's group incorporated PDI into a metal-organic polymer (Zn-PDI) revealing its highly photoactive nature for the oxidation of benzyl alcohols and amines.[98b] However, it was a great challenge to self-assemble the PDI with non-substituents at the parent skeleton (PDINH) due to its insolubility in most common organic solvents. Non-covalent supramolecular self-assembly prepared through independently pure soft molecules was reported for the first time in 2016 by Yongfa Zhu and coworkers where purely organic small-PDINH (non-substituents at the parent skeleton) building blocks were utilized to fabricate willow leaf-shaped supramolecule-engineered nano-assemblies *via* rapid solution dispersion method.[97] In this specific method, non-covalent interactions assisted self-assembly of the molecules, consequencing in solid-state precipitate in presence of poor solvent with superior morphology tunability and structural flexibility in contrast to the conducting polymers. This self-assembled supramolecular system expressed outstanding photocatalytic activity for the degradation of phenolic compounds, anionic dyes (methyl orange/MO), cationic dyes (Rhodamine B/RhB, methylene blue/MB), and $O_2$ evolution through water splitting in the absence of co-catalyst under visible light irradiation (Figure 9). The strong cofacial π–π stacking present in PDINH molecules facilitated efficient delocalization of long-range π-electrons, leading to the augmentation of charge mobility and electronic coupling, further contributing to its noteworthy photocatalytic properties.

## 5.3. *$CO_2$ reduction reaction*

$CO_2$ reduction reactions nowadays have gained much attention to the scientific community with increasing environmental issues like 'Global warming'. A significant amount of effort has been given to increase the

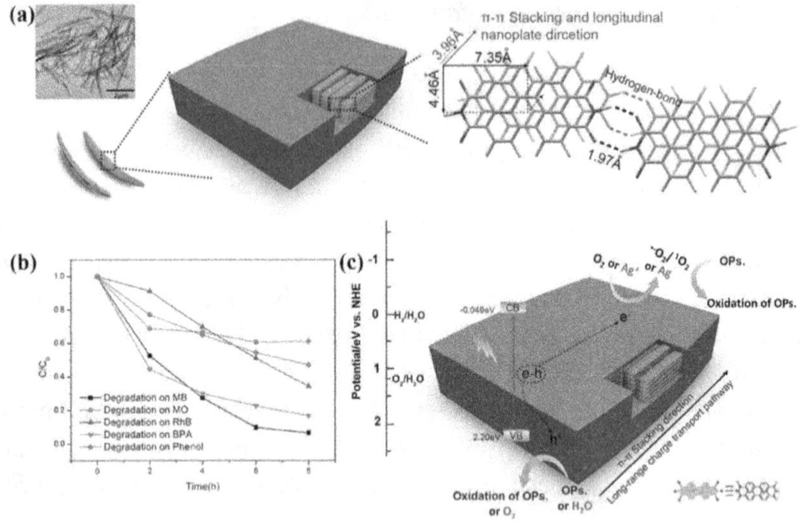


Figure 9. (a) Illustration of a single PDINH molecule showing the intermolecular arrangement within the nanoplate displaying the cofacial π–π stacking distance. (b) Photodegradation performance of self-assembled PDINH supramolecular system on various probe pollutants under visible light irradiation (λ > 420 nm). (c) Graphical representation of the self-assembled PDINH supramolecular system electronic energy level structure showing the photocatalytic reaction's mechanism under visible light irradiation (λ > 420 nm). Reproduced with permission from Ref. 97. Copyright 1989, WILEY-VCH.

catalytic activity, stability and efficiency of the photocatalytic systems for $CO_2$ reduction and it has been observed that high performance occurs mostly in non-aqueous mediums or with organic solvent-water mixture medium.[99] The main reason for that is the competitive reaction of hydrogen evolution along with the $CO_2$ reduction, as the former reaction has a comparable reduction potential value with the latter.[99] Thus stable, efficient, and aqueous photocatalytic systems are very rare in organic photosystems. To get such stable systems that will work under pure water medium, we need rare metal complexes like Ruthenium or Cadmium complexes. Thus, construction of an organic supramolecular photocatalytic system is very challenging.

A very recent report appeared from Stupp's group utilizing a stable dye called ullazine, which is known for its efficient donor ability in dye-sensitized solar cells to reduce $CO_2$ in an aqueous-organic solvent mixture.[100] They have synthesized an amphiphilic carboxylic acid derivative of illazine dye and made them self-assembled in a gel system (Figure 10). Using dicobalt complexes as the molecular catalyst they have

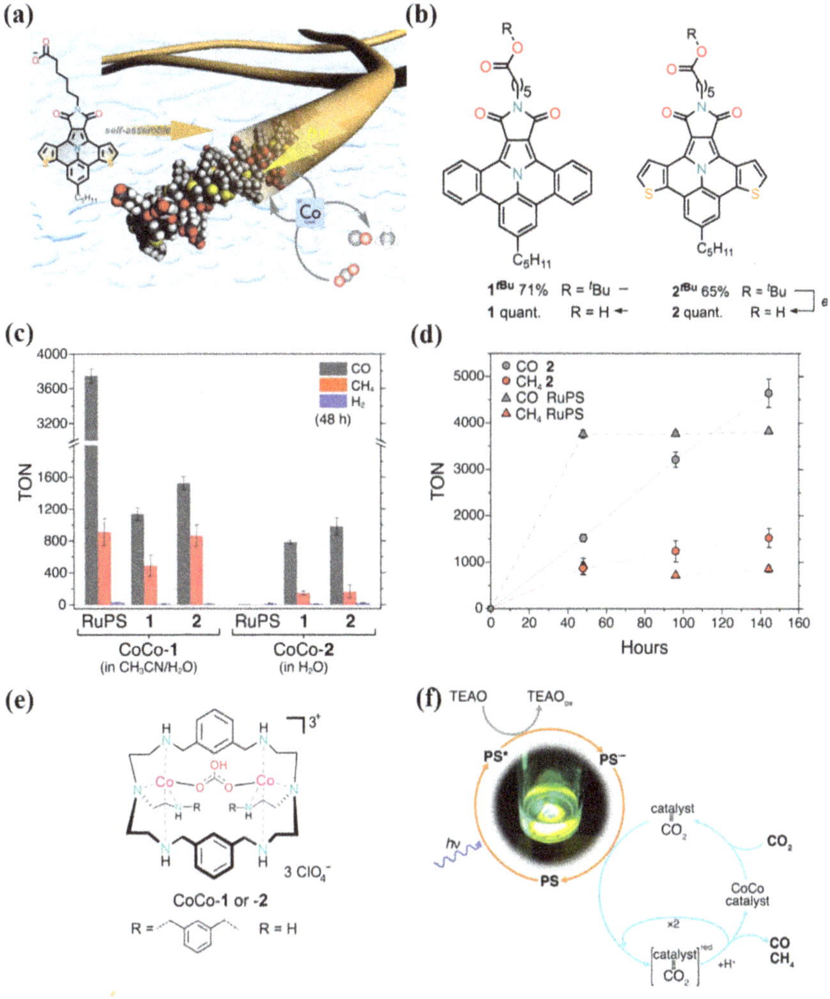


Figure 10. (a) Schematic representation of self-assembled system of Ullazine amphiphilic derivative performing $CO_2$ reduction reaction. (b) Structures of sensitizer monomers. (c) Catalytic turnovers in $H_2$ (blue), CO (gray), and $CH_4$ (red) for consuming as photosensitizers $Ru(phen)_3(PF6)_2$ (RuPS) or either 1 or 2, with dinuclear cobalt complexes 1. (d) Generation of CH4 (red symbols) and CO (gray symbols) with time (48, 96, and 144 h) for diareno-fused ullazine 2 and CoCo-1. (e) Structure of catalyst and (f) Proposed reaction mechanism for the $CO_2$ reduction reaction using TEOA as the sacrificial electron donor. Reproduced with permission from Ref. 100. Copyright 2022, ACS.

achieved an efficient photocatalytic system to reduce $CO_2$ to CO and $CH_4$ with $TON_{CO} = 1525 \pm 84$ and $TON_{CH4} = 865 \pm 137$ which even show better performance than a comparable system sensitized with $Ru(phen)_3(PF_6)_2$ in the pure aqueous medium.

## 6. Perspective and Future Outlook

The aforementioned examples of energy conversion techniques using supramolecular structures indeed show great potential in enhancing fuel production. The very next challenge in this field will be to enhance the efficiency, robustness, recyclability, and device processability of these self-assembled nanostructures. Despite a considerable amount of progress using various assembled dye systems to perform photocatalysis in fuel production, efficiency remains a great challenge till now. Enhancement of efficiency largely depends on the system's ability to utilize the whole solar spectrum and the ability to increase exciton lifetime considerably. Similar to nature, efficient energy transfer through successive light-harvesting dyes not only utilizes various ranges of the solar spectrum, but also enhances the excited state electron's lifetime to harvest the solar energy gracefully. Thus, to mimic nature to its best, a multicomponent supramolecular system can make a great impact in this regard. It is well known that a donor-acceptor type hetero junction is way more efficient for the charge transport properties compared to a single-component stack.[33] Thus using a multicomponent block copolymer system for photocatalysis can be an advanced system to look forward to. But constructing a multicomponent system is a great challenge in the field of supramolecular chemistry. In a pioneering work from Manner group, the use of a multicomponent-blend structure to perform photocatalytic hydrogen evolution reaction comes out to show the highest turnover number in organic systems reported till now. This eye-opening report shows the importance of a multi-component system in photocatalysis.[96] Recently, works from our group on the synthesis of multicomponent systems opened the possibility to utilize their properties in a wide range of applications.[37,38] These systems are fundamentally very important for the photocatalysis process where we can have a vast tunability of the HOMO-LUMO levels of the materials, which not only will cover the whole visible range spectrum in a single system but essentially will help to modulate the redox properties to perform photocatalytic reactions selectively. Playing with various sequences in a multi-component system can enhance the selectivity of $CO_2$ reduction over $H_2$, which can immensely solve the problem of low efficiency of $CO_2$ reduction in an aqueous solution. Hence, we believe multicomponent supramolecular systems are the most potent candidates to revolutionize the field of photocatalysis in the near future. These strategies are the closest way to mimic nature successfully.

## 7. Conclusion

Supramolecular materials have emerged recently as a potential candidate for energy conversion which can closely mimic nature in the best possible way. The non-covalent interactions provide them the freedom to utilize various functions in a reversible way which includes an additional benefit of self-healing, similar to natural systems. Also, these systems can simultaneously utilize the properties of various components in a single system. Non-covalent interactions overcome the recyclability issues of many other covalent systems. Along with that, $\pi$-$\pi$ interactions in the assembled state of the supramolecular systems is an added advantage compared to other systems. As discussed earlier, $\pi$-chromophoric systems behave very differently in the assembled state compared to their monomeric systems. A high lifetime of exciton can easily be generated through the assembly of those $\pi$-chromophores which is an essential requirement for photocatalysis. The scope of tunability of these supramolecular systems makes them distinct from other systems when it comes to harvesting light energy or performing various redox reactions. Combining these two properties opens up the horizon for supramolecular systems in the energy conversion field. In conclusion, here we have provided an overview of various supramolecular systems reported recently for energy conversion functions and also presented our perspective on the way forward with these materials.

## References


1. F. Wohler. Ueber künstliche Bildung des Harnstoffs. *Ann. Phys. Chem.* **88** (2), 253 (1828).
2. J. M. Lehn. Supramolecular Chemistry-Scope and Perspectives Molecules, Supermolecules, and Molecular Devices (Nobel Lecture). *Angew Chem. Int. Ed.* **27** (1), 89 (1988).
3. J. M. Lehn. From Supramolecular Chemistry Towards Constitutional Dynamic Chemistry and Adaptive Chemistry. *Chem. Soc. Rev.* **36** (2), 151 (2007).
4. F. Hof, S. L. Craig, C. Nuckolls, J. Rebek and R. Julius. Molecular Encapsulation. *Angew. Chem., Int. Ed.* **41** (9), 1488 (2002).
5. E. Persch, O. Dumele and F. Diederich. Molecular Recognition in Chemical and Biological Systems. *Angew. Chem. Int. Ed.* **54** (11), 3290 (2015).
6. E. J. Dale, N. A. Vermeulen, M. Juricek, J. C. Barnes, R. M. Young, M. R. Wasielewski and J. F. Stoddart. Supramolecular Explorations: Exhibiting the Extent of Extended Cationic Cyclophanes. *Acc. Chem. Res.* **49** (2), 262 (2016).
7. M. De Loos, B. L. Feringa and J. H. Van Esch. Design and Application of Self-Assembled Low Molecular Weight Hydrogels. *Eur. J. Org. Chem.* (17), 3615-3631 (2005).
8. R. Orbach, L. Adler-Abramovich, S. Zigerson, I. Mironi-Harpaz, D. Seliktar and E. Gazit. Self-Assembled Fmoc-Peptides as a Platform for the Formation of Nanostructures and Hydrogels. *Biomacromolecules.* **10** (9), 2646 (2009).
9. M. Zhou, A. M. Smith, A. K. Das, N. W. Hodson, R. F. Collins, R. V. Ulijn and J. E. Gough. Self-assembled Peptide-based Hydrogels as Scaffolds for Anchorage-dependent Cells. *Biomaterials.* **30** (13), 2523 (2009).
10. D. J. Abdallah and R. G. Weiss. Organogels and Low Molecular Mass Organic Gelators. *Adv. Mater.* **12** (17), 1237 (2000).
11. A. Ajayaghosh and V. K. Praveen π-Organogels of Self-Assembled *p*-Phenylenevinylenes: Soft Materials with Distinct Size, Shape, and Functions. *Acc. Chem. Res.* **40** (8), 644 (2007).
12. A. Ajayaghosh, V. K. Praveen and C. Vijayakumar. Organogels as Scaffolds for Excitation Energy Transfer and Light Harvesting. *Chem. Soc. Rev.* **37**, 109 (2008).
13. O. M. Yaghi, M. O'Keeffe, N. W. Ockwig, H. K. Chae, M. Eddaoudi and J. Kim. Reticular Synthesis and the Design of New Materials. *Nature.* **423**, 705 (2003).
14. A. P. Côté, A. I. Benin, N. W. Ockwig, M. O'Keeffe, A. J. Matzger and O. M. Yaghi, Porous, Crystalline, Covalent Organic Frameworks. *Science.* **310** (5751), 1166 (2005).
15. H. Furukawa, K. E. Cordova, M. O'Keeffe and O. M. Yaghi. The Chemistry and Applications of Metal-organic Frameworks. *Science.* **341** (6149), 1230444 (2013).
16. M. A. Green, Commercial Progress and Challenges for Photovoltaic. *Nat. Energy* 1, 15015 (2016).

17. A. Polman, M. Knight, E. C. Garnett, B. Ehrler and W. C. Sinke. Photovoltaic materials: Present Efficiencies and Future Challenges. *Science.* **352** (6283), aad44241-442410 (2016).
18. P. K. Nayak, S. Mahesh, H. J. Snaith and D. Cahen. Photovoltaic Solar Cell Technologies: Analysing the State of the Art. *Nat. Rev. Mater.* ***4***, 269-285 (2019).
19. A. S. Tayi, A. K. Shveyd, A. C. H. Sue, J. M. Szarko, B. S. Rolczynski, D. Cao, T. J. Kennedy, A. A. Sarjeant, C. L. Stern, W. F. Paxton, W. Wu, S. K. Dey, A. C. Fahrenbach, J. R. Guest, H. L. Mohseni, X. Chen, K. L. Wang, J. F. Stoddart and S. I. Stupp. Room-temperature Ferroelectricity in Supramolecular Networks of Charge-transfer Complexes. *Nature.* **488**, 485-489 (2012).
20. N. Robertson, Optimizing Dyes for Dye-Sensitized Solar Cells. *Angew. Chem. Int. Ed.* **45** (15), 2338 (2006).
21. J. A. A. W. Elemans, A. E. Rowan and R. J. M. Nolte, Mastering Molecular Matter. Supramolecular Architectures by Hierarchical Self-assembly. *J. Mater. Chem.* **13** (11), 2661 (2003).
22. K. Liu, Y. Kang, Z. Wang and X. Zhang. 25th Anniversary Article: Reversible and Adaptive Functional Supramolecular Materials: "Noncovalent Interaction" Matters. *Adv. Mater.* **25** (11), 5530-5548 (2013).
23. A. S. Weingarten, R. V. Kazantsev, L. C. Palmer, M. McClendon, A. R. Koltonow, A. P. Samuel, D. J. Kiebala and M. R. Wasielewski. S. I. Stupp, Self-assembling Hydrogel Scaffolds for Photocatalytic Hydrogen Production. *Nat. Chem.* **6**, 964-970 (2014).
24. F. Tantakitti, J. Boekhoven, X. Wang, R. V. Kazantsev, T. Yu, J. Li, E. Zhuang, R. Zandi, J. H. Ortony, C. J. Newcomb, L. C. Palmer, G. S. Shekhawat, M. O. de la Cruz, G. C. Schatz and S. I. Stupp. Energy Landscapes and Functions of Supramolecular Systems. *Nat. Mater.* **15**, 469-476 (2016).
25. G. D. Scholes, T. Mirkovic, D. B. Turner, F. Fassioli and A. Buchleitner. Solar Light Harvesting by Energy Transfer: From Ecology to Coherence. *Energy Environ. Sci.* **5** (11), 9374-9393 (2012).
26. C. Train, R. Gheorghe, V. Krstic, L. M. Chamoreau, N. S. Ovanesyan, G. L. J. A. Rikken, M. Gruselle and M. Verdaguer. Strong Magneto-chiral Dichroism in Enantiopure Chiral Ferromagnets. *Nature Mater.* **7**, 729–734 (2008).
27. P. Wang, S. Guo, H. J. Wang, K. K. Chen, N. Zhang, Z. M. Zhang and T. -B. Lu, A Broadband and Strong Visible-light-absorbing Photosensitizer Boosts Hydrogen Evolution. *Nat. Commun.* **10**, 3155 (2019).
28. Y. Guo, Q. Zhou, J. Nan, S. Wenxin, F. Cui and Y. Zhu, Perylenetetracarboxylic Acid Nanosheets with Internal Electric Fields and Anisotropic Charge Migration for Photocatalytic Hydrogen Evolution. *Nat. Commun.* **13**, 2067 (2022).
29. A. T. Haedler, K. Kreger, A. Issac, B. Wittmann, M. Kivala, N. Hammer, J. Köhler, H.-W. Schmidt and Hildner, R. Long-range Energy Transport in Single Supramolecular Nanofibres at Room Temperature. *Nature.* **523**, 196 (2015).

30. F. Würthner, Z. Chen, F. J. M. Hoeben, P. Osswald, C. You, J. Pascal, J. V. Herrikhuyzen, A. P. H. J. Schenning, P. P. A. M. Van der Schoot, E. W. Meijer, H. A. Edwin Beckers, S. C. J. Meskers and R. A. J. Janssen. Supramolecular p–n-Heterojunctions by Co-Self-Organization of Oligo(p-phenylene Vinylene) and Perylene Bisimide Dyes. *J. Am. Chem. Soc.* **126** (34), 10611 (2004).
31. X. Wang, G. Guerin, H. Wang, Y. Wang, I. Manners and M. Winnik. Cylindrical Block Copolymer Micelles and Co-micelles of Controlled Length and Architecture. *Science* **317** (5838), 644 (2007).
32. J. B. Gilroy, T. Gädt, G. R. Whittell, L. Chabanne, J. M. Mitchels, R. M. Richardson, M. A. Winnik and I. Manners. Monodisperse Cylindrical Micelles by Crystallization-driven Living Self-assembly. *Nat. Chem.* **2**, 566–570 (2010).
33. W. Zhang, W. Jin, T. Fukushima, A. Saeki, S. Seki, T. Aida, Supramolecular Linear Heterojunction Composed of Graphite-like Semiconducting Nanotubular Segments. *Science*. **334** (6054), 340 (2011).
34. (a) W. Wagner, M. Wehner, V. Stepanenko and F. Würthner. Supramolecular Block Copolymers by Seeded Living Polymerization of Perylene Bisimides. *J. Am. Chem. Soc.* **141** (30), 12044 (2019). (b) T. Aida, W. Meijer and S. I. Stupp. Functional Supramolecular Polymers. *Science*. **335** (6070), 813 (2012).
35. B. Adelizzi, A. Aloi, A. J. Markvoort, H. M. Ten Eikelder, I. K. Voets, A. R. Palmans and E. W. Meijer. Supramolecular Block Copolymers under Thermodynamic Control. *J. Am. Chem. Soc.* **140** (23), 7168 (2018).
36. A. Sarkar, R. Sasmal, C. Empereur-mot, D. Bochicchio, S. V. K. Kompella, K. Sharma, S. Dhiman, B. Sundaram, S. S. Agasti, G. M. Pavan and S. J. George. Self-Sorted, Random, and Block Supramolecular Copolymers via Sequence Controlled, Multicomponent Self-Assembly. *J. Am. Chem. Soc.* **142** (16), 7606 (2020).
37. A. Sarkar, T. Behera, R. Sasmal, R. Capelli, C. Empereur-mot, J. Mahato, S. S. Agasti, G. M. Pavan and S. J. George. Cooperative Supramolecular Block Copolymerization for the Synthesis of Functional Axial Organic Heterostructures. *J. Am. Chem. Soc.* **142** (26), 11528 (2020).
38. A. Sarkar, R. Sasmal, A. Das, A. Venugopal, S. S. Agasti, S. J. George, Tricomponent Supramolecular Multiblock Copolymers with Tunable Composition via Sequential Seeded Growth. *Angew. Chem. Int. Ed.* **60** (33), 18209 (2021).
39. B. Adhikari, Y. Yamada, M. Yamauchi, K. Wakita, X. Lin, K. Aratsu, T. Ohba, T. Karatsu, M. J. Hollamby and N. Shimizu. Light-induced Unfolding and Refolding of Supramolecular Polymer Nanofibres. *Nat. Commun.* **8**, 15254 (2017).
40. K. K. Kartha, N. K. Allampally, A. T. Politi, D. D. Prabhu, H. Ouchi, R. Q. Albuquerque, S. Yagai and G. Fernández. Influence of Metal Coordination and Light Irradiation on Hierarchial Self-assembly Processes. *Chem. Sci.* **10** (3), 752 (2019).
41. O. Szczepankiewicz, B. Linse, G. Meisl, E. Thulin, B. Frohm, C. Sala Frigerio, M. T. Colvin, A. C. Jacavone, R. G. Griffin, T. Knowles, D. M. Walsh and S. Linse.

N-Terminal Extensions Retard Aβ42 Fibril Formation but Allow Cross-Seeding and Coaggregation with Aβ42. *J. Am. Chem. Soc.* **137** (46), 14673 (2015).
42. M. örnquist, T. C. T. Michaels, K. Sanagavarapu, X. Yang, G. Meisl, S. I. A. Cohen, T. P. J. Knowles and S. Linse. Secondary Nucleation in Amyloid Formation. *Chem. Commun.* **54** (63), 8667 (2018).
43. S. I. A. Cohen, S. Linse, L. M. Luheshi, E. Hellstrand, D. A. White, L. Rajah, D. E. Otzen, M. Vendruscolo, C. M. Dobson and T. P. J. Knowles. Proliferation of Amyloid-B42 Aggregates Occurs through a Secondary Nucleation Mechanism. *Proc. Natl. Acad. Sci. U. S. A.* **110** (24), 9758 (2013).
44. S. Datta, Y. Kato, S. Higashiharaguchi, K. Aratsu, A. Isobe, T. Saito, D. D. Prabhu, Y. Kitamoto, M. J. Hollamby, A. J. Smith, R. Dalgliesh, N. Mahmoudi, L. Pesce, C. Perego, G. M. Pavan and S. Yagai. Self-Assembled Poly-Catenanes from Supramolecular Toroidal Building Blocks. *Nature*. **583**, 400 (2020).
45. N. Sasaki, M. F. J. Mabesoone, J. Kikkawa, T. Fukui, N. Shioya, T. Shimoaka, T. Hasegawa, H. Takagi, R. Haruki and N. Shimizu. Supramolecular Double-Stranded Archimedean Spirals and Concentric Toroids. *Nat. Commun.* **11**, 3578 (2020).
46. N. Singh, B. Lainer, G. J. M. Formon, S. De Piccoli and T. M. Hermans. Re-Programming Hydrogel Properties Using a Fuel-Driven Reaction Cycle. *J. Am. Chem. Soc.* **142** (9), 4083 (2020).
47. N. Singh, A. Lopez-Acosta, G. J. M. Formon and T. M. Hermans. Chemically Fueled Self-Sorted Hydrogels. *J. Am. Chem. Soc.* **144** (1), 410 (2022).
48. S. Sarkar, A. Sarkar, A. Som, S. S. Agasti and S. J. George. Stereoselective Primary and Secondary Nucleation Events in Multicomponent Seeded Supramolecular Polymerization. *J. Am. Chem. Soc.* **143** (30), 11777 (2021).
49. B. Daum, D. Nicastro, J. Austin, J. R. McIntosh and W. Ku¨hlbrandt, Arrangement of Photosystem II and ATP Synthase in Chloroplast Membranes of Spinach and Pea. *Plant Cell.* **22** (4), 1299 (2010).
50. M. Li, W. Ni, X. Wan, Q. Zhang, B. Kan and Y. Chen. Benzo[1,2-b:4,5-b']dithiophene (BDT)-based Small Molecules for Solution Processed Organic Solar Cells. *J. Mater. Chem. A* **3** (9), 4765 (2015).
51. A. Tang, C. Zhan and J. Yao, E. Zhou. Design of Diketopyrrolopyrrole (DPP)-Based Small Molecules for Organic-Solar-Cell Applications. *Adv. Mater.* **29** (2), 1600013 (2017).
52. R. Freeman, M. Han, Z. Álvarez, J. A. Lewis, J. R. Wester, N. Stephanopoulos, M. T. McClendon, C. Lynsky, J. M. Godbe, H. Sangji, E. Luijten and S. I. Stupp. Reversible Self-assembly of Superstructured Networks. *Science* **362** (6416), 808 (2018).
53. E. R. Zubarev, M. U. Pralle, E. D. Sone. and S. I. Stupp. Self-Assembly of Dendron Rodcoil Molecules into Nanoribbons. *J. Am. Chem. Soc.* **123** (17), 4105 (2001).

54. X.-H. Jin, M. B. Price, J. R. Finnegan, C. E. Boott, J. M. Richter, A. Rao, S. M. Menke, R. H. Friend, G. R. Whittell and I. Manners. *Science*. **360** (6391), 897 (2018).
55. N. J. Hestand, R. V. Kazantsev, A. S. Weingarten, L. C. Palmer, S. I. Stupp and F. C. Spano. *J. Am. Chem. Soc.* **138** (36), 11762 (2016).
56. R. V. Kazantsev, A. J. Dannenhoffer, T. Aytun, B. Harutyunyan, D. J. Fairfield, M. J. Bedzyk and S. I. Stupp. Molecular Control of Internal Crystallization and Photocatalytic Function in Supramolecular Nanostructures. *Chem.* **4** (7), 1596 (2018).
57. Z. Han, F. Qiu, R. Eisenberg, P. L. Holland and T. D. Krauss. Robust Photogeneration of $H_2$ in Water Using Semiconductor Nanocrystals and a Nickel Catalyst. *Science*. **338** (6112), 1321 (2012).
58. S. W. Boettcher, E. L. Warren, M. C. Putnam, E. A. Santori, D. Turner-Evans, M. D. Kelzenberg, M. G. Walter, J. R. McKone, B. S. Brunschwig, H. A. Atwater and N. S. Lewis. Photoelectrochemical Hydrogen Evolution Using Si Microwire Arrays. *J. Am. Chem. Soc.* **133** (5), 1216 (2011).
59. P. Poddutoori, D. T. Co, A. P. S. Samuel, C. H. Kim, M. T. Vagnini and M. R. Wasielewski. Photoinitiated Multistep Charge Separation in Ferrocene–zinc Porphyrin–diironhydrogenase Model Complex Triads. *Energy Environ. Sci.* **4** (11), 244 (2011).
60. B. S. Veldkamp, W.-S. Han, S. M. Dyar, S. W. Eaton, M. A. Ratner and M. R. Wasielewski. Photoinitiated Multi-step Charge Separation and Ultrafast Charge Transfer Induced Dissociation in a Pyridyl-linked Photosensitizer–cobaloxime Assembly. *Energy Environ. Sci.* **6** (6), 1917 (2013).
61. S. Berardi, S. Drouet, L. Francas, C. Gimbert-Surinach, M. Guttentag, C. Richmond, T. Stoll and A. Llobet Molecular Artificial Photosynthesis. *Chem. Soc. Rev.* **43** (22), 7501 (2014).
62. A. Bassegoda, C. Madden, D. Wakerley W., E. Reisner and J. Hirst. Reversible Interconversion of CO2 and Formate by a Molybdenum Containing Formate Dehydrogenase. *J. Am. Chem. Soc.* **136** (44), 15473 (2014).
63. K. Li, B. Peng and T. Peng. Recent Advances in Heterogeneous Photocatalytic CO2 Conversion to Solar Fuels. *ACS Catal.* **6** (11), 7485 (2016).
64. T. Reda, C. M. Plugge, N. J. Abram and J. Hirst. Reversible Interconversion of Carbon Dioxide and Formate by an Electroactive Enzyme. *Proc. Natl. Acad. Sci. U. S. A.* **105** (31), 10654 (2008).
65. F. C. Spano. Excitons in Conjugated Oligomer Aggregates, Films, and Crystals. *Annu. Rev. Phys. Chem.* **57**, 217 (2006).
66. T.-S. Ahn, A. M. Müller, R. O. Al-Kaysi, F. C. Spano, J. E. Norton, D. Beljonne, J.-L. Brédas and C. J. Bardeen. Experimental and Theoretical Study of Temperature Dependent Exciton Delocalization and Relaxation in Anthracene Thin Films. *J. Chem. Phys.* **128** (5), 054505 (2008).

67. N. J. Hestand and F. C. Spano. Molecular Aggregate Photophysics beyond the Kasha Model: Novel Design Principles for Organic Materials. *Acc. Chem. Res.* **50** (2), 341 (2017).
68. A. Fujishima and K. Honda. Electrochemical Photolysis of Water at a Semiconductor Electrode. *Nature*. **238**, 37 (1972).
69. A. M. Appel, J. E. Bercaw, A. B. Bocarsly, H. Dobbek, D. L. DuBois, M. Dupuis, J. G. Ferry, E. Fujita, R. Hille and P. J. Kenis. Frontiers, Opportunities, and Challenges in Biochemical and Chemical Catalysis of CO2 fixation. *Chem. Rev.* **113**, 6621 (2013).
70. M. Aresta, A. Dibenedetto and A. Angelini. Catalysis for the Valorization of Exhaust Carbon: from CO2 to Chemicals, Materials, and Fuels. Technological Use of CO2. *Chem. Rev.* **114** (3), 1709 (2014).
71. J. Barber. Photosynthetic Energy Conversion: Natural and Artificial. *Chem. Soc. Rev.* **38** (1), 185 (2009).
72. G. W. Crabtree and N. S. Lewis. Solar Energy Conversion. *Phys. Today* **60** (3), 37 (2007).
73. M. K. Nazeruddin, A. Kay, I. Rodicio, R. Humphry-Baker, E. Mueller, P. Liska, N. Vlachopoulos, M. Graetzel, Conversion of Light to Electricity by cis X2-Bis(2,28-bipyridyl-4,48-dicarboxalate) Ruthernium(II) Charge Transfer Sensitizers. *J. Am. Chem. Soc.* **115**, 6382 (1993).
74. M. Grätzel. The Artificial Leaf, Bio-mimetic Photocatalysis. *Cattech.* **3**, 3 (1999).
75. C. Santato, M. Ulmann and J. Augustynski. Photoelectrochemical Properties of Nanostructured Tungsten Trioxide Films. *J. Phys. Chem. B.* **105** (5), 936 (2001).
76. S. U. M. Khan and J. Akikusa. Photoelectrochemical Splitting of Water at Nanocrystalline n-Fe2O3 Thin Film Electrodes. *J. Phys. Chem. B* **103** (34), 7184 (1999).
77. N. A. Romero and D. A. Nicewicz. Organic Photoredox Catalysis. *Chem. Rev.* **116** (17), 10075 (2016).
78. K. E. Dalle, J. Warnan, J. J. Leung, B. Reuillard, I, S. Karmel and E. Reisner. Electro- and Solar-Driven Fuel Synthesis with First Row Transition Metal Complexes. *Chemical Reviews* **119** (4), 2752 (2019).
79. R. Gueret, L. Poulard, M. Oshinowo, J. Chauvin, M. Dahmane, G. Dupeyre, P. P. Lainé, J. Fortage and M. N. Collomb. Challenging the [Ru(Bpy)3]2+ Photosensitizer with a Triazatriangulenium Robust Organic Dye for Visible-Light-Driven Hydrogen Production in Water. *ACS Catal.* **8**, 3792 (2018).
80. A. Garci, J. A. Weber, R.M. Young, M. Kazem-Rostami, M. Ovalle, Y. Beldjoudi, A. Atilgan, Y. Jue Bae, W. Liu, L. O. Jones, C. L. Stern, G. C. Schatz, O. K. Farha, M. R. Wasielewski and J. F. Stoddart. Mechanically interlocked pyrene-based photocatalysts. *Nat. Catal.* **5**, 524 (2022).
81. L. S. Herculano, L. C. Malacarne, V. S. Zanuto, G. V. Lukasievicz, O. A. Capeloto and N. G. Astrath, Investigation of the Photobleaching Process of Eosin Y in Aqueous Solution by Thermal Lens Spectroscopy. *J. Phys. Chem. B.* **117** (6), 1932 (2013).

82. Y.-H. Chiu, T.-F.M. Chang, C.-Y. Chen, M. Sone and Y.-J. Hsu. Mechanistic Insights into Photodegradation of Organic Dyes Using Heterostructure Photocatalysts. *Catalysts*. **9** (5), 430 (2019).
83. E. Jelly. Spectral Absorption and Fluorescence of Dyes in the Molecular State. *Nature* **138**, 1009 (1936).
84. G. Scheibe. Über die Veränderlichkeit der Absorptionsspektren in Lösungen und die Nebenvalenzen als ihre Ursache. *Angew. Chem.* **50** (11), 212 (1937).
85. N. J. Hestand and F. C. Spano. Molecular Aggregate Photophysics Beyond the Kasha Model: Novel Design Principles for Organic Materials. *Acc. Chem. Res.* **50** (2), 341 (2017).
86. F. Würthner, T. E. Kaiser and C. R. Saha-Möller. J-Aggregates: From Serendipitous Discovery to Supramolecular Engineering of Functional Dye Materials. *Angew. Chem. Int. Ed.* **50** (15), 3376 (2011).
87. N. J. Hestand and F. C. Spano. Expanded Theory of H- and J-Molecular Aggregates: The Effect of Vibronic Coupling and Intermolecular Charge Transfer. *Chem. Rev.* **118** (15), 7069 (2018).
88. A. T. Haedler, K. Kreger, A. Issac, B. Wittmann, M. Kivala, N. Hammer, J. Köhler, H.-W. Schmidt and R. Hildner. Long-range Energy Transport in Single Supramolecular Nanofibres at Room Temperature. *Nature*. **523**, 196 (2015).
89. A. S. Weingarten, A. J. Dannenhoffer, R. V. Kazantsev, H. Sai, D. Huang and S. I. Stupp. Chromophore Dipole Directs Morphology and Photocatalytic Hydrogen Generation. *J. Am. Chem. Soc.* **140** (15), 4965 (2018).
90. A. S. Weingarten, R. V. Kazantsev, L. C. Palmer, D. J. Fairfield, A. R. Koltonow and S. I. Stupp. Supramolecular Packing Controls H2 Photocatalysis in Chromophore Amphiphile Hydrogels. *J. Am. Chem. Soc.* **137** (48), 15241– 15246 (2015).
91. K. Kong, S. Zhang, Y. Chu, Y. Hu, F. Yu, H. Ye, H. Ding and J. A Hua. Self-Assembled Perylene Diimide Nanobelt for Efficient Visible-Light-Driven Photocatalytic H2 Evolution. *Chem. Commun.* **55** (56), 8090 (2019).
92. P. Liao., Y. Hu, Z. Liang, J. Zhang, H. Yang, L.-Q. He, Y.-X Tong, J.-M. Liu, L. Chen and C.-Y Su. Porphyrin-based Imine Gels for Enhanced Visible-light Photocatalytic Hydrogen Production. *J. Mater. Chem. A*. **6** (7), 3195 (2018).
93. P. Wang, S. Guo, H. J. Wang, K. K. Chen, N. Zhang, Z. M. Zhang and T.-B. Lu. A Broadband and Strong Visible-light-absorbing Photosensitizer Boosts Hydrogen Evolution. *Nat Commun.* **10**, 3155 (2019).
94. Y. Liu, L. Wang, H. Feng, X. Ren, J. Ji, F. Bai and H. Fan. Microemulsion-Assisted Self-Assembly and Synthesis of Size-Controlled Porphyrin Nanocrystals with Enhanced Photocatalytic Hydrogen Evolution. *Nano Lett.* **19** (4), 2614 (2019).
95. J. Tian, Y. Zhang, L. Du, Y. He, X.-H. Jin, S. Pearce, J.-C. Eloi, R. L. Harniman, D. Alibhai, R. Ye, D. L. Phillips and I. Manners. Tailored Self-assembled Photocatalytic Nanofibres for Visible-light-driven Hydrogen Production. *Nat. Chem.* **12**, 1150 (2020).

96. Y. Jiao, L. Đorđević, H. Mao, R. M. Young, T. Jaynes, H. Chen, Y. Qiu, K. Cai, L. Zhang, X. Y. Chen, Y. Feng, M. R. Wasielewski, S. I. Stupp and J. F. Stoddart. A Donor–Acceptor [2]Catenane for Visible Light Photocatalysis. *J. Am. Chem. Soc.* **143** (21), 8000 (2021).
97. D. Liu, J. Wang, X. Bai, R. Zong and Y. Zhu. Self-Assembled PDINH Supramolecular System for Photocatalysis under Visible Light. *Adv. Mater.* **28** (23), 7284 (2016).
98. (a) Y. Guo, Q. Zhou, J. Nan, W. Shi, F. Cui and Y. Zhu. Perylenetetracarboxylic Acid Nanosheets with Internal Electric Fields and Anisotropic Charge Migration for Photocatalytic Hydrogen Evolution. *Nat. Commun.* **13**, 2067 (2022), (b) L. Zeng, T. Liu, C. He, D. Shi, F. Zhang and C. Duan. Organized Aggregation Makes Insoluble Perylene Diimide Efficient for the Reduction of Aryl Halides via Consecutive Visible Light-Induced Electron-Transfer Processes. *J. Am. Chem. Soc.* **138** (12), 3958 (2016).
99. H. Takeda, C. Cometto, O. Ishitani and M. Robert. Electrons, Photons, Protons and Earth-Abundant Metal Complexes for Molecular Catalysis of CO2 Reduction. *ACS Catal.* **7** (1), 70–88 (2017).
100. O. Dumele, L. Đorđević, H. Sai, T. J. Cotey, M. H. Sangji, K. Sato, A. J. Dannenhoffer and S. I. Stupp. Photocatalytic Aqueous CO2 Reduction to CO and CH4 Sensitized by Ullazine Supramolecular Polymers. *J. Am. Chem. Soc.* **144** (7), 3127 (2022).

CPSIA information can be obtained
at www.ICGtesting.com
Printed in the USA
JSHW060139200223
37688JS00007B/11

9 789811 270949